FIG. 68.—Negritoids; natives of western part of Southern India.

Fig. 69.—Negritoid statuettes from Palembong District, Sumatra. (Photographs donated to Dr. Hrdlička by Mr. L. C. Westenanck, former Resident of Palombong.)

bong, Sumatra, there still prevail in the island among the whites as well as the natives, beliefs in the existence of wild men. There are two varieties. The Orang Pandak (orang=man, pandak=short) is said to live in the almost impenetrable mountain forests of the central and southern parts of the island. The natives describe him as black, short, long-haired, and wild, but not unsurmountably shy—will ask the Malay natives for tobacco. The second form is the Orang Sedapak. He is said to live in the unhealthy lowlands of the southeastern parts of Sumatra. He is said to have the body of a child of about 12 and to have long red hair on head and body. He is very shy and runs but does not climb. On June 20 an expedition was to leave, in charge of Captain Bor, for the determination of what this creature really is. In addition Sumatra has been yielding for some time peculiar stone sculptures, including heads that seem to represent the Negrito.

In the mountainous region of the upper parts of the Malay Peninsula, according to information given to Dr. Hrdlička, there still live thousands of negritoid people, and there are many caves waiting to be explored.

The visit to Java was made chiefly for the purpose of inspecting the site of the Pithecanthropus, but Dr. Hrdlička also desired to satisfy himself as to any possible cultural traces of early man, and as to the present population. As a result of the generous assistance given by the authorities,[1] he was able to see the natives in practically the whole of the island and especially to examine that important region which gave the precious remains of the Pithecanthropus—the valley of the Bengawan or Solo River, a fairly large river, beginning in the south of the island and running north and then east to Soerabaya. Here exists a veritable treasure-house for anthropology and paleontology, where nothing has been done since the Selenka expedition of 1910, which was the only one since the work of Dr. Dubois in 1891-'93. The lower deposits along the river are full of the fossil bones of Tertiary and Quaternary mammals, but among them at any time may be remains of greater value. Many of the fossils fall out of exposed strata every year and lie in the mud, where the natives occasionally gather them and take them to their homes.

---

[1] Dr. Hrdlička wishes to thank especially Dr. B. Schrieke of Veltevreden; Mr. J. Th. Jarman, the Assistant Resident at Ngawi; Mr. and Mrs. S. H. Pownall, at Banjoewangi; and Messrs. C. P. Kuykendall and R. R. Winslow, U. S. Consuls respectively at Batavia and Soerabaya.

FIG. 70.—The Dubois Monument opposite the site of the Pithecanthropus. (Photograph taken for Dr. Hrdlička by Mr. J. Th. Jarman, Assistant Resident, Ngawi.)

FIG. 71.—Natives opposite the site of the Pithecanthropus bringing in fossils found about that site. (Photograph by Dr. Hrdlička.)

FIG. 72.—The Pithecanthropus site from the opposite bank. (Photograph by Dr. Hrdlička, May 26, 1925.)

FIG. 73.—The Pithecanthropus site from the opposite bank. (Photograph taken for Dr. Hrdlička by Assistant Resident Jarman later in the summer.)

When the actual site of the Pithecanthropus was reached by Dr. Hrdlička, under the guidance of the Assistant Resident of Ngawi and his Chief of Police, a whole gang of natives advised by the police were already waiting there, bringing each a smaller or larger pile of fossils gathered from the muddy ledges of the river as these were exposed by the receding water. These fossils were eagerly examined but they included no remains of any Primate. A selection was made, to which the boys added a few specimens collected at that moment from about the site which gave the Pithecanthropus. On the top of the opposite bank stands a cement monument erected by Dubois and pointing to the spot which yielded what are probably the most precious remains in existence.

Further excavation here and in other localities along the river would be relatively easy and a few years of sustained work here is one of the great needs of Anthropology.

After the site of the Pithecanthropus and its neighborhood were examined, a little dug-out with two natives took Dr. Hrdlička down the river to Ngawi, a distance by river of perhaps 15 miles. During this trip both of the banks could be closely examined. They and their prolongation south-eastward are of interest geologically and there may be spots of paleontological value, but there are no other sites as promising in the latter respect as that near Trinil.

From Madioen Dr. Hrdlička's journey led to the eastern portion of Java, where it was interesting to find in spots traces of the pre-Malay Hindoo population which peopled the island in early historic times. In the central part of Java these people evidently reached a high degree of culture and left imposing ruins.

From Java Dr. Hrdlička traveled by boat along the northwestern and western Australian coast, stopping at all the little ports from Derby to Perth. This gave the opportunity to see numerous pure-blood Australians, and also some of their impressive nocturnal ceremonies. Here was encountered an exceptional type of the Australian from the Wyndham district, differing considerably from the rest of the natives. Here also were seen for the first time full-blood and otherwise full-colored Australians with tow hair; more were seen later on the Trans-Australian Railway at Ooldea. In southern Australia other aborigines were seen, particularly on the lower Murray River.

The principal Australian Museums of interest to Anthropology are located at Perth, Adelaide, Melbourne, Sydney and Brisbane. They were found to contain astonishingly rich collections, ethnological, archeological and anthropological, from Australia, Tasmania,

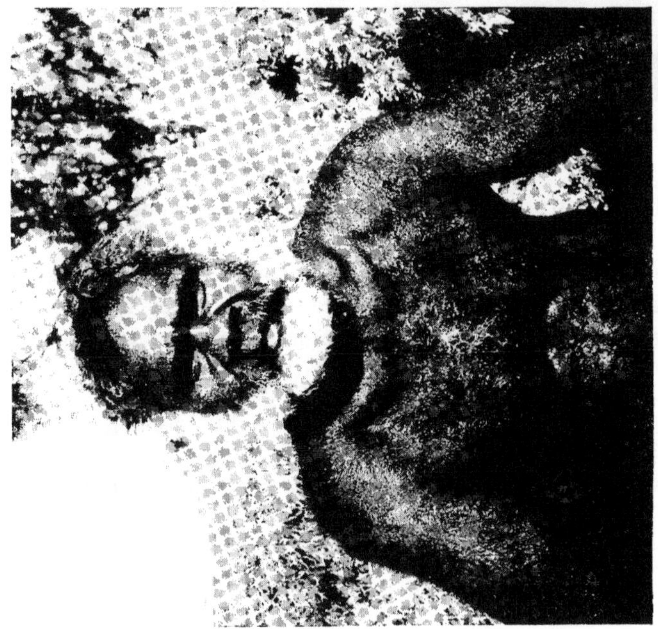

FIG. 74.—A full-blood Australian man from the lower Murray River. (Photograph by Dr. Hrdlička.)

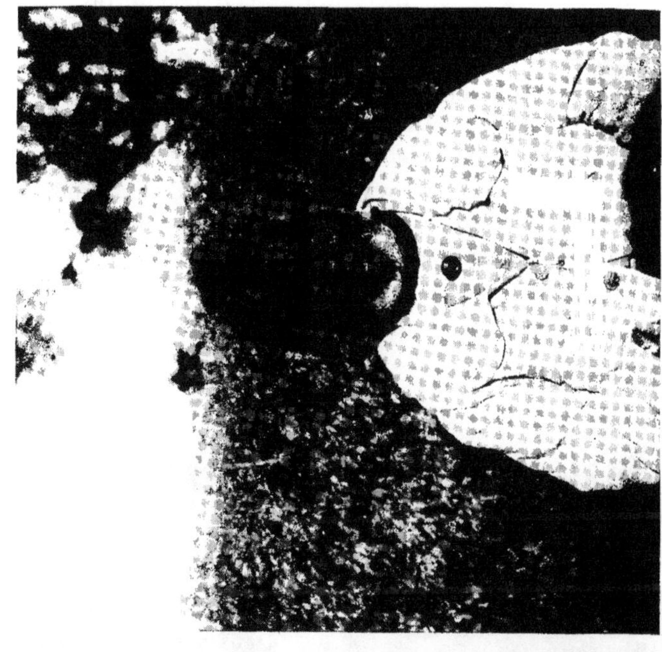

FIG. 75.—A full-blood Australian woman from the lower Murray River. (Photograph by Dr. Hrdlička.)

and Melanesia. In addition there are several noteworthy private collections of this nature, two of which (Dr. Basedow's and Dr. Pulleine's, Adelaide), were seen; and important somatological collections are being built up at the Anatomical Departments in the principal cities.[1] The greatest collection of human skeletal material is that of the Museum of Adelaide. It consists of over 600 skulls of the Australian aborigines, with numerous skeletons, and it is being constantly added to under a beneficial law which obliges all the police officials of the State of which Adelaide is the capital to

FIG. 76.—Three of the tow-haired full-blood Australians of the Ilgarene Tribe. (Photograph donated to Dr. Hrdlička.)

forward to the Museum any aboriginal skeletal remains that may be found.

These precious somatological collections Dr. Hrdlička was permitted to utilize and nearly five weeks were spent in the work, resulting in securing essential measurements on 1,000 well-identified skulls of Australians, and on such of the Tasmanians as are preserved in the institutions visited.

[1] For aid given in connection with his work in Australia, Dr. Hrdlička is particularly indebted and thankful to the following: Dr. I. S. Battye, Director of the Perth Museum; Mr. A. E. Morgan, U. S. Consular Agent at Perth; Dr. A. E. Waite, Director of the Museum, Adelaide; Dr. J. A. Kershaw, Curator of the National Museum, Melbourne; Dr. C. Anderson, Director of the Australian Museum, Sydney; Professors of Anatomy, R. J. A. Berry (Melbourne), F. Wood Jones and A. N. Burkitt; Drs. Herbert Basedow and R. H. Pulleine at Adelaide; and the U. S. Consul General at Melbourne.

The data obtained in Australia, supplemented by those on the Tasmanian material in the College of Surgeons, London, throw a very interesting and to some extent new light on the moot questions of both the Australian and the Tasmanian aborigines. According to these observations, the Australian aborigines deserve truly to be classed as one of the more fundamental races of mankind, and yet it is a race which shows close connections with our own ancestral stock—not with the negroes or Melanesians (except through admixture), but with the old white people of postglacial times. They carry, however, some admixtures of the Melanesian blacks, which is more pronounced in some places than in others.

As to the Tasmanians, the indications are that they are in all probability but a branch of the Australians, modified perhaps a little in their own country. Both peoples have lived, and the Australians of the northwest live largely to this day, in a paleolithic stage of stone culture. They are still making unpolished stone tools, which in instances resemble the Mousterian implements or later European paleolithic types. But they are also capable of a much higher class of work. Today, about Derby, bottles are used in making beautifully worked spear heads.

In the Anatomical Department of the University of Sydney, with the kind aid of Professor Burkitt, Dr. Hrdlička had the chance to examine several times the Talgai skull, believed to be of geological antiquity. The specimen was seen to bear undeniable affinities with the Australian cranial type, but the very large palate and the teeth need further consideration.

.From Australia Dr. Hrdlička's journey led to South Africa and disembarking at Durban, Natal, the first task was to see as many as possible of the Zulu, about whose exact blood affinities there was some doubt. Large numbers were seen, and the conclusion was reached that they are unquestionably true negroes, though now and then as in other negro tribes, showing a trace of Semitic (Arab?) type due probably to old admixtures.

The two main objects of the visit to South Africa were the investigation on the spot of the important find of the Rhodesian skull, and of the recent discovery of the skull of a fossil anthropoid ape at Taungs, which had been reported as being possibly a direct link in the line of man's ascent. South Africa is a land full of anthropological interest. There is the disappearing old native population of Bushmen, Strandloopers, and Hottentots; the newer negro population which amounts already to over 7,000,000 and is steadily increasing; the almost stationary population of 1,500,000 South African whites of Dutch and English derivation, who are blending together and producing a type of their own (as is also happening on a larger scale in Australia); and there are abundant remains of "paleolithic" cultures. Of equal interest are the great finds of the

FIG. 77.—Mr. Zwigelaar, the miner who with his "boy" discovered the Rhodesian skull, with the specimen shortly after the find was made. (Photograph loaned Dr. Hrdlička by Mr. Zwigelaar.)

Broken Hill mine, Northern Rhodesia, 2,015 miles north of Cape Town, and of the Buxton quarry, 1,000 miles further southward.

The discovery in 1921 at Broken Hill in Southern Rhodesia of the skull of the so-called "Rhodesian Man" was an event of much scientific importance. The find, moreover, is still enigmatic. The skull shows a man so primitive in many of its features that nothing like it has been seen before. The visit to the Broken Hill mine in which the skull was discovered proved a good demonstration of the necessity of a prompt following up by scientific men of each such accidental discovery. The impracticability of such a following up in this case has resulted in a number of errors and uncertainties on important aspects of the case, some of which have already misled students of the find. It was possible to clear up some of the mooted points, but others remain obscure and can be definitely decided only by further discoveries.

As one of the results of the present visit, it was possible to save and bring for study a collection of bones of animals from the cave,

FIG. 78.—Animal and human bones secured by Dr. Hrdlička at the Broken Hill Mine; all from the Bone Cave. (Photograph by Dr. Hrdlička.)

the lower recesses of which gave the Rhodesian skull, and also two additional mineralized human bones belonging to two individuals; all of which, to facilitate the study of the whole subject, were deposited with the earlier relics in the British Museum. The mine is by no means exhausted, and since the interest of everybody on the spot is now fully aroused to these matters, there is hope that more of value may yet be given to science from this locality.

A visit to the Taungs or rather Buxton quarry which yielded, late in 1924, the high-class anthropoid ape announced in February of this year by Professor Dart (*Nature,* Feb. 7), revealed also most interesting conditions from the standpoint of geology, paleontology, and anthropology. Here are remnants of a vast plateau, eroded in the middle by a river to a shallow valley with an escarpment of long cliffs on each side. In the western escarpment, in ferruginous shales, is an ancient basin filled with remarkably pure limestone. This limestone in turn, through water action, had become honeycombed with crevices and caves, and in these caves lived, and especially went to die, ancient baboons and also the recently found anthropoid ape the existence of which, so far south, has never been suspected. These remains became covered with sand blown in from the Kalahari Desert. This sand was in turn permeated with water carrying lime in solution, forming hard rock in which the remains of the ancient creatures are enclosed; and here they appear in the stone as this is blasted. This site is by no means exhausted, at least as far as the smaller apes are concerned. But to get to the fossils, a man must climb with the help of a rope a 60-foot vertical cliff, and thrusting his foot into crevices, must hammer off piece after piece of the hard rock which contains the remains. In this manner Dr. Hrdlička found five baboon skulls, only one of which however could be preserved. Other fossils besides those of baboons have been found in this quarry—turtles, crabs, large eggs and bones.

Dr. Hrdlička examined the large fossil skull at Johannesburg University where it is deposited in Professor Dart's laboratory.[1] It belongs to a species of anthropoid ape of about the size of a chimpanzee and evidently related to this form, though there are certain differences, especially in the brain. These differences suggest that this ape may possibly have been somewhat superior to the chimpanzee and nearer to the human. But it is not necessarily a form that stood in the direct line of the human phylum.

In "paleoliths," South Africa is rich. They may be found in favorable spots along the sea shore; in the gravels, banks, and vicinity of rivers; and they are common in caves. They present forms rather more like those of India than those of old western Europe; but here and there, are also close resemblances to the earlier or later European types. The question of the antiquity of these implements has not yet been satisfactorily worked out as a great many are found on the surface and are plainly recent; others may be ancient. That not all the sites where such implements occur and have hitherto been regarded as ancient, are of that nature, was seen along the Zambesi

---

[1] Those in South Africa whose aid in Dr. Hrdlička's work is hereby specially and thankfully acknowledged are: Professor Raymond A. Dart and many of his colleagues at the Johannesburg University; the officials of the Broken Hill Development Company, Northern Rhodesia; those of the Northern Lime Company, Bechuanaland; Mr. Neville Jones of the London Mission, near Bulawayo; Professor M. R. Drennan at the Cape Town University; and Mr. Dewitt C. Poole, U. S. Consul General at Cape Town.

Fig. 79.—The western wall of the Taungs quarry. The dark spot in the center is the opening to what remains of the stalagmite cave. On the left is the 60-foot wall of limestone, from about the middle of which came the anthropoid skull. On the extreme right is seen the semi-consolidated filling of a great old cavity in the rock.

Fig. 80.—Northern wall of the Taungs quarry. A darker patch slightly to the right of the center and mid-way between the face and top of the cliff shows the filled-in tunnel in which Dr. Hrdlička found five fossil baboon skulls.

on both sides of the river at Victoria Falls. Here stone implements were reported as occurring in the ancient gravels of the river, deposited along the sides of the stream before the formation of the falls. A three-days' examination of conditions, in company with two Americans, a South African engineer and some negroes was sufficient to show that the cultural remains here extend over a considerable distance along both sides of the river, are numerous, superficial, and in all probability not very ancient. A good-sized collection of the worked stones was secured for the National Museum.

The Bushmen and the Strandloopers whose remains are being found in shell heaps and in caves along the southern coast of Cape Colony, were apparently identical, judging from the osteological evidence that could be seen, and both show a strong affinity with the Hottentots. And all the essential characteristics of the three, outside of stature and muscular development, appear to be radically connected with the negro.

Dr. Hrdlička has returned deeply impressed with the opportunities for and the need of anthropological research offered by all these distant parts of the world, and the openings everywhere for American cooperation. The story of man's origin, differentiation, spread and struggle for survival, is evidently greater, far greater than ordinarily conceived, and a vast amount of work remains for its satisfactory solution.

A brief stop on the return journey was made in England, where, thanks to the courtesy of Sir Arthur Keith, of the Royal College of Surgeons, the precious Tasmanian cranial collection of that institution could be examined. Here also, thanks to those in charge of the Department of Geology, British Museum (Natural History), it was possible to examine once more the Rhodesian originals.

ARCHEOLOGICAL INVESTIGATIONS AT PUEBLO BONITO AND PUEBLO DEL ARROYO, NEW MEXICO

Under the auspices of the National Geographic Society, Mr. Neil M. Judd, curator of American archeology, U. S. National Museum, continued, during the summer months of 1925, his exploration[1] of Pueblo Bonito, a prehistoric communal village in northwestern New Mexico. The extensive excavations inaugurated at this particular ruin in 1921 had been concluded by the autumn of 1924. In course of the

[1] Smithsonian Misc. Coll., Vol. 72, Nos. 6 and 15; Vol. 74, No. 5; Vol. 76, No. 10; Vol. 77, No. 2.

FIG. 81.—The methods employed in excavating Pueblo Bonito varied as local conditions changed. Rooms enclosed by high walls offered greatest difficulty. In the above view, taken in August, 1924, debris first turned with hand shovels is being dragged by horse-drawn scrapers to waiting tram cars for transfer to, and deposition in, the deep arroyo visible just beyond the Expedition's camp. (Photograph by O. C. Havens. Courtesy of the National Geographic Society.)

work, however, traces of earlier structures underlying the floors of the great pueblo were frequently disclosed by the explorers. Such discoveries naturally required thorough understanding and, to this end, the studies planned for Pueblo Bonito in 1925 were intended to be mainly chronological and stratigraphical. While these technical investigations—by their very nature, slow and tedious—were in progress, most of the 25 Indian laborers employed by the expedition were put to work in the neighboring ruin, Pueblo del Arroyo.

This latter structure is severely rectangular except that its two ends are connected by an eastward curving series of low, one-story rooms. Since the south wing and the extramural habitations adjoining it had been excavated during the two previous field seasons, the efforts of the 1925 party were centered on the middle portion of the village. Some of the outward results of this investigation are apparent in the accompanying illustrations.

In accord with the original plan of procedure, the north wing of Pueblo del Arroyo and the curved series of rooms on the east have been left undisturbed. This decision was made deliberately and with realization that modern archeological research, no matter how thorough, is rarely conclusive; that the National Geographic Society or some other institution might, at some future time, wish to confirm the deductions of the current expedition.

In seeking last summer to establish a chronology for Pueblo Bonito several outstanding discoveries were made. Not all of these could have been anticipated. It was learned, for example, that the site occupied by this largest of all prehistoric villages in Chaco Cañyon National Monument had been utilized long prior to construction of the great pueblo itself—a pueblo whose massive architecture has won the admiration of American antiquarians generally. Vast quantities of wind-blown detritus, floor sweepings and other refuse had accumulated during the centuries throughout which Pueblo Bonito was inhabited—accumulations on which the more recent dwellings were constructed. Ten feet below the foundations of these latter houses were the broken remains of two primitive structures erected by pre-Pueblo peoples; that is, by Indians who had not yet learned the benefits of such community enterprise as is represented by the complex dwellings and obvious civic organization of later house-building tribes. The deposits which covered these primitive structures—deposits consisting of successive layers of ash, blown sand and rubbish from razed and rebuilt dwellings—gave the long-sought stratigraphic evidence by means of which the inhabitants of Pueblo Bonito could be separated.

Fig. 82.—In May, 1877, Mr. W. H. Jackson, of the Hayden Geological Surveys, visited and described the Chaco Canyon ruins. Forty-eight years later, as a guest of the Pueblo Bonito Expedition, he examined those same ancient pueblos with Mr. Judd. (Photograph by O. C. Havens. Courtesy of the National Geographic Society.)

Fig. 83.—Two venerable Navaho who, risking serious illness through telling stories before the first snow of winter, related to Mr. Judd their observations and experiences in Chaco Canyon, 70 years ago. (Photograph by O. C. Havens. Courtesy of the National Geographic Society.)

Pueblo Bonito was occupied contemporaneously by two distinct groups of prehistoric agriculturists. The explorations of the preceding four summers had established this fact beyond question. Their dwellings, their ceremonial chambers, their cultural remains were alike divisible into two classes. One of these was early; the other, late. And yet, throughout a considerable period the two were co-existent. As the evidence accumulated with each successive expedition it became the more certain that Pueblo Bonito, as it stood

FIG. 84.—From the north rim of Chaco Canyon one looked down upon a veritable maze of foundation walls that emerged from beneath the outer rooms of Pueblo Bonito and continued eastward more than 500 feet. (Photograph by O. C. Havens. Courtesy of the National Geographic Society.)

at the time of its abandonment, represented the individual and yet cooperative efforts of these two distinct peoples. Their clan organizations were obviously similar; their daily activities were probably identical; the utensils they used daily in their several households, while exhibiting no marked difference to the untrained eye, showed to the archeologist dissimilarities that could only have resulted from the industry of separate groups, each trained to its own mode of thought and self expression. The researches of the expedition during 1925 fixed beyond reasonable conjecture the truth of these earlier deductions.

Two members of Mr. Judd's scientific staff devoted their undivided attention last summer to examination of the pottery fragments collected from the individual rooms explored during the previous four seasons. These sherds were separable into various types, based on the stratigraphic evidence above mentioned. As an indication of the vast number of vessels fashioned by prehistoric potters, final tabulation of the fragments from Pueblo Bonito alone, after eliminating all possible duplicates, shows over two hundred thousand

Fig. 85.—Not the least puzzling of the many riddles connected with the foundations above pictured was the purpose served by this group of seven huge, ash-filled ovens. (Photograph by O. C. Havens. Courtesy of the National Geographic Society.)

individual pieces. These represent successive periods of occupancy as well as variations in design and technique for each such period. In assembling the information conveyed by these fragments of broken pottery more than two million sherds were handled at least twice. Lesser ceramic collections from Pueblo del Arroyo were studied with equal devotion to detail.

By a series of four great trenches the site occupied by Pueblo Bonito, together with its immediate surroundings, was laid bare. Each trench provided a cross section by means of which the Expedition staff was enabled to visualize the successive changes which

Fig. 86.—A great trench through the west court of Pueblo Bonito revealed deeply buried walls, a partially razed kiva over 50 feet in diameter and stratified accumulations of household refuse. Herein was pictured the domestic history of this ancient village. (Photograph by O. C. Havens. Courtesy of the National Geographic Society.)

Fig. 87.—The Expedition's third exploratory trench bisected the east refuse mound and ended against the outer south wall of Pueblo Bonito. More than 5 feet beneath the latter and overlain by successive strata of man-made rubbish was a huge fireplace. (Photograph by O. C. Havens. Courtesy of the National Geographic Society.)

occurred throughout the entire period of human occupancy. Hearths were encountered eight feet, and more, below the surface; evidence of frequent reconstruction, of alterations in the grouping of dwellings, of destructive conflagrations were disclosed by these cross sections.

In addition, a former channel for the diversion of flood waters—mud-laden floods that refused to stay within the man-made banks intended for them—to cultivated fields below the village was brought to light for the first time since abandonment of Pueblo Bonito a thousand years or more ago. No trace of small irrigation ditches has yet been found in Chaco Canyon and the diversion channel just mentioned tends, therefore, to confirm the opinion of the Expedition's staff and technical advisers that the ancient Bonitians practiced a system of flood irrigation not unlike that employed today by certain of the more successful Navaho farmers of northwestern New Mexico or, on a larger scale, by the Hopi, Pima and Papago Indians of Arizona.

During the course of the National Geographic Society's explorations at Pueblo Bonito and Pueblo del Arroyo, indisputable evidence of the former presence of human beings, semi-sedentary in habit, has been observed as much as 20 feet below the present valley surface. These ancient folk were far less civilized than the Bonitians but, owing to the depth at which their infrequent remains lie buried, we may not hope to learn much about them. Fragmentary remains are there but the limits of the crude culture to which they belong are still indeterminable.

There are factors that point to certain changes in the geophysical appearance of Chaco Canyon since the period of Pueblo Bonito's greatest prestige; there is accumulative evidence in support of possible climatic changes during or since that same period. The former existence of a prehistoric arroyo, 18 feet deep immediately in front of Pueblo Bonito, was finally established by the Expedition of 1925. The data already assembled suggest that this ancient water course, now completely filled and leveled over, may have rendered the cultivated fields of the Bonitians unproductive, thus forcing abandonment of this, at one time the most influential and powerful of all prehistoric pueblos in the southwestern United States.

Fig. 88.—In excavating the central portion of Pueblo del Arroyo, debris from the individual rooms was wheeled in barrows to accessible points and dragged thence to an elevated trap through which it was poured into steel tram cars. (Photograph by O. C. Havens. Courtesy of the National Geographic Society.)

Fig. 89.—By September, 1925, exploration of Pueblo del Arroyo had been completed but the original court pavement still lay several feet below the last level of human occupancy. (Photograph by O. C. Havens. Courtesy of the National Geographic Society.)

ARCHEOLOGICAL AND ANTHROPOMETRICAL WORK IN MISSISSIPPI

Henry B. Collins, Jr., assistant curator of ethnology, U. S. National Museum, was engaged during the summer of 1925 in an archeological reconnoissance and exploration of the ancient Choctaw territory in Mississippi for the Bureau of American Ethnology. In cooperation with the Bureau, the Mississippi Department of Archives and History detailed Mr. H. H. Knoblock, assistant in the department, to participate in the work.

The region selected for investigation was the eastern part of the state, the former center of the Choctaw tribe. Here are found not only the village sites known to have been occupied by the Choctaw within historic times, but also a number of prehistoric mounds similar to those found throughout the Mississippi Valley and in other parts of the South and East, denoting a still earlier occupancy of this region by either the Choctaw themselves or by related tribes.

At the time of first contact with Europeans, the Choctaw were the most numerous of all the southern Indians. They are also generally regarded as a basic type, culturally and physically, of the great Muskhogean linguistic stock. In any consideration of the ethnic problems of the South, therefore, the Choctaw must assume a place of importance, but as yet very little work has been done among them. It was decided, therefore, that operations for the summer should be confined to definitely-known Choctaw territory, devoting part of the time to exploration of historic village sites and part to the excavation of prehistoric mounds, in an attempt to establish as far as possible the relation of the two.

Mr. Collins left Jackson, May 24, to make a reconnoissance of the field and to select sites for exploration. Mounds were examined in the counties of Hinds, Lowndes, Winston, Neshoba, Kemper, Newton, Lauderdale, Clarke, and Wayne. Perhaps the most important earthwork examined was Nanih Waiya, the sacred mound of the Choctaws, in the southern part of Winston County, shown in figure 90. This famous mound, one of the largest and best preserved in Mississippi, is regarded by the Choctaw as their place of origin, and figures prominently in their legendary history.

The historic village sites visited were Holihtasha, Yanabi, East Yazoo, Shomo Takali, and Ibetap Okla Iskitini, all in Kemper County; Halunlawasha and Kastasha in Neshoba County; Coatraw in Newton; Coosha in Lauderdale; and Yowanne in Wayne County.

The work of reconnoissance was concluded early in June and excavations were begun on a group of eight small burial mounds on the farm of Mr. Lawrence Slay near Crandall, in Clarke County. These mounds averaged about 30 feet in diameter and were for the most part unstratified. Skeletal remains, ranging in number from one individual to fifteen or more, were found in each mound. Evidences of cremation were observed in several of the mounds and in one of them, resting on a thick layer of charcoal, was found a compact mass of broken and calcined bones, representing the remains of a dozen or more individuals. As is usual in mounds of this type, the bones were in a very poor state of preservation even

Fig. 90.—Nanih Waiya, sacred mound of the Choctaw, in Winston County, Mississippi. According to Choctaw tradition, Nanih Waiya is the birth place of the tribe, the first man and woman having come up from the underworld through the mound.

where they had not been subjected to fire. Figure 91, showing the bone layer in one of the mounds, will afford some idea of the condition in which the bones were found. From the few accounts we have of the early explorations into the Choctaw country, it is known that the burial customs of this tribe during the eighteenth century and earlier were different from those practiced at a later date. The dead were formerly placed on a platform erected for the purpose, from which, some months later, they were removed by the "bone-pickers," whose official duty it was to carefully scrape and clean the bones. These were then placed in cane hampers and deposited in the bone house, one or more of which was to be found in each Choctaw village. The bones were later taken from the bone house and carried some distance from the village, where they were finally buried in a promiscuous heap and a mound of earth erected over them This custom was not confined to the Choctaw alone but was quite widespread as is evidenced by the presence of similar mounds or ossuaries in many localities of the eastern United States and Canada.

Following the excavation of the Crandall mounds, work was begun on a much larger mound of a different type on the property of Dr. B. J. and Mr. R. L. McRae, near the town of Increase. Approximately one third of the mound was excavated by trenching, and while no skeletal material and only a few artifacts were found,

Fig. 91.—Layer of human bones in small sand mound near Crandall, Clarke County, Mississippi.

the peculiar stratification seemed to warrant as thorough an examination as was made (see fig. 92). This stratification consisted of a series of brilliantly colored sand layers, yellow, brown, orange, blue-gray, and pure white, from which, at the center of the mound, there suddenly arose a dome-shaped structure of compact yellow clay. This clay dome and the succession of colored sand strata probably had a ceremonial significance, having been placed on the floor of what had very likely been a temple, the site of which was later covered over with a mound of earth, on the top of which, still later, there probably stood a temple or council house. Colored sand strata in much the same arrangment have also been found in the effigy mounds of Wisconsin.

Within this small inner mound or clay dome was found a rectangular ornament of sheet copper and silver enclosing a core of wood, shown *in situ* in figure 93. Both copper and silver are shown by analysis to be native American, probably from the Lake Superior region. Silver and copper ornaments practically identical to this have been found in small numbers in Florida, Tennessee, Ohio, and Michigan.

Thin, flaked knives, struck with a single blow from flint cores, were found both in the mound and in the adjoining field. These

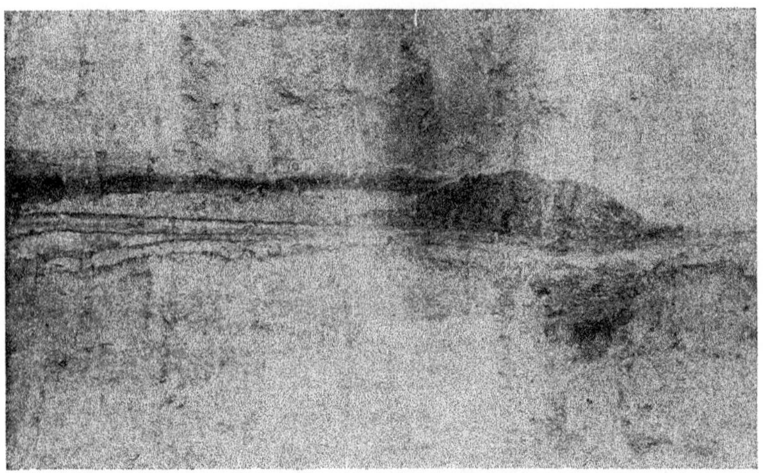

FIG. 92.—Stratification of colored sand and clay in McRae mound, near Increase, Clarke County, Mississippi.

are identical in every respect with the flaked knives from Flint Ridge in Ohio which, while abundant in the Ohio mounds, are rarely found in other localities.

With the most significant features of the McRae mound so strongly suggesting northern influence, we must conclude that the builders of this Mississippi mound maintained at least a close trade relationship with the northern tribes. While undoubtedly the many mounds and various other earthworks of North America were built by Indian tribes of diverse stocks, there are certain resemblances between even the most distant of them which suggest a contact something more than sporadic.

Work was next begun on a group of seven small mounds on the property of Mr. J. M. Kettler near Hiwannee, in Wayne County.

In size, construction, and contents they were similar to those previously excavated at Crandall.

With the completion of the mounds at Hiwannee in early July, explorations were started on the site of the historic Choctaw village of Coosha, near Lockhart in Lauderdale County, on the property of Mr. W. E. Frederickson. The cemetery was located and ex-

FIG. 93.—Ornament of sheet copper and silver in McRae mound, Clarke County, Mississippi.

amined but the burials were comparatively recent, dating probably between 1800 and 1830, by which time the Choctaw had lost much of their native culture. Except for the presence of mortuary offerings, consisting for the most part of beads, porcelain, and cooking utensils, the burials were typical of those of the whites.

It is believed that the archeological work described above reveals something in the nature of chronological cross section of the three most important phases of Choctaw culture history. At the Coosha village, conditions were found which were typical of the last phase of Choctaw history, when their native culture was rapidly breaking down and assuming the essential features of the dominant white civilization. The small burial mounds at Crandall and Hiwannee, dating perhaps from the first half of the eighteenth century, belong in all probability to the period immediately preceding, when mounds were still built, but only for purposes of burial. The earliest of the three stages, perhaps antedating by centuries the smaller burial mounds, is represented in the McRae mound.

FIG. 94.—Choctaw Indians engaged in a game of native ball at Philadelphia, Mississippi. There are at the present time about 1,100 Choctaws in the state.

The explorations at Coosha brought to a close the archeological work of the summer, and Mr. Collins and Mr. Knoblock proceeded to Philadelphia, Miss., where measurements and observations were made on the living Choctaw, of whom there are still about a thousand in the state. Measurements were secured on 58 adults, some of which undoubtedly have white or negro blood.[1] The cooperation of Mr. T. M. Scott, of the Indian Agency at Philadelphia, was very helpful in this work. Particular thanks are also due to Mr. Weaver Bridges of Philadelphia and to Mr. H. B. Cole, of Quitman, both of whom rendered valuable assistance in many ways.

[1] The results of this study have been published in the American Journal of Physical Anthropology, Vol. 8, No. 4, 1925.

Fig. 95.—Group of Choctaw Indians assembled for a game of native ball at Philadelphia, Mississippi, July, 1925. Measurements and physiological observations were taken on a number of the Indians in this group. (Photograph by Knoblock.)

ARCHEOLOGICAL STUDIES OF THE WUPATKI NATIONAL MONUMENT

So little was known in 1853 of the physical features of the region between the Zuñi and the Colorado Rivers, that in that year Captain Sitgreaves, of the U. S. Corps of Topographical Engineers, was sent by the Secretary of War to follow the Zuñi River to its junction with the Little Colorado and on to Yuma. One object of this trip, which now seems almost ludicrous, was to determine the "navigable properties" of the two rivers. It did not take long to determine this. Captain Sitgreaves records that between Camps 13 and 14, near the Great Falls of the Little Colorado, "all the prominent points [were] occupied by the ruins of stone houses of considerable size, and in some instances of three stories in height. They are evidently," he writes, "the remains of a large town, as they occurred at intervals for an extent of eight or nine miles, and the ground was thickly strewed with fragments of pottery in all directions."

In 1900 these ruins, then locally called the Black Falls ruins, were described and first figured by the present Chief of the Bureau of American Ethnology, Dr. J. Walter Fewkes, who was so much impressed by their magnitude that he recommended they should be preserved by the National Government.[1] In 1925, his hope was realized and they were declared a National Monument by Presidential proclamation.

Other literature on this monument is meagre, but the archeology of the Flagstaff region has lately been studied by Professor and Mrs. Colton, who have published valuable material on the "small houses" of this region.[2] The construction of a road recently laid bare a prehistoric cemetery at Young's Canyon, 18½ miles east of Flagstaff, and the objects there brought to light have been acquired and a description of them published.[3]

At least two types of large ruins occur in this area, the former of which is represented by the Citadel, figures 96 and 97. The second

---

[1] A cluster of Arizona ruins which should be preserved. Records of the Past, Vol. III, Pt. 1, pp. 3-19, Washington, 1904.

[2] The little known small house ruins in the Coconino forest. Mem. Am. Anthrop. Assn., Vol. V, No. 4, 1918.

Did the so-called cliff dwellers of central Arizona also build hogans? Am. Anthrop., Vol. 22, 1920.

[3] An archeological collection from Young's Canyon, near Flagstaff, Arizona. Smithsonian Misc. Coll., Vol. 77, No. 10.

FIG. 96.—The Citadel and terraced gardens, and small ruins near them.

FIG. 97.—Rooms and fallen walls on top of Citadel.

type, figure 101, contains rectangular ruins of slabs of stone masonry erected on the rims of shallow canyons, which were apparently used as habitations.

The most imposing ruin is the so-called Citadel, which from a distance resembles a volcanic peak, the cone of which is capped with walls made of sandstone and lava rocks, the sides of the elevation being strewn with these blocks which have fallen from the walls at the apex. Figure 96 shows this Citadel from the side, and figure 97 indicates the broken-down walls and the general character of the fallen masonry on the surface. The Citadel had many rooms apparently built around a central open space or plaza, the masonry being composed of lava blocks and sandstone slabs. The numerous stones down the sides of the Citadel now and then are arranged in rows, calling to mind retaining walls or trincheras so common on hillsides of southern Arizona and the northern states of Mexico.

Tcuaki (Snake House). The ruin here first called Tcuaki (figs. 98, 99), is situated about 30 miles from Flagstaff, Arizona, and is in reality composed of two clusters of rooms connected by a long wall, possibly modern, which follows the summit of a sandstone ridge.[1] The corners of the walls have fallen down the sides of the ridge, covering the roofs of the subterranean rooms at the base of the cliffs. In the year 1900, rafters and remains of these roofs were well shown and a few fragments of roofs still remain *in situ,* as if lately abandoned. As in all these ruins, there were numerous basal rooms from which many specimens were obtained, but the best objects were found in the rooms above the surface or those on the summit of the ridge. The rooms were two stories high, the floors of the lowest story being generally buried under fallen debris.

Alaki (Horn House). The ruin called Alaki or Horn House, also of the Great House dormitory type, is characteristic of the Wupatki monument. It is shown in figure 101. There are massive buildings above and the subterranean basal walls below. In the neighborhood of Alaki there are several small stone subterranean houses, each generally with a single room, possibly pit dwellings.

The canyon on the rim of which Alaki stands becomes "blind" a few hundred feet from the ruin and is exceptional in having excavated in its walls small rectangular cubby holes formerly closed by slabs of stone, all of which have been broken and many have been destroyed. These recesses were apparently used for storage cysts, and are not confined to this canyon.

---

[1] Records of the Past, *op. cit.*

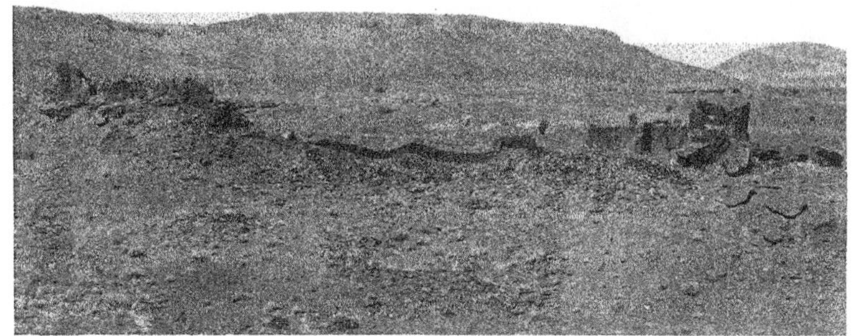

FIG. 98.—Tcuaki, showing the two large buildings and connecting wall, which was possibly constructed by sheep herders. These and neighboring buildings are called Wupatki in the President's proclamation.

FIG. 99.—Tcuaki.

Fig. 100.—Typical one-room house.

Fig. 101.—Alaki.

Fig. 102.—Ruin near Alaki.

These ruins are referred to in the map in the President's proclamation as belonging to the Citadel group.

Wukoki. Wukoki, figure 103, the tallest building of the Black Falls buildings, belongs to the second type and is the highest and best preserved of these buildings. It presents some of the finest masonry on the monument, and almost entirely covers the top of the small mesa. From a distance across the sands, its prominent walls resemble a castle, its base rising about 15 feet above the neighboring plain.

General features of the walls. The walls of the large buildings that form the most striking feature of the Wupatki monument

FIG. 103.—Wukoki (near Wupatki Monument). It is hoped that Wukoki may later be included in the Monument.

belong to the fine masonry of the Southwest. They are constructed of a filling faced on both surfaces with slabs of dressed stone, often more or less artificially worked and laid in courses. But, as in the construction of all masonry in the Southwest, some of the first principles of masonry are neglected. For instance, the corner binding of stones is neglected and there is seldom an overlapping of the same in order to tie the component stones of the wall together. The corners especially show weakness in this particular for they are seldom fitted together. The stones are sometimes roughly hewn, the alternating courses being often pitted on the surfaces, but also occasionally decorated with incised geometrical designs. The doors and entrances to the rooms are generally lateral and the few windows are small, circular or rectangular. Two or three of these buildings contain but a single room, others are multi-chambered, and in the largest buildings three stories can be distinguished.

The roofs of all the rooms are simple, flat coverings, supported by several large logs extending from one wall to its opposite, across which are laid willow twigs supporting a layer of cedar bark, the whole being covered with clay. There are evidences of fireplaces centrally placed, and in a few instances ashes were found in one corner. In one room there is an elaborate chimney, possibly modern, communicating through the walls with a fireplace, extending a few feet above the highest masonry. Although the walls are generally constructed of sandstone, they also show alternate layers of lava rock which, contrasted with the intervening courses of red sandstone, add a picturesque appearance to the whole. As is customary in pueblo construction the stones are laid in adobe in which often occur imbedded pottery shards and fragments of stones. The marks of human fingers also appear in the mud plastering, especially where the walls are protected by overhanging cliffs. The walls of the rooms were seen in one case to be plastered on the inside with adobe decorated with red pigment, like the dadoes of the modern Hopi houses.

At the base of the cliffs there are several walled-in shrines or rude rock inclosures, generally containing water-worn pebbles or fossil shells which were no doubt religious offerings. These also contained prayer sticks, figure 105, *b, b'*, and in one instance a wooden image of human form, figure 105, quite unlike, however, any idols which have previously been described. Clusters of pictographs generally of geometrical forms adorn the walls of the canyons on which the larger buildings have been built.

The characteristic features of the Wupatki buildings are the small rectangular stone-walled rooms of good masonry and the subterranean rooms at the base of the low cliffs on which they stand. No kivas have yet been excavated; those shown on the map are doubtful ceremonial rooms.

Typical specimens of Wupatki Monument pottery resemble those found in the cemetery at Young's Canyon.[1] They belong to a prepuebloan type found throughout Arizona underlying the more brilliant Sikyatki ware and the related Homolobi-Chevlon forms. The

[1] Smithsonian Misc. Coll., Vol. 77, No. 10.

FIG. 104.—Type designs on inside of bowls from Wupatki, southern Tokonabi ceramic culture.

a. Key figures covering whole inside area.

b. Alien designs.

c. Central unfilled space surrounded by serrated bands.

d. Central unoccupied space and peripheral border. Quadrate design in which square white figures with black dots predominate.

e. Typical quadrate design with Greek fret border.

f. Quadrate design in white on a black background. The square central figure is characteristic of Wupatki ware.

prevailing ceramic types are corrugated, black on white, and polychrome, dull red with shiny black interiors. The food bowls often have out-curve rims and single loop handles. The colander form, or food bowl perforated with holes, occurs both at Wupatki and at Marsh Pass.

The designs on the well decorated pottery, both black on white and polychrome, figure 104, are similar to those from Tokonabi. These decorations cover the whole interior of the bowls except the very center, and have a quadrate arrangement, in some examples of which rows of white squares with black dots predominate. The flaring rims of food bowls are adorned with special geometrical patterns forming the framework of the main interior design.

The dead were inhumated or cremated. A typical human interment found in the sands near the ruin, Wukoki, was enclosed in a cyst made of stone slabs set on edge, covered with another slab of rock, resembling those in the Marsh Pass region. In one of these graves was a skeleton with two pendants made of lignite and turquoise mosaic like Hopi ear rings, figure 105, *c*, as fine specimens of jewelry as any known from the Little Colorado region. Several mortuary vessels were found with the dead, and also remains of fabrics, apparently kilts or garments.

Additional facts of a comparative nature are much needed in order to explain the difference between decorations on the Wupatki pottery and the beautiful designs on the Homolobi-Chevlon ceramics from ruins higher up on the Little Colorado. There is also a difference between the pottery and architecture of this region and those of Young's Canyon a few miles away. So far as known it would appear that the Homolobi-Chevlon culture, which is allied to that of the Hopi ruins, Awatobi and Sikyatki, and to that of the Jeddito and Bitarhootci valleys, although late prehistoric, was more modern than that of Wupatki.

As elsewhere suggested, both architecture and ceramic designs from Wupatki are practically the same as those from Marsh Pass, probably indicating that the latter is an identical and synchronous culture area which in Hopi Snake legends is called Tokonabi (Kayenta). This far-flung culture area is quite unlike that which occurs higher up on the Little Colorado at Homolobi and Chevlon, the artifacts of which are more closely related to the Sikyatki and Awatobi and probably is more modern, as Hopi migration legends state.

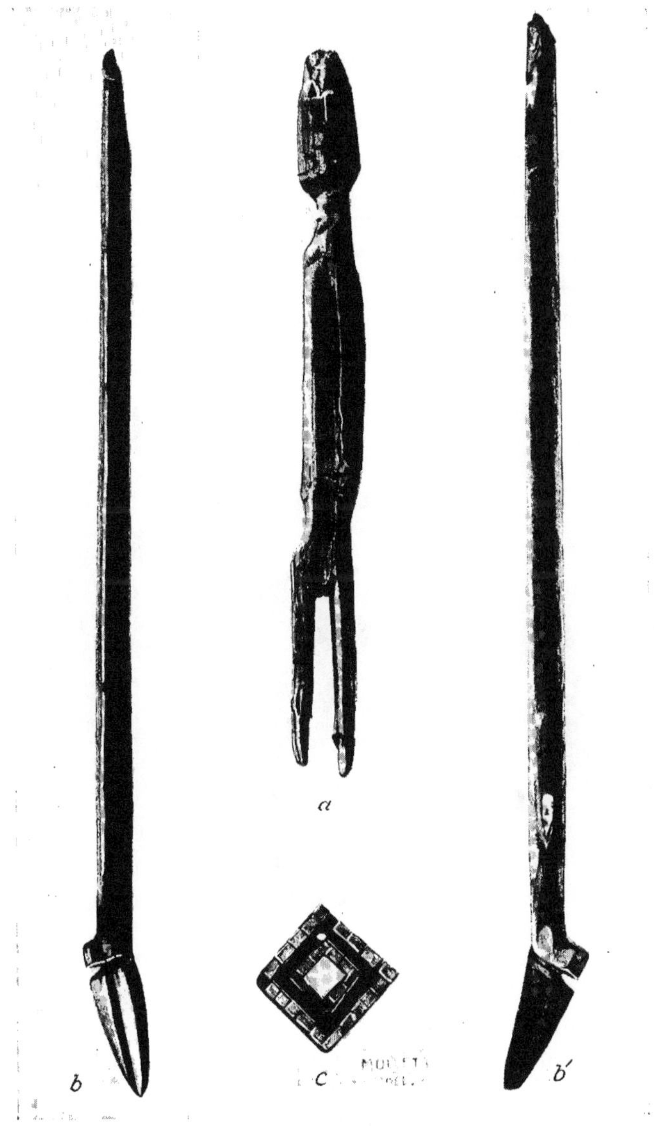

FIG. 105.—*a*, Wooden image from shrine near Tcuaki. *b*, *b'*, Possible shaft of throwing stick from near Tcuaki. *c*, One of two mosaic ear pendants made of turquoise and lignite set on lignite base from stone grave near Wukoki. The grave contained a large decorated black and white vase and fragments of cloth.

RESEARCHES ON THE ARCHEOLOGY OF SOUTHERN CALIFORNIA

At the close of June, 1925, Mr. J. P. Harrington, ethnologist of the Bureau of American Ethnology, proceeded to Santa Barbara, California, to continue his researches on the Mission Indians. Many Indian rancheria sites of this region were visited in the course of his work. In the Santa Ynez valley alone, more than 40 ruined villages were inspected, and in the Otay and Simi valleys some 30 more, at several of which extensive excavations were made.

Pictographs were discovered and photographed; also many rocks which represent mythological personages or form the crucial land-

Fig. 106.—Medicine rock surrounded with cacti.

mark of ancient legends. Spirit footprints on the rocks, of gigantic size, were said to have been made by the "first people" when the earth was still soft and muddy.

At San Marcos the boulders on a hillside represent the warriors of a mythic battle; some are standing with the blood from wounds running down their sides, seen as stains on the rock. A site was also visited where two boulders are situated six feet apart. Indian boys used to attempt to jump from one to the other, and if they succeeded it was a sign that they would be able to jump around the mountains in later years without skinning their legs. A medicine rock (fig. 106) was also visited, a symmetrical pinnacle of stone surrounded with a circle of cacti in such a way as quite effectually to keep away intruders.

At the Rincon, Mr. Harrington discovered the ruins of a medicine house formerly used by the island wizards for secret ceremonies. This remarkable structure consists of a natural cave chamber, blackened with the smoke of former fires. On the east side is a circular enclosure 18 feet in diameter, the walls of which, made by piling up rocks to a height of about three feet, are still partly standing. Pine trees were formerly laid across from the top of this wall to the roof of the cave, and on these thatch was placed. An extended study was made of this temple, and the cave floor was excavated. There was a tradition that if a common Indian came on this place by mistake he would be struck dead and that if an innocent person happened to stroll near, thunder, lightning and rain would immediately result.

Since little is known of the California Indian house of this section, it was thought desirable to construct a replica of a native wigwam and to photograph step by step the building of it (fig. 107). In all more than 40 exposures were made as the work progressed, thus securing a record for future description. The wigwam is circular and about 13 feet in diameter and 7 feet high. The site for the house was selected and the ground cleared and leveled. Using a short bar of willow for digging the post holes and the hand for scooping out the earth, eight slender willow poles, 15 to 20 feet high, were erected in the form of a Greek cross. The pairs of poles opposite each other were then lashed together to form arches 7 feet high. Other uprights were then added until the poles formed a complete wall and were only "a short step apart." Smaller poles, called "latas" in Spanish, were then lashed on the outside of the uprights at intervals of about a foot (fig. 108). The thatching material—tule, carrizo, brakes, or grass—was laid on in tiers, the lowest tier standing upright with the butt end in the ground to form a firm base for the wall and subsequent tiers upside down, as the inverted leaves shed water better. The material used for lashing this house was mescal fiber. Outer "latas," opposite the inner ones, hold the thatch firm, and it is sewed in place by means of a great needle of willow wood which is poked through the thatch, two workers being required at this stage of the construction. The thatching, when compressed, is only four inches thick, and is impervious to wind and rain. At the top of the house an ample hole is left for the exit of smoke. A fire place with pot-resting stone occupies the center of the earth floor.

Fig. 107.—Modern wigwam of Southern Californian Indians; constructed under Mr. Harrington's direction.

Fig. 108.—Framework of wigwam of Southern Californian Indians; constructed under Mr. Harrington's direction.

The doorway of the house is only a yard high and about two feet wide. It is closed by a door made by tying together an oblong frame of willow poles and weaving small willow twigs to fill this frame. Another type of door is a tule mat, stood on end. The Indian word "to lock the door" really says "to tie the door," for the only protection from intruders when Indians went away and left the house was such a mat or frame, carelessly tied—and Indian etiquette, which said that a stranger should not enter. Similar mats were used for sleeping and for sitting on the floor, and are surprisingly warm when used for such purposes. The Indians slept with nothing between themselves and the cold earth but one of these "petates de tule."

Some of the Indians had their houses lined on the inside with similar tule mats, much as we use wallpaper, but in the poorer houses, thatch and twigs showed on the inside. Between the poles and thatching, all kinds of Indian utensils and furnishings were stored out of the reach of children and in sight when needed for use.

On completing the Indian house, Mr. Harrington started in the latter part of October on an expedition to the Cañada de las Uvas. This trip proved rich in discovery along several different lines. Many of the archeological sites visited had not been touched since Indian times, and Mr. Harrington found without difficulty the old hut circles on the surface of the ground, either marked by rings of rocks, formed by the Indians clearing the surface for the hut, or by rings of raised earth which mark the former walls.

The first work was at the village of Sikutip, where the Indian huts were formerly clustered at the southwest border of the cienega. There are many interesting rocks and caves in the neighborhood, and four minor springs were located. Only a mile away is Choriy, another large village.

The largest village discovered was that known as Milyahu (fig. 109). This differed from the other sites in being located on a detached rocky hill which has the appearance of a great towering citadel when seen from the arroyo. The little Indian wigwam circles, varying in diameter from 12 to 20 feet, were found all over the summit of this hill. The water used by the people of Milyahu is supplied by a spring which gushes forth from the sandy bed of the dry arroyo opposite the middle of the Indian town. All the water had to be carried up a cliff 75 feet or more in height, reminding one of the practice at some of the Pueblos. Figure 110

Fig. 109.—The great citadel of Milyahu, viewed from downstream. The spring is in the bed of the arroyo at the left. The entire top of the hill was covered with thatched jacales, the pits for which have not been disturbed to any extent since the time the village was occupied.

Fig. 110.—A wigwam circle at Milyahu, perfectly preserved except that a rock or two has rolled inside. These circles were easy to trace and will throw much light on the average size of the California Indian jacal.

shows a typical wigwam circle on top of Milyahu. The Milyahu cemetery has unfortunately been washed away by the arroyo.

On a hill on the Santa Maria ranch, Mr. Harrington discovered an Indian fortification wall (fig. 111) which is evidently in much the same condition as when it was abandoned. The rocks have been piled to form a parapet five feet or more in height, the wall forming a corral around the top of the hill. It was used for outlook purposes.

FIG. 111.—Old Indian fortification works on the crest of the hill west of the Santa Maria ranch house. The wall of piled up rocks forms a parapet around the top of the hill, and was used as a lookout by the ancient Indian inhabitants.

Besides obtaining many unique traditional songs, Mr. Harrington devoted special attention to pictographs, both photographing them and tracing them off full size on thin paper, so that they can be reproduced in their natural colors.

STUDIES OF THE FOX AND OJIBWA INDIANS

Dr. Truman Michelson, ethnologist, Bureau of American Ethnology, left Washington about the middle of June, 1925, to renew his researches among the Fox Indians near Tama, Iowa. There he verified a number of texts previously obtained on various ceremonials; he also procured additional information on several sacred packs, as that of Pyätwäyā, belonging to the Thunder gens, and that called ke'tcimī'cam^{mi} in Tepashit's care belonging to the Thunder gens; and the one called Sāgimā 'kwäwA, which belongs to the Bear gens, formerly in the possession of Pushitonequa, the last recognized chief of the Foxes. He obtained also much information on two other sacred packs, one belonging to the Thunder gens, the other to the Bear gens, the existence of which were hitherto unknown. The information on all these packs was obtained mainly

FIG. 112.—Da lo tti wa, a Fox woman, gathering flags to make mats. (Photograph by Grace Scott.)

in the current Fox syllabary and English paraphrases secured from Horace Poweshiek and Harry Lincoln. During his visit, a Winnebago enrolled among the Foxes was injured in an automobile accident, and Dr. Michelson had the rare opportunity of witnessing a sweat-lodge performance and listening to the Winnebago prayers and songs accompanying it.

Dr. Michelson proceeded on August 21 to Odanah, Wisconsin, to obtain some first-hand information on Ojibwa gentes. The following live at or near Odanah: Marten, Loon, Eagle, Bull Head, Bear, Sturgeon, Great Lynx, Crane, Lynx, and Chicken. The chief belongs to the Loon gens; a "head-man" to the Bear gens; and the messenger to the Marten gens. In sharp contrast with the Foxes, the gentes apparently have no special rituals. Exogamy is still practiced for the most part. It may be noted that cross-cousins as well as parallel cousins are not allowed to marry. It should be stated that, as with the Menomini, the offspring of a white man and an Ojibwa woman belongs to the Chicken gens. Although almost all of the Ojibwa at Odanah are Christians, Dr. Michelson found

FIG. 113.—Some birch-bark dwellings of the Ojibwa. From a postal card purchased at Ashland, Wisconsin. The scene obviously is at the pageant held at Apostles' Islands.

that they have a vivid recollection of their ancient religious rites, and he obtained detailed information on some of these. On September 2, Dr. Michelson proceeded to the vicinity of L'Anse, Michigan, where he located a family of Stockbridge Indians, but none spoke their own language. He observed that the Ojibwa dialects spoken at Odanah and in the vicinity of L'Anse differs markedly in some respects from the western dialects, to judge from Jones' Ojibwa Texts. On September 12, Dr. Michelson went to Mt. Pleasant, Michigan, where he began a preliminary survey of the Ojibwa, Ottawa, and Potawatomi of the neighborhood.

ETHNOLOGICAL RESEARCHES AMONG THE IROQUOIS AND CHIPPEWA

Mr. J. N. B. Hewitt, ethnologist, Bureau of American Ethnology, left Washington in May, 1925, for field duty and resumed his studies among the Six Iroquois Nations or Tribes, namely, the Mohawk, Seneca, Onondaga, Oneida, Cayuga, and the Tuscarora, all dwelling on the Haldimand Grant on the Grand River in Ontario, Canada.

In previous years Mr. Hewitt had recorded with great care from the dictation of the most intelligent living statesmen, ritualists, and counsellors, voluminous texts relating to the complex institutions of the League. Because the war of the American Revolution had badly disrupted the tribes of the League and the League itself, Mr. Hewitt inevitably encountered variant versions of many portions of the traditions, rituals, chants, and addresses relating to the organization, constitution, ordinances, and regulations of the League, and recorded these variant versions. In the furtherance of this task, Mr. Hewitt again took up the literary study, interpretation, and translation of the texts embodying the laws, ordinances, and the regulations, the chants and the rituals of condolence for the dead *rotiyaner* and *koñtiyaner* (the native name of the federal counsellors), and the installation of the *rotiyaner* and the *koñtiyaner* (elect) (who constituted the councils of the tribe and of the League, in addition to the chiefs). The first is the masculine, and the second the feminine, form of the noun.

The organic institutions of the League of the Iroquois for over one hundred and fifty years have been subject to the action of various destructive external and internal forces, and so it is that many of the most distinctive institutions of the League have long been inoperative through the failure of the leaders to execute them.

The Governor General in Council by an Order in Council on September 17, 1924, abrogated the organic institutions of the Canadian part of the League. This crisis in the affairs of these tribes arose because the government of the League of the Iroquois had become such a travesty of the complex institution established by the great prophet-statesman, Deganawida, and his astute collaborators, that it failed to function organically.

By the aid of Mohawk informants, Mr. Hewitt was enabled to resolve the lexical and the grammatic difficulties of the Mohawk texts of certain important rituals of the Council of Condolence for deceased *rotiyaner* and the installation of the *rotiyaner* elect and to translate into free English speech one of these rituals and to discover the reason for its most peculiar name. The title of this ritual is *Ka'rhawe^{n}' hrā'to^{n}'* in Mohawk, and *Ga' hawe^{n}' hä' dï* in Onondaga, meaning "Cast or Thrown Over the Grand Forest." To learn the cause of giving so peculiar a name is to learn one of the processes of constructing rituals.

Legislative or ceremonial action is taken by the tribe only through the orderly cooperation of the two constitutive Sisterhoods of clans, commonly called Phratries in early ethnologic writings. This dualism in the highest units of organization was based originally on definite mythic concepts. Briefly, the one Sisterhood of tribes symbolized the Female Principle or Motherhood in Nature, and the other, the Male Principle or Fatherhood in Nature.

The Sisterhood of tribes functioned by the independent action of its constituent institutional units—every several tribe. In turn, every tribe functioned through the units of its own internal organization—each several clan, to execute its prescribed part in the larger federal action, which otherwise could not be authentic and authoritative; so that a clan, or an individual in a clan, in special cases involving personal rights, might prevent vital federal action. So personal rights were abundantly safeguarded.

In addition to the chant called "Cast Over the Grand Forest" mentioned above, the most distinctive one of the Council of Condolence and Installation of the League of the Iroquois is that which is designated as "The Seven Songs of Farewell." This is intoned in behalf of the deceased member of the Federal Council which, as a Council of Condolence and Installation, has met to condole his death with his kinsmen and to install his successor. These two chants are respectively divided into two portions. The first six of the "Seven Songs of Farewell" are followed by the first five paragraphs of the chant "Cast Over the Grand Forest." A veil of skins divides the Mother from the Father Side during the chanting of the "Farewell Chant."

By a searching study of all symbolic terms and phrases occurring in the chants of these rituals, Mr. Hewitt was able to identify the phrase, "the veil of skins" with the other phrase "the Grand Forest." The "Grand Forest" represents ritualistically the totality of the forests which intervene between the lands of the Mourning Side of the League and those of the other side. Mr. Hewitt also made a free English translation of the chants, "Cast Over the Grand Forest," and "The Seven Songs of Farewell."

Mr. Hewitt made a reconnaissance trip to the Chippewa of Garden River, Canada, for the purpose of expanding his knowledge of certain Chippewa texts, recorded in 1921 by him from the dictation of Mr. George Gabaoosa of Garden River, Canada, and also to obtain data in regard to the derivation of two very important proper names, namely, Chippewa and Nanabozho (appearing in literature also as Nenabojo, Menaboju, and Wenaboju).

The name Chippewa is the generic designation of a historically important group of Algonquian tribes of the northwestern United States. Various unsatisfactory derivations have been given to it, and it appears in literature with no less than 97 variant spellings.

For two years Mr. Hewitt has had in mind a definition of the name Chippewa which brings out one of the distinctive arts of these people, just as the Ottawa received their name of "The Traders" because for the moment the business of trading was then ethnically distinctive. To those who first gave the name Chippewa to these people, picture-writing was their preeminent characteristic. And the birch bark records of the Chippewa are sufficiently prominent in their culture to be noteworthy. The stem of the term may be found in the Chippewa expression, *nind ojibiwa,* meaning "I mark, write, on some object." The form *ojibiwa* used as an appellative in the plural would become *ojibweg,* which used as an ethnic appellation signifies "those who make pictographs." Mr. George Gabaoosa, of Garden River, Canada, a most intelligent Chippewa, collaborated in the derivation of this tribal name.

The present writer is not aware that any consistent meaning has been given by any other student to the proper name Nanabozho (Wenaboju, Menaboju, etc., being other spellings of it) of the Algonquian biogenetic myth. Briefly, it is the Myth of Mudjikewis, the First Born on Earth, commonly called The Story of Inābi"-oji'o' (*i. e.,* Nanabozho.) This story, which is remarkable for beauty and comprehensiveness, relates that on the shore of the great primal sea dwelt Misakamigokwe (*i. e.,* The Entire Earth Mother) and her Daughter. This Entire Earth Woman is the impersonation of the inert earth, while the daughter is the life-increasing power of the earth—the Life Mother—the Mother of all Living Things. These two personages were of the super-race of the "first people" who lived when the earth was yet new.

The Entire Earth Woman cautioned her daughter, saying, "Daughter, bend not yourself against the sun at noon-tide, because the Great Father Spirit at that time looks on you. Remember,

I command you, not to forget my words, for surely if you do, evil will befall us; since our time to increase the number of living things on the earth on which we live, is not yet fully come."

But there came an evil day when, very busy with her mat-making, and with her back unconsciously turned sunward at noon-tide, she dropped her mat-gauge on the ground and unwittingly stooped forward to pick it up. Instantly, she was seized with exhausting pains after the manner of women.

So, in due time, the daughter gave birth to a son, whom she named Inābi″oji'o' (*i. e.*, Formed, Created, by a Look). She continued bearing offspring until four other sons were born to her—all brothers of Inābi″oji'o'. In order of their birth these brothers were named Ningabeon (The West), Kiwedinese (The North), Wabanese (The East), and Shawanese (The South). After this time Misakamigokwe became Nokomis, the grandmother of all living things.

In this highly condensed and abbreviated recital of the common Algonquian myth of the Beginnings is given the key to the literal signification of the name Nanabozho (Wenabozho, etc.), or Inābi″-oji'o'. This name is based on the common Algonquian verb *w ā b*, "to see, to look," which with associated elements, expressed and understood, gives the literal meaning "created, conceived, made, through the look, the gaze (of the Great Father Spirit).

ETHNOLOGICAL WORK AMONG THE OSAGE OF OKLAHOMA

In May, 1925, Mr. Francis La Flesche, ethnologist, of the Bureau of American Ethnology, visited the Osage Reservation, Oklahoma, to continue his work among the Indians of that reservation. He was met at Nelogany station by his friend, Ku-zhi-si-e, in his big automobile and driven to his house in the Indian village about a mile from the town of Pawhuska.

The afternoon was passed by Mr. La Flesche and Ku-zhi-si-e in planning the work they were to do together. They agreed to do first that part which promised to be the most laborious, that of recording and classifying the personal names of the members of the tribe according to gentes and sub-gentes, and of giving where possible, the meaning of each name. To avoid confusion in the performance of the work, Ku-zhi-si-e suggested the use of the early annuity pay rolls which contained the names of every man, woman, and child.

Early the following morning, Mr. La Flesche and his assistant appeared before Superintendent J. George Wright of the Agency and made known to him the purpose of their visit. The Superintendent immediately put one of his clerks to work to find the record wanted. The clerk found a roll for the year 1877, which he placed in Mr. La Flesche's hands. Then he and his friend sped back to the little cottage where they began this work.

FIG. 114.—1, Coneflower, life emblem of the Peacemaker gens. 2, Haircut of the children of that gens to resemble the flower.

The two put about three weeks of steady work on the revision of the roll. The spelling of nearly all of the 1991 names found on the roll had to be corrected, and to the name of each annuitant was added the name of the gens to which he or she belonged, as, for example, Ku-zhi-si'-e, Wa-tse-tsi gens, They Who Came from the Stars.

When the work on the gentile personal names of the members of the tribe was completed, Mr. La Flesche and Ku-zhi-si-e took up the task of recording the names of the wild plants known to the Osage and their uses where they could be ascertained. This was out-door work and more agreeable than the other. One day as the two were driving over the hills, Ku-zhi-si-e called to his driver to stop. He pointed to a patch of yellow flowers which he said were called "ba-shta," hair-cut (Cone-Flower). This was the flower chosen by the people of the Peacemaker gens to be their life and peace emblem. The hair of the children of this gens was cut to resemble the sacred flower as the sign of a petition to the Power that brought it into existence, to grant the little one a long and fruitful life (see fig. 114). They drove on to a small-wooded stream where there were many aquatic plants. The *ci*^{n} *(Sagittaria latifolia)*, Ku-zhi-si-e pointed out as a food plant; the *Mi-ke-the* (cattail), the leaves of which he said were used to thatch houses as well as for medicine; the *ça-zhi*^{n}*-ga*, rush, *(Eleocharis interstincta)*, which was used for making mats to sit upon in the house, and for making the inner shrine of the sacred hawk. As the two drove along the border of a small lake, Ku-zhi-si-e also pointed out the *tse'wa-the* (water chinkapin), as a food plant.

On the last day of the exploration, the two drove to the top of a high hill where they stopped and got out of the car. After a moment's silence, Ku-zhi-si-e gave a wide sweep of his right arm and said, "All the grasses, weeds, shrubs, and trees that we see around us as far as the eye can reach, are medicine, but we know the qualities of only a few. Some plants like the *çta-in-ga-hi* (persimmon tree) serve as medicine and food. There is life in all plants and all are active. There are medicines for horses as well as for human beings."

STUDIES OF INDIAN MUSIC AMONG THE MENOMINI OF WISCONSIN

In July and August, 1925, Miss Frances Densmore, collaborator of the Bureau of American Ethnology, recorded the songs and studied the musical customs of the Menomini Indians in Wisconsin. Three classes of songs recorded among the Menomini have not been found in other tribes. These are the songs connected with games played as a "dream obligation," the songs of "adoption dances," and the songs connected with the use of packs or bundles, by individuals, for the purpose of securing success on the hunt or war path.

Fig. 115.—James Pigeon-hawk. (Photograph by Miss Densmore.)

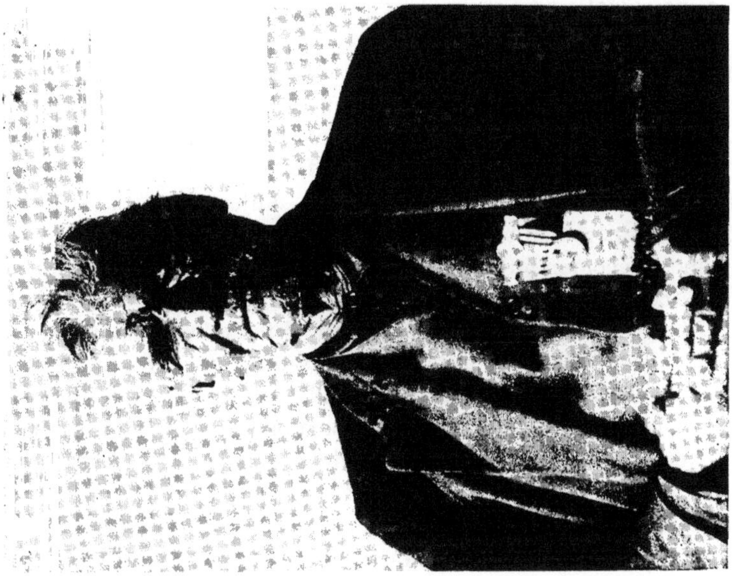

Fig. 116.—John Shawunopinas. (Photograph by Miss Densmore.)

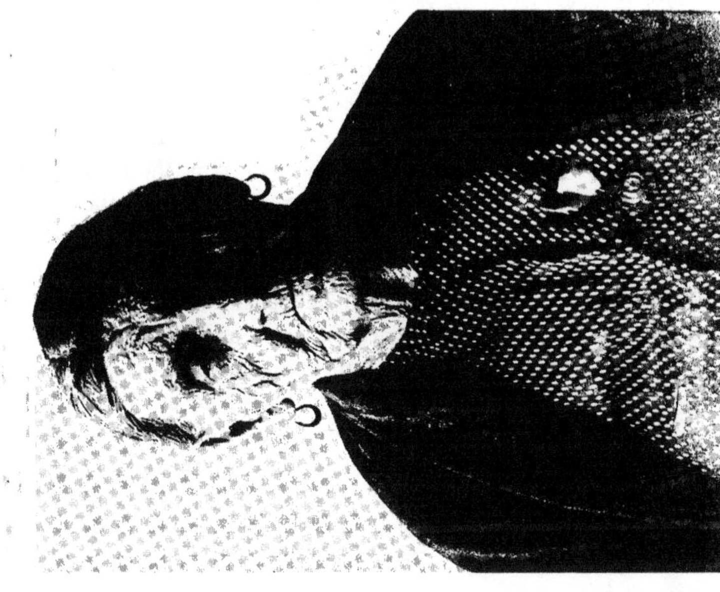

Fig. 117.—Agnes Sullivan. (Photograph by Miss Densmore.)

The two games played in order to secure benefits promised in dreams are the "bowl and dice game," played by women who have dreamed of the "four spirit women in the east," and the lacrosse game, played by men who have dreamed of the "thunderers." The manner of playing the first game was demonstrated by Catherine Laughery, whose dream requires that she play the game once every summer. She also recorded the song given her by the spirit women when, in her dream, she visited their lodge. A lacrosse game was thus played at a gathering attended by Miss Densmore. It is believed that a sick person will be benefited by attending a lacrosse game played in a ceremonial manner, and a song used on such an occasion was recorded by James Pigeon-Hawk (fig. 115). An adoption dance is held when a family wishes to adopt someone in place of a member of the family who has died. Two sets of songs are used at these dances, one attributed to the east and the other to the south god. The family select the songs to be used, the songs from the east god being chosen if the adoption is to be an important, dignified occasion. These are the more highly regarded as the east god is supposed to be the greater. Songs of both sorts were recorded, together with a description of the ceremonial action attending the adoption.

The individual war and hunting packs of the Menomini consist of certain articles wrapped in tanned hide. Under certain conditions the owner opens the pack in a ceremonial manner and exposes the contents, singing certain songs. A description of the bundles and manner of their use was obtained, together with the songs which are sung to make them effective.

The songs of the Dream dance formed a subject of special study, as Miss Densmore had attended this dance when held by the Chippewa and Menomini in 1910. (See Bull. 53, Bur. Amer. Ethn., pp. 142-180.) A few songs were sung by both tribes at certain points in the ceremony, and a comparison of the two renditions was desired. The "Pipe song" showed the same rhythmic phrase in the two renditions, and a somewhat similar duplication was found in the "Drum song." Thirty-three Menomini songs of the Dream dance were recorded, and the place where the dance is held was visited and photographed.

Songs used in the treatment of the sick were, as usual, an important phase of the research. Among those recording such songs was James Pigeon-Hawk, who, in treating the sick, uses songs which he received from his uncle and grandfather. One of these songs is sung when digging medicinal roots in order to make them effective. The Menomini differ from the Chippewa and Sioux in that they employ a "diagnostician" who decides whether a sick person shall be treated by a doctor giving material remedies or shall be taken to an exhibition of magic power by a "juggler." A description of the performances of the latter, with several of their songs, was obtained from a man familiar with their practices.

War songs received attention, about 25 of this class being recorded. Several war songs were connected with the Black Hawk war, which took place about 1832, while others were connected with the enlistment and service of Menomini Indians in the Civil War. John Shawunopinas (fig. 116) is a member of the G. A. R.

FIG. 118.—Exterior of medicine lodge.

and recorded a song with the words "The white man points his pipe at me," meaning "The white man asks me to join him in war." Agnes Sullivan recorded five old, vigorous war songs and asked to be photographed wearing her badge as a member of the Auxiliary to the American Legion. Among the miscellaneous songs obtained were those of the moccasin game, several social dances, the legends of Manabus, and the lullaby sung to little children, the latter being a variant of the Chippewa lullaby. Four melodies played on a cedar flute were recorded.

The Menomini Medicine Society, corresponding to the Chippewa Mĭde′ wĭwĭn, held a meeting in July which was attended by Miss Densmore. This was an interesting opportunity, as the society meets only once or twice a year to initiate members, and was particularly valuable, as Miss Densmore had recorded the songs of the

Fig. 119.—Interior of medicine lodge.

Fig. 120.—Women sitting in medicine lodge, their medicine bags hanging on the framework behind them.

Mĭde'wĭwĭn. Remaining at the Menomini ceremony about four hours she found that the songs used were chiefly Chippewa songs, with words in that language. She witnessed the "shooting with spirit power," in which members of the society thrust their medicine bags toward others with deep ejaculations. These persons "become unconscious" for a time and the performance is continued until all the members have thus "received spirit power." The meeting was held in a long lodge (fig. 118), and the members sat on the ground or danced in a line around the lodge, the songs being accompanied by a "water drum" and three gourd rattles (fig. 119). At the right of this illustration may be seen the top of the "water drum;" the dancers, including a little child; and the gifts for the leaders which are hung from a horizontal pole. The man near the drum is carrying his medicine bag, probably made of weasel skin. Many medicine bags were made of otter and were elaborately decorated, the material indicating the owner's rank in the society (fig. 120).

Continuing her study of Chippewa customs, especially those connected with the treatment of the sick, Miss Densmore visited the Cass Lake and Mille Lac reservations in Minnesota, in June, 1925, obtaining additional specimens of medicinal plants with descriptions of their uses. At Mille Lac she witnessed the making of two native dwellings of bark and rushes, and took photographs at various stages of the construction. Specimens of native implements made of wood were also obtained.

INVESTIGATION OF SHELL AND SAND MOUNDS ON PINELLAS PENINSULA, FLORIDA

Mr. David I. Bushnell, Jr., collaborator of the Bureau of American Ethnology, while on the west coast of Florida during the present year, visited various shell and sand mounds on Pinellas Peninsula. The peninsula extends southward into Tampa Bay and is of irregular form. On the northeast it is bounded by Old Tampa Bay, on the southeast and south by Tampa Bay, and on the west by Boca Ciega Bay, the latter being separated from the Gulf of Mexico by low, sand keys. Part of the peninsula is quite low and the entire region is infested with mosquitoes and other insects in vast quantities. Much of the shore is bordered by a broad stretch of marsh, with a dense growth of semi-tropical vegetation, but some of this marshy expanse has recently been reclaimed in the endeavor to make it suitable for building sites.

The shores of Tampa Bay—Espiritu Santo Bay of earlier maps—are interesting as having been the landing place of DeSoto and his numerous party in the year 1539, when "On Friday the 30th of May they landed in Florida, two leagues from a towne of an Indian lord, called Ucita." Unfortunately, the exact position of this ancient settlement is not known, but from the manner in which it was approached from the Gulf, as told in the Spanish narratives, it must have stood near the deeper channel which is found on the east and south sides of Tampa Bay, away from Pinellas Peninsula.

FIG. 121.—Large mound, formed of shells and sand, standing about midway between Maximo Point and Point Pinelos. The graded way, or approach to the summit, is on the left. Camera pointed northeast.

Three classes of works can be distinguished on the peninsula, all of which were erected near the shore. First are the large mounds, of a definite form, composed of shells and sand; second are the sand mounds; and third are the shellheaps, of no clearly defined shape, which resulted from the extensive use of shellfish for food.

Nearly a half century has passed since these mounds were visited by S. T. Walker, of Clearwater, Florida, at that time connected with the U. S. Fish Commission, and by whom they were briefly described in the Report of the Smithsonian Institution for the year 1879. The most important work is in the extreme southern part of the region, about midway between Maximo Point on the west and Point Pinelos a little south of east. It stands about four hundred yards from the water and the greater part of the intervening area is a marsh with a thick, matted growth of low vegetation. This mound, viewed from the southwest, is shown in figure 121. It was visited by Walker in 1879 who referred to it as "the most beautiful mound that I have seen in South Florida." And he continued, "The mound is situated in a 'rosemary scrub,' and rises to an imposing height above the low trees in its vicinity. Its outlines are beautifully regular . . . . with a beautiful inclined roadway leading up to its western side." During past years excavations have been made in the structure by persons seeking hidden treasure, and consequently the surface is now quite irregular but it is possible to follow the original lines. The mound is about two hundred feet in length and twenty feet in height. It is composed of shells and sand, and the following shells were collected from the several excavations: *Fasciolaria tulipa; Macrocallista gigantea; Busycon canaliculata; Pecten irradians; Strombus pugilis; Venus mercenaria mortoni; Fulgur perversa.*[1] No fragments of pottery were discovered although known to occur in the mound.

A symmetrical mound stands near the shore of Big Bayou, some two and one half miles east of north of the work just described. It appears to be formed entirely of sand, is a hundred feet or more in diameter and rises about ten feet above the original surface. It is owned by Mr. Glen Taylor, to whom I am indebted for much assistance during visits made to the scattered sites on the peninsula. About four hundred yards northeast of this mound were two large shell mounds, now destroyed. They are said to have been formed for the greater part of large oyster shells, and fragmentary pottery was found throughout the mass. When the site was examined large shells of the *Ostrea virginica* were discovered, some being seven inches in length. These were in the lowest and consequently oldest part of the mound or mounds.

Next to be considered are works on the west side of the peninsula, south of the railroad bridge across Four Mile Bayou. On the shore, opposite the mouth of the bayou, is an extensive shellheap, part of which is shown in figure 122, a view looking southwest, over the waters of Boca Ciega Bay to the low keys beyond. About one thousand feet inland from the shellheap is a large shell and

[1] All shells have been identified by Dr. Paul Bartsch, of the U. S. National Museum.

Fig. 122.—Shellheap on the west side of the peninsula at the mouth of Long or Four Mile Bayou. The view is looking southwest showing the waters of Boca Ciega Bay.

Fig. 123.—Compact mass of shells exposed near the summit of a large mound standing about three hundred yards inland, east, from the mouth of Four Mile Bayou.

sand mound, quite similar in appearance to the great mound in the southern part of the peninsula. A small part of the southern section of the work has been removed and the upper portion, thus exposed to view, is shown in figure 123. This reveals a compact mass of shells, usually of small size. The shells collected from this exposed portion of the mound were: *Venus mercenaria mortoni; Cardium robustum; Fulgur perversa; Fasciolaria tulipa; Arca ponderoso; Ostrea virginica; Pecten irradians; Melongena corona; Chione cancellata; Cardita floridana.* Fragments of pottery were mingled with the shells, all having stamped designs over the entire surface. Many bits of human bones, indicating burials, were found near the summit. One fragment of a human skull revealed traces of a red pigment with which it had probably been covered.

Southward from the preceding site, at the end of the electric line running to St. Petersburg, and on the shore of Boca Ciega Bay, is a shellheap having a diameter of a hundred yards and an elevation of approximately ten feet. It extends eastward from the edge of the water to the road which has cut through the eastern portion, exposing the interior of the mass as shown in figure 124. The shells gathered here were: *Fasciolaria gigantea; Fulgur perversa; Strombus pugilis; Fasciolaria tulipa; Venus mercenaria mortoni; Pecten irradians; Ostrea virginica; Spicula similis; Cardium isocardia.* A closer view of the mass of shells in this exposure is given in figure 125. All fragments of pottery discovered mingled with the shells and decomposed vegetal matter were smooth on both surfaces, entirely undecorated.

On the eastern side of the peninsula, north of east from the several mounds just mentioned, is another ancient site. It is near the narrowest part of the bay, south of the bridge over Old Tampa Bay, and exactly west of Port Tampa on the opposite shore. However, the water is not visible from the site on account of the very dense vegetation. Here are shellheaps, low and spreading and less clearly defined than others already mentioned. They extend in a general course from southeast to northwest and terminate abruptly at a sand mound. The latter is about one hundred and twenty feet in diameter, although it is difficult to distinguish where the artificial work actually begins; its elevation is approximately seven feet. This appears to have been the burial place which belonged to the nearby settlement. Many burials were discovered by Dr. J. W. Fewkes in the eastern part of this mound during the winter of 1923-1924, at which time about one quarter of the structure was examined. Fragments of pottery, decorated with designs in incised

FIG. 124.—Section of a spreading shell mound in a region known as "The Jungle," on the west side of Pinellas Peninsula, at the end of the electric line running to St. Petersburg.

FIG. 125.—A closer view of the mass of shells forming the mound in "The Jungle."

Fig. 126.—Fragment of a large pottery vessel from the burial mound on Weeden Island. One-half natural size.

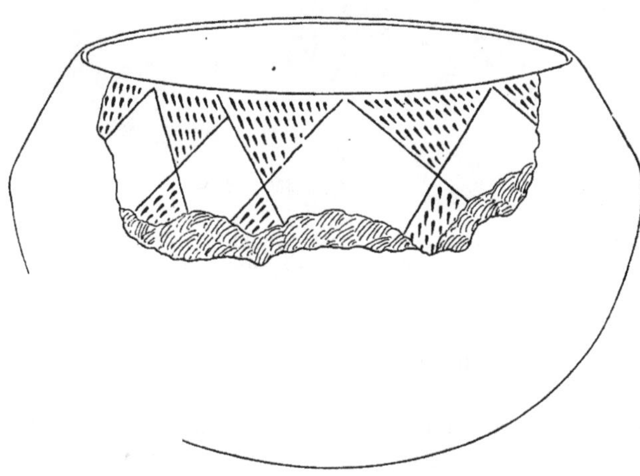

Fig. 127.—The probable form of the vessel, suggested by the shape of the fragment. (Drawn by DeLancey Gill.)

lines, were recovered from parts of the excavation, and although this is considered typical of the west coast of Florida no examples were found by the writer on other sites on the peninsula. In addition Dr. Fewkes encountered many fragments bearing the usual stamped design, and exactly like that recovered from the large mound on the west side of the peninsula, but it was not determined conclusively whether the two types were intermingled or whether they occurred in two distinct layers one above the other. A careful examination of the mound, especially the central portion, would undoubtedly make it possible to discover which of the two forms of pottery

FIG. 128.—The low, spreading burial mound on Weeden Island. The large cabbage palmetto, to which the sign is attached, stands very near the center of the mound.

is the older, if a difference actually exists. A photograph of the mound is reproduced in figure 128. On the right is the end of the excavation made by Dr. Fewkes, the large marker being attached to a tree near the center of the work. A fragment of the rim of a large earthen vessel, found by the writer in the excavation, is shown in figure 126, and a sketch of the probable shape of the vessel in figure 127. It was a well-made vessel, thin, and of a yellowish-brown tint. The opening was about ten inches in diameter.

Other large mounds occur on other parts of the peninsula, and all are of the greatest interest at this time. As yet it is not possible to identify the tribe or tribes by whom the ancient settlements were occupied.

SMITHSONIAN MISCELLANEOUS COLLECTIONS
VOLUME 78, NUMBER 2

MEXICAN MOSSES COLLECTED BY BROTHER ARSÈNE BROUARD

BY
I. THÉRIOT
Fontaine La Mallet, France

(PUBLICATION 2867)

CITY OF WASHINGTON
PUBLISHED BY THE SMITHSONIAN INSTITUTION
JUNE 15, 1926

The Lord Baltimore Press
BALTIMORE, MD., U. S. A.

MEXICAN MOSSES COLLECTED BY BROTHER ARSÈNE BROUARD[1]

By I. THÉRIOT

FONTAINE LA MALLET, FRANCE

Brother Arsène (F. S. C.), who resided in Mexico previous to 1914, made important collections of mosses, more especially in the states of Puebla and Michoacán. Compelled by war and revolution to leave the country, he saved what he could of his collections and tendered the mosses to the U. S. National Museum at Washington.

The specimens had still to be named. Himself of French descent, Brother Arsène preferred that his collections be entrusted to a French bryologist for identification. My friend Cardot was the one fitted for this undertaking, but unfortunately, having suffered much during the war, he had given up bryological study completely. Brother Arsène asked me to take M. Cardot's place. At first I demurred, since the Mexican flora was almost unknown to me; then I considered that this flora had been studied more especially by the French bryologists (as witness the number of species bearing the authority Bescherelle or Cardot), and it seemed to me a duty to receive their legacy and to carry on their labor.

Further, the determination of Brother Arsène's mosses would not require of me an all-inclusive knowledge of the Mexican flora. Though the collection totaled more than 1,000 numbers, the number of species would scarcely exceed 200, less than one-third of the moss flora known at present.

I have found the study of the material very attractive and very profitable, since each species is represented by numerous specimens from different localities. The bryologist, elaborating in his study the flora of a distant region, can not but be most grateful to collectors as wise as Brother Arsène, whose generous gatherings permit one the better to appreciate the range of variation in each species, the stability or instability of distinguishing characters, and the worth of species already described, and hence to bring about clearer understanding and to make reductions.

---

[1] With M. Thériot's permission, the comments and critical notes, written in French, were translated by the late Edward B. Chamberlain.

The larger proportion of the collections comes from the states of Michoacán and Puebla. To avoid frequent repetitions, I give below a list of the localities most frequently cited and their altitudes. The reader will surely be willing, if he deems it worth while, to refer to this tabulation.

STATE OF MICHOACÁN, NEIGHBORHOOD OF MORELIA

Jesús del Monte, 2,000 meters.
Campanario, 2,200-2,300 meters.
Cerro Azul, 2,300-2,400 meters.
Cerro San Miguel, 2,200 meters.
Loma Santa Maria, 2,000 meters.
Loma del Zapote, 1,950 meters.
Loma de la Huerta, 2,000 meters.
Carríndapaz, 2,100 meters.
Rincón, 1,950-2,100 meters.
Bosque San Pedro, 1,950 meters.
Andameo, 2,100 meters.

STATE OF PUEBLA

Huejotzinco, 2,200 meters.
Esperanza, 2,400 meters.
Xúchitl, 2,400-2,800 meters.
Vicinity of Puebla: Cholula road, 2,180 meters.
Hacienda Alamos, 2,170 meters.
Hacienda Batán, 2,120 meters.
Rancho Guadalupe, 2,150 meters.
Cerro Tepoxúchitl, 2,370 meters.
Rancho Posadas, 2,170 meters.
Cerro Guadalupe, 2,200 meters.

The present contribution covers but a portion, around a third, of Brother Arsène's rich collections. Others will follow in accordance with the progress of my identification of the specimens.

DICRANACEAE

CERATODON STENOCARPUS Bry. Eur. (29-30:) Cerat. 4. 1846 (*nom.*); C. Müll. Syn. 1: 647. 1849

Campanario (7459); Jalapa, Veracruz, 1400 m. (7970a).

AONGSTROEMIA PUSILLA Thér., sp. nov.

(Fig. 1)

Jesús del Monte (7605 p. p.), among tufts of *Pogonatum toluccense* var.

Pusillus. Caulis simplex, interdum ramosus, 2-3 mm. altus. Folia appressa, inferiora ovato-acuminata, costa percurrente, ceteris sensim majora, costa excurrente, haud attenuata, basi 40-50 μ crassa, marginibus planis, superne sinuolatis; rete pellucido, cellulis basilaribus breviter rectangularibus, ceteris hexagonis vel rhombeis, 24-30 μ longis, 8 μ latis, parietibus tenuibus flexuosis. Folia perichaetialia longiore cuspidata, pedicellus brevissimus, vix 1 mm. longus, capsula immersa, subglobulosa, 0.6 mm. longa, 0.5 mm. crassa, calyptra minuta, cucullata, operculum tantum obtegens. Caetera desunt.

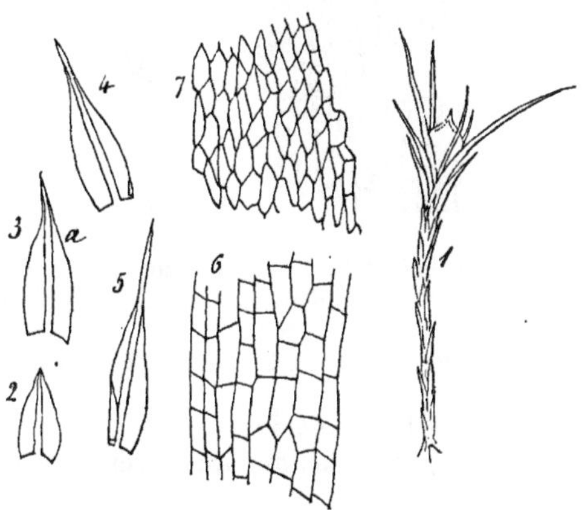

FIG. 1.—*Aongstroemia pusilla* Thér. 1, entire plant, × 12; 2, 3, 4, leaves, × 17; 5, perichaetial leaf, × 17; 6, basal areolation, × 200; 7, areolation at point *a*, × 200.

This species is scarcely to be compared with any save *Aongstroemia jamaicensis* C. M. (*A. brevipes* Hpe., fide R. S. Williams in N. Amer. Flora 15: 79. 1913), which is distinguished by greater robustness and taller stems (up to 6 cm.); by longer, perfectly entire, more gradually acuminate leaves; by the median and upper leaf cells being longer and narrower, and with the walls very incrassate; by the costa being thin basally and widened above; and by the thicker, more elongate capsule. Since the plant is abundantly distinct from all its congeners, it seemed rather superfluous to destroy the solitary capsule available in order to study the peristome.

AONGSTROEMIA ORIENTALIS Mitt. Trans. Linn. Soc. II. 3: 154. 1891

(FIG. 2)

Mixed with other mosses, especially *Campylopus Arsenei* and *Anomobryum* sp.: Rincón (s. n.); Campanario (7559 p. p., 7561 p. p., 7577 p. p.).

Statura reteque *A. julaceae* (Hook.) Mitt. sat similis, differt foliis apice patulo-arcuatis, parum secundis, subacutis, marginibus sinuolatis, costa percurrente, angustiore, vix 30 μ crassa.

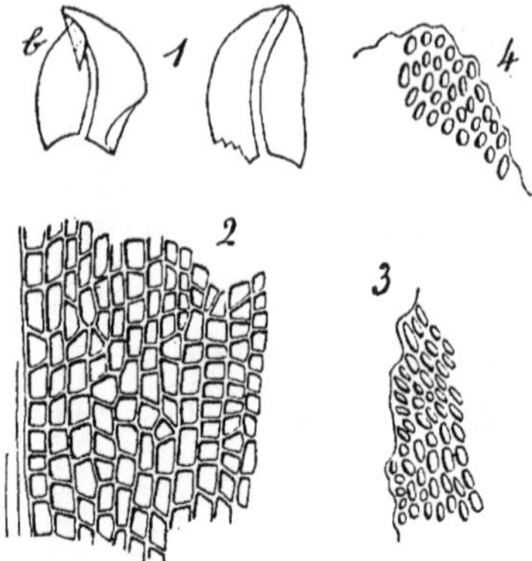

Fig. 2.—*Aongstroemia orientalis* Mitt. 1, leaves, × 30; 2, basal areolation, × 200; 3, areolation at point *b*, × 200; 4, apical cells, × 200.

DICRANELLA VARIA (Hedw.) Sch. Corol. Bry. Eur. 13. 1855, forma

Hacienda Alamos (4632, 4633).

A form with short, short-acuminate leaves, which are at most 1 to 1.2 mm. long.

ORTHODICRANUM FLAGELLARE (Hedw.) Loesk. Studien 85. 1910

(Fig. 3)

Cerro Azul (4989).

Sterile. Caespites compacti. Caulis 1-2 cm. altus, apice ramulis numerosis, gracilis, flagelliformis, folia parva, remota, obtusa emittens. Folia caulina sicca crispato-circinata, humida erecta parum patula, flexuosa, e basi oblonga sensim longe et tenuiter acuminata, canaliculato-tubulosa, superne denticulata, dentibus minutis et remotis, marginibus haud incrassatis, 3-3.8 mm. longa, 0.5 mm. lata, folia inferiora breviora, basi longiora, marginibus saepe integris, omnia fragilia, frequenter effracta; costa percurrente vel breviter excurrente, basi 90 μ lata, dorso remote et obtuse dentata, cellulis alaribus (7–9-seriatis) hexagonis, fuscis, 15-20 μ latis, auriculas parum excavatas et totam basin laminae efficientibus, cellulis suprabasilaribus linearibus, firmis, ceteris quadratis vel breviter rectangularibus, valde chlorophyllosis, diam. 10 μ.

The species has the habit and color of *H. arboreum* Mitt., to which it seems to be related, though more slender. It is distinguishable at once by the flagelliform branches covering the upper part of the stems, by the more fragile, often broken leaves which are shorter, narrower, and only slightly dentate, by the smaller, more numerous alar cells, and by the flattened costa. *H. proliferum* Mitt. also differs decidedly, according to description, in the ligulate, obtuse, strongly dentate leaves, etc.

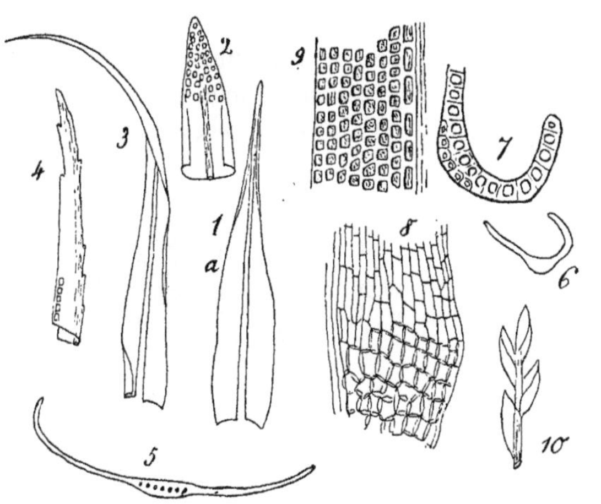

FIG. 3.—*Orthodicranum flagellare* (Hedw.) Loesk. 1, leaf from midstem, × 17; 2, its acumen, × 90; 3, upper leaf, × 17; 4, its acumen, × 90; 5, transverse basal section of a leaf, × 90; 6, section of leaf in upper third, × 90; 7, the same, × 200; 8, basal areolation, × 90; 9, areolation at point *a*, × 200; 10, portion of a flagelliform branch, × 17.

DICRANUM FRIGIDUM C. Müll. Bot. Zeit. 17: 219. 1859

Vicinity of Puebla (4957).

CAMPYLOPUS (Pseudocampylopus) ARSENEI Thér., sp. nov.

(FIG. 4)

Rincón (s. n.); Cerro Azul (4981); Campanario (7573).

Caespites densissimi. Caulis valde radiculosus, 1-3 cm. altus, dense foliosus. Folia erecto-appressa, apice plus minus flexuosa, lanceolato-linearia, longe et tenuiter subulata, superne denticulata, e basi valde concava, deinde involuto-tubulosa, alis angustissimis (e medio ad summum e cellulis 1-seriatis compositis), 3.5-4 mm. longa, 0.4-0.5 mm. lata; auriculis minutis, vix excavatis, cellulis alaribus (2-ser.) laxis, hyalinis vel fuscis; rete suprabasilari hyalino, cellulis marginalibus (5–6-ser.) linearibus, internis rectangularibus, cellulis laminae anguste rectangularibus vel rhomboidalibus, 35-40 μ longis, 6 μ latis, parum chlorophyllosis, parietibus haud incrassatis; costa basi 0.35 mm. crassa, excurrente, dorso superne sulcata, in sectione transversali e 3-4 stratis cellularum formata, ventrali e cellulis amplis, inanibus, interno e cellulis (eurycystis) incrassatis, dorsali (1-2) e cellulis parvis, incrassatis (substereidis). Pedicellus 5 mm. altus, calyptra ciliata, capsula oblonga, symmetrica, sicca suberecta, sulcata, sporae 12-15 μ crassae.

The structure of the costa is a little ambiguous. If a section is made in the lower third, among the substereid cells of the dorsal epidermis, there are a few of smaller lumen which may be considered stereids. A section from the acumen, however, shows no stereids at

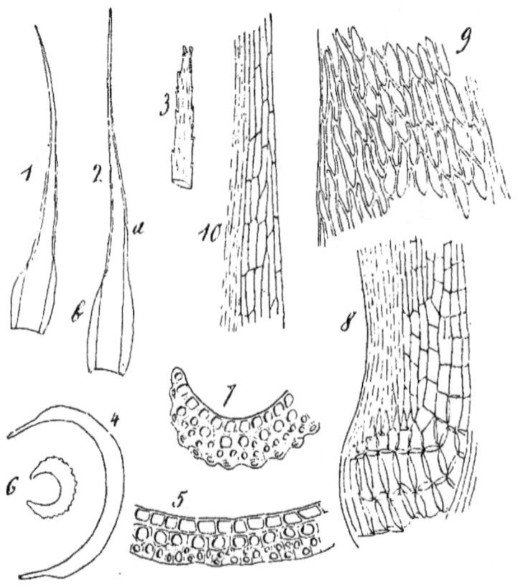

Fig. 4.—*Campylopus Arsenei* Thér. 1, 2, leaves, × 12; 3, acumen, × 90; 4, basal cross section of a leaf, × 90; 5, portion of same, × 300; 6, cross section of leaf towards apex, × 90; 7, portion of same, × 300; 8, auricle and suprabasal areolation, × 130; 9, cells of lamina at point *b*, × 200; 10, cells at point *a*, × 200.

all. The species is surely one of the links in the chain connecting *Pseudocampylopus* and *Eucampylopus*. In the shape and size of the leaves it bears an odd resemblance to *Campylopus subturfaceus* Card., but the stems of the latter are very short and the auricles of the leaves, above the base, are composed of subquadrate or rhomboidal, chlorophyllose cells of large diameter. *C. Chrismari* (C. M.) Mitt. has a non-ciliate calyptra, a pedicel 10 mm. long, and the leaves 5 to 7 mm. long.

CAMPYLOPUS SUBTURFACEUS Card. Rev. Bryol. 37: 119. 1910

Campanario (7561 p. p.).

Young, incompletely developed plants. It is necessary to place here also a moss collected by Liebmann and labeled *Campylopus pusillus* by Schimper (Herb. Mus. Paris); it is totally different from the Orizaban *C. pusillus* (collected by F. Müller), which, as Mr. R. S. Williams has already observed, belongs to *Campylopodium*.

CAMPYLOPUS (Atrichi) MEXICANUS Thér., sp. nov.

(Fig. 5)

Campanario (7576, type); Cerro Azul (4544 p. p., 4781, 4797 p. p.).

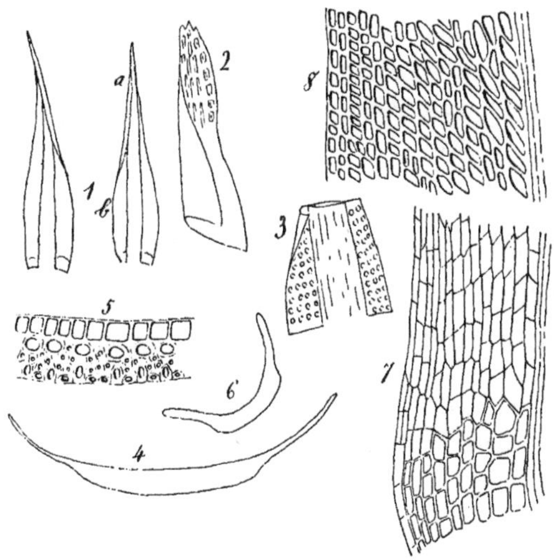

Fig. 5.—*Campylopus mexicanus* Thér. 1, leaves, × 12; 2, acumen, × 90; 3, portion of acumen at point *a*, × 90; 4, basal cross-section of leaf, × 90; 5, portion of same, × 300; 6, apical cross-section, × 90; 7, auricle and suprabasal areolation, × 130; 8, areolation of the lamina at point *b*, × 200.

Mollis, caespites densi. Caulis 1.5-2 cm. altus, radiculosus, basi terra obrutus, dense foliosus, saepe in axillis foliorum superiorum ramuli flagelliformes. Folia mollia, erecto-appressa, apice flexuosa, humida parum patula, lanceolata, breviter acuminata, integra vel superne denticulata, parum concava, marginibus planis superne leviter involutis, 3 mm. longa, 0.55 mm. lata; costa 0.25-0.30 mm. crassa, percurrente, dorso laevi, in sectione transversali structura normali, auriculis planis male limitatis; cellulis suprabasilaribus hyalinis vel parum chlorophyllosis, marginalibus (4-5 ser.) linearibus, juxtacostalibus rectangularibus, cellulis laminae quadratis vel rhomboidalibus, magnis, diam. 10 μ, sed valde inaequalis, recte seriatis, parietibus parum incrassatis, Pedicellus 6 mm. altus, calyptra ciliata, capsula immatura.

I can compare this moss only with *Campylopus Roellii* R. & C., but that species has the leaves longer, with a more slender, involute-tubulose acumen, and the nerve proportionately broader.

CAMPYLOPUS INTROFLEXUS (Hedw.) Brid. Bryol. Univ. 1: 472. 1826

Loma Santa Maria (7862, 7871, 7873, 7877); Andameo (4817, 4837); Rincón (4563); Loma de la Huerta (4835); Tlaxcala, Santa Ana Chiautempan (4853).

Nearly all these specimens differ one from the other, showing once more the extreme variability of this species: The leaves may be auricled or not; the hair point long or short; the costal lamellae, of 2, 3, or 4 cells, may begin at the very base or appear only in the lower third of the leaf; the capsule may be more or less elongate, and more or less costate.

METZLERELLA LEPTOCARPA (Sch.) Card. Rev. Bryol. 38: 100. 1911

Leptotrichum leptocarpum Besch. Prodr. Bryol. Mex. 34. 1871.

Cerro Azul (4544, 4547, 4555, 4774); Campanario (4771, 7511, 7551, 7574, 7928).

I notice a rather wide variation in the height of the plants, in the direction and length of the leaves (erect or falciform-secund), in the length of the seta, and also in the capsule, which is sometimes a little arcuate (no. 7551).

FISSIDENTACEAE

FISSIDENS PRINGLEI Card. Rev. Bryol. 36: 69. 1909

Hacienda Alamos (4724); Camino de Cholula (4848, 4860 p. p.).

FISSIDENS HERIBAUDI Broth. & Par.; Card. Rev. Bryol. 40: 33. 1913

Rancho Guadalupe (4604, 4609); Morelia, Parque de San Pedro (4920).

FISSIDENS ARSENEI Broth. & Par.

Cerro Guadalupe (685, 803); Hacienda Alamos (4768).

An unpublished species; I furthermore believe that it is simply a form of the preceding species. The leaves are smaller, crisped when dry, and not readily flattening out when moist, with a hyaline margin that is much widened at base, and with shorter, usually erect capsules. Further, there are transitional forms, as for example one received from Brother Héribaud (under name of *Fissidens Heribaudi*) which matches exactly neither *F. Arsenei* nor *F. Heribaudi,* but is intermediate between the two extreme forms. No. 4768 is another form, with a narrower margin (4 to 6 cells), that tends to approach the next species.

FISSIDENS TORTILIS Hpe. & C. M. Bot. Zeit. 22: 340. 1864

Puebla, without definite locality (695); Rancho Guadalupe (4610); Huejotzinco (4858); Andameo (4823, 4825, 4839); Loma Santa Maria (4903, 5064, 7859, 7863); Querétaro, Jurica (11003).

This species, which Bescherelle lists in his Prodromus (Mém. Soc. Sci. Nat. Cherbourg **16**: 170. 1872), seems to have been lost sight of or misunderstood since the publication of that work, yet it is not rare in Mexico. I have come to the conclusion that the moss distributed in the Pringle collection under the name *Fissidens reclinatulus* C. M. should be referred to *F. tortilis,* and further am of the opinion that the Costa Rican *F. reclinatulus* should not be kept specifically distinct from *F. tortilis.* At the same time it should be stated that the specimens of Tonduz's collecting which I have studied have been polyoicus (the male flowers sometimes terminal on a special branch, sometimes on a short branch at the base of the fruiting stem), and that I have seen nothing like this condition in the Mexican plants studied.

According to my observations, *Fissidens tortilis* Hpe. & C. M. is characterized as follows: Leaves unequal, crisped when dry, not readily flattening out when moist, with the dorsal lamina a little decurrent, the margin usually of the same coloration as the rest (but sometimes hyaline) and reaching the apex of the leaf, the margin formed of 2 or 3 cells in the apical lamina, and of 3 or 4 cells in the true lamina. The capsule is inclined.

FISSIDENS TORTILIS Hpe. & C. M. var. BREVIFOLIUS (Card.) Thér., comb. nov.

Fissidens reclinatulus var. *brevifolius* Card. Rev. Bryol. **36**: 69. 1909.
Fissidens pennaeformis Par. MS.

Morelia, Bosque San Pedro (4575); Puebla, in horto archiepiscopali (ex hb. Paris sub nom. *F. pennaeformis*).

Cardot established his variety *brevifolius* upon no. 10699 of Pringle's collection, but the latter subsequently distributed under no. 10699 another plant that does not differ from *F. tortilis.*

FISSIDENS (Bryoidium) LONGIDECURRENS Thér., sp. nov.

(Fig. 6)

Morelia, Loma Santa Maria (4892, 4906).

Dioicus? Caulis 7-12 mm. altus. Folia 12–15-juga, inaequalia, sicca crispata, difficile emollentia, oblongo-lanceolata, late acuminata, subacuta, marginata, 1.2-2 mm. longa, 0.4-0.5 mm. lata, lamina apicali ¼ ad ⅓ folii aequante, lamina dorsali longe decurrente, limbo angusto hyalino e 1-2 seriebus cellularum angustarum composito, limbo

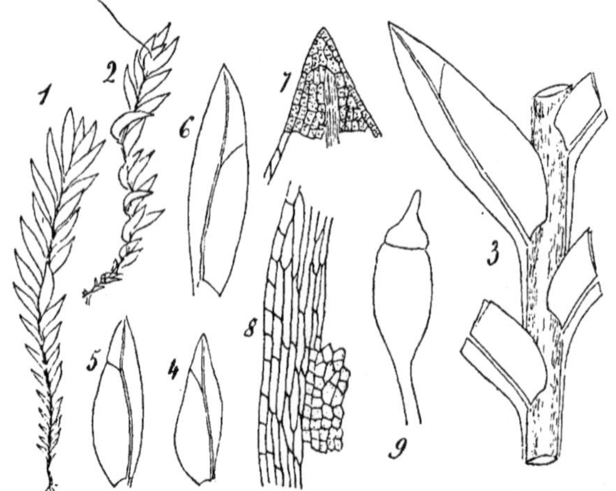

Fig. 6.—*Fissidens longidecurrens* Thér. 1, 2, entire plants, × 4; 3, portion of a stem, × 17; 4, 5, median stem leaves, × 17; 6, upper stem leaves, × 17; 7, apical cells, × 200; 8, margin of the true lamina, × 200; 9, moist capsule, × 17.

laminae verae dilatato e 6-8 seriebus cellularum composito; costa concolore, basi 40 μ lata, sub apice evanescente; rete obscuro (alis pellucido), cellulis minutis, irregularibus, parietibus tenuibus, diam. 6-7 μ. Pedicellus geniculato-flexuosus, 3 mm. longus, capsula suberecta, oblonga, basi attenuata, operculum conicum, brevirostratum.

This species is surely close to *F. Pringlei* Card., but differs in the longer stems, the greater decurrence of the leaves (they unite with the leaf below), and the smaller cells, especially those in the true lamina. It is also close to *F. aequalis* Salm. in size and in the measurements and shape of the leaves and in their areolation, but is nevertheless distinct in the unequal leaves with a shorter apical lamina and a more decurrent dorsal lamina.

FISSIDENS (Bryoidium) FLEXUOSUS Thér., sp. nov.

(FIG. 7)

Morelia, Andameo (4824).

Caulis erectus, flexuosus, 1.5 cm. longus. Folia 15–20-juga, inferne minuta, cetera sensim majora, sicca applanata, haud crispata, marginata, 2-2.5 mm. longa, 0.5-0.6 mm. lata, oblongo-lanceolata, acuminata, acuta, lamina apicali ⅓ folli aequante, lamina dorsali parum decurrente, limbo lamina apicali hyalino e 1–2 seriebus cellularum angustarum composito, limbo lamina vera dilatato, e 4–5 seriebus cellularum composito; costa concolore, basi 36 μ crassa, sub apice finiente; rete obscuro, cellulis hexagonis, parietibus tenuibus, diam. 6 μ, basilaribus vix majoribus. Caetera ignota.

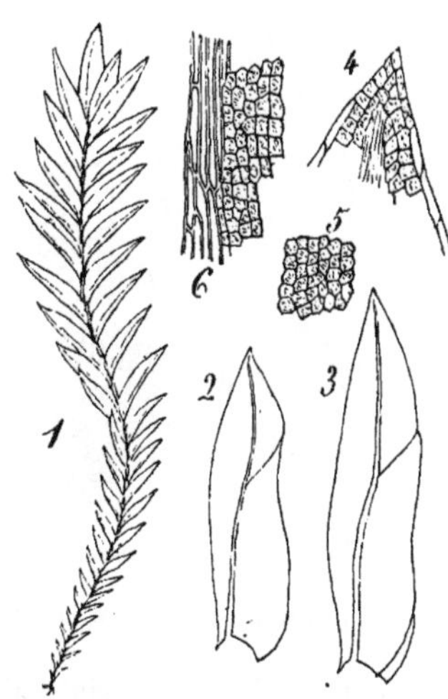

FIG. 7.—*Fissidens flexuosus* Thér. 1, whole plant, × 4; 2, medium stem leaf, × 17; 3, upper stem leaf, × 17; 4, apical cells, × 200; 5, median cells, × 200; 6, margin of true lamina, × 200.

This differs from *F. aequalis* Salm. in its greater size and its unequal leaves, which are longer and neither secund nor crisped when dry. From all Mexican species of the section *Bryoidium* it differs in the elongate stems having large flat leaves with a compact obscure areolation.

FISSIDENS (Semilimbidium) BROUARDI Thér., sp. nov.

(FIG. 8)

Morelia, Loma Santa Maria (7858).

Rhizautoicus, pusillus, terrestris. Caulis brevis, simplex, vix 2 mm. altus. Folia 5–7-juga, inferiora remota minuta, sensim majora, oblongo-lanceolata, breviter acuminata, acuta, 0.8-1.2 mm. longa, 0.25-0.35 mm. lata, alis vaginantibus, limbo pellucido, lutescente, sinuato, e cellulis 4-seriatis composito marginatis, laminis apicali et dorsali elimbatis, crenulatis, lamina dorsali attenuata haud decurrente; costa sub apice evanida, basi 20–30 μ lata; rete obscuro, cellulis rotundato-hexagonis, papillosis, diam. 6 μ. Pedicellus erectus, 2 mm. altus. Caetera ignota.

This species differs from *Fissidens Ravenelii* Sull. in the smaller size, shorter leaves with the apical lamina reaching hardly one-third of the leaf, etc. It is distinguished from *F. Nicholsoni* Salm. and

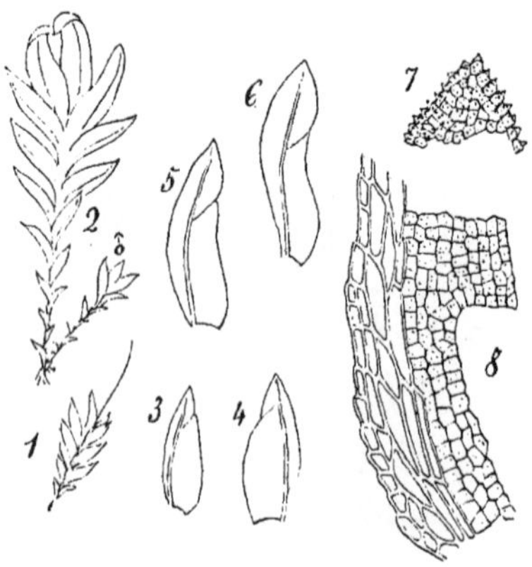

FIG. 8.—*Fissidens Brouardi* Thér. 1, plant with seta, × 4; 2, another plant with male and female flower, × 12; 3, 4, 5, stem leaves, × 17; 6, perichaetial leaf, × 17; 7, apical cells, × 200; 8, margin of true lamina, × 200.

F. hemicraspedophyllus Card. by the terrestrial habitat and the shape of the leaves; from the first also by the broader margin, and from the second in not having the margin extended to the apex of the vaginant lamina.

FISSIDENS (Semilimbidium) MICHOACANUS Thér., sp. nov.

(FIG. 9)

Morelia (7892).

Dioicus? Terrestris, pusillus. Caulis 2 mm. altus. Folia 5–8-juga, inaequalia, media 0.7 mm. longa, 0.25 mm. lata, lamina apicali brevi, immarginata, lamina dorsali anguste basi attenuata vel rotundata, immarginata, lamina vera ad ⅔ vel ¾ folii producta, marginata tantum in foliis perichaetialibus, limbo e cellulis 2–3-seriatis superne et inferne evanescente; costa ante apicem dissoluta angusta, basi 12–18 μ lata; rete obscuro, cellulis rotundato-hexagonis, chlorophyllosis, papillosis indistinctis. Pedicellus geniculatus, vix 2 mm. longus, capsula erecta, oblonga, e basi breviter attenuata, operculum conicum longirostratum capsulam aequante, peristomii dentes sub ore inserti, 0.27 mm. longi, usque fere ad basin fissi. Caetera desunt.

This species is near the preceding in the sum of its characters, but is readily distinguishable by its appearance. The fertile stems are more densely leafy, and are accompanied by sterile stems that have more numerous, shorter, subequal leaves. The leaves also, save the perichaetial ones, are marginless, the costa is only half as wide, and

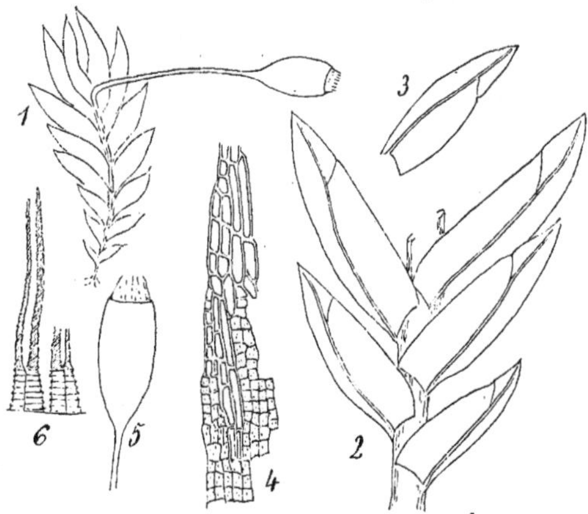

FIG. 9.—*Fissidens michoacanus* Thér. 1, entire fertile plant, × 12; 2, portion of fertile stem, × 30; 3, leaf from a sterile stem, × 30; 4, margin of true lamina (perichaetial leaf), × 200; 5, moist capsule, × 17; 6, teeth of peristome, × 90.

the pedicel is geniculate. The capsule of *F. Brouardi,* which I have not seen, will doubtless furnish other distinguishable marks. The plant is close also to *F. Ravenelii* Sull., but differs in having a short apical lamina, no dentation in the true lamina, the costa always terminating below the apex, the areolation more opaque, the pedicel shorter, and the peristome teeth divided almost to the base.

FISSIDENS (Heterocaulon) PSEUDO-EXILIS Thér., sp. nov.

(FIG. 10)

Morelia, Loma Santa Maria (4899, 4928).

Rhizautoicus. Caulis fertilis 2 mm. altus. Folia 3-4-juga, inaequalia, integra, lamina apicali breviter acuminata, acuta, ⅓ folii aequante, lamina dorsali immarginata, attenuata, lamina vera tantum e medio marginata (cellul. 2-3-seriatis); rete pellucido, cellulis mediis et superioribus hexagonis, laevibus, parce chlorophyllosis, diam. 7–9 μ, basilaribus (lamina vera) majoribus, breviter rectangularibus; costa valida, 30–36 μ lata, ante apicem evanescente. Caulis sterilis dense foliosus, 3 mm. altus; folia 12–15-juga, subaequalia, duplo minora, subobtusa, immarginata, lamina dorsali ad insertionem evanescens. Pedicellus pallidus, gracilis, 3-4 mm. longus, capsula erecta vel parum inclinata, ovata, brevicollis, leptoderma, operculum conicum brevirostratum, peristomii dentes usque ad ⅔ fissi, sporae laeves, 18 μ crassae.

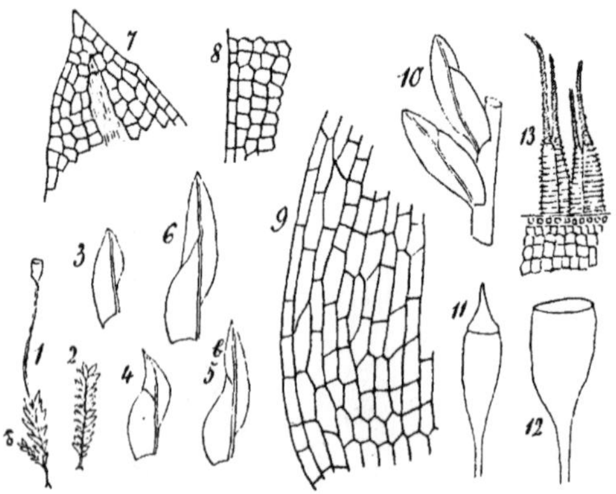

FIG. 10.—*Fissidens pseudo-exilis* Thér. 1, fertile plant, × 4; 2, sterile plant, × 4; 3, 4, 5, 6, leaves from a fertile stem, × 17; 7, apical cells, × 200; 8, marginal cells at point *b*, × 200; 9, basal areolation of true lamina, × 200; 10, leaves from a sterile stem, × 30; 11, very young capsule, × 17; 12, deoperculate capsule, × 17; 13, teeth of peristome, × 90.

In habit and size the fertile stems recall *F. exilis* Hedw., but our species is easily distinguished by its dimorphous stems, its entire leaves with smaller cells and the true lamina partially marginate, its larger spores, etc. This species is the first member of the section *Heterocaulon* reported from North America.

FISSIDENS ASPLENIOIDES (Sw.) Hedw. Musc. Frond. 3: 65. *pl. 28.* 1801

Cerro Azul (5097); Morelia (7888, 7898, 7899, 7910, 7919); Campanario (7942a).

TRICHOSTOMACEAE (in part)

PLEUROCHAETE LUTEOLA (Besch.) Thér., comb. nov.

Trichostomum luteolum Besch. Mém. Soc. Sci. Nat. Cherbourg **16**: 178. 1872.
Pleurochaete mexicana Broth. MS. in hb. Levier, no. 7458.

Puebla: Esperanza (4661).

A stouter plant than *P. squarrosa* (Brid.) Lindb. of Europe. The basal areolation of the leaves is composed of much longer rectangular cells, and the margin, which is formed of more elongate cells, extends sensibly farther up the leaf.

LEPTODONTIUM EXASPERATUM Card. Rev. Bryol. 36: 74. 1909

Esperanza (4728).

LEPTODONTIUM HELICOIDES Card. Rev. Bryol. 36: 75. 1909

Cerro Azul (4538, 4549, 4776 p. p., 4790); Campanario (7444, 7517, 7929).

These specimens which I have referred to *L. helicoides* have scarcely more in common than the helicoid arrangement of the leaves on the stem. They differ among themselves in respect to the length of the leaves, in the width of the acumen or revoluteness of the border, and in regard to the development of the papillae. On the other hand, I am unable to refer them to any other American species, and am forced to the opinion that *L. helicoides* is an extremely variable species, perhaps an extreme form of *L. ulocalyx* (C. M.) Mitt. Thus no. 7444, from Campanario, with elongate, less dentate leaves, indicates a tendency toward this latter species, while nos. 4549 and 4776, from Cerro Azul, with more incrassate cells and long, salient, sometimes bifurcate papillae, approach *L. exasperatum*. Many species, often based on a single specimen, have been proposed in this genus; I am confident that when it proves possible to study these in the field, or with ample material, the number will have to be greatly reduced.

BRYACEAE (in part)

LEPTOBRYUM PIRIFORME (L.) Wils. Bry. Brit. 219. *pl. 28.* 1855

Puebla: Rio San Francisco (922, 5002).

EPIPTERYGIUM MEXICANUM (Besch.) Broth. in E. & P. Nat. Pflanzenfam. 1^3: 555. 1903

Loma Santa María (4900).

BRYUM (Argyrobryum) CINEREUM Thér., sp. nov.

(Fig. 11)

Finca Guadalupe (736).

Sterile, pusillum, viride cinereum, innovationibus gracilibus, 0.5 cm. longis, dense foliosis, julaceis, arcuatis. Folia minuta, imbricata, cordato-ovata, subobtusa vel acuta, 0.36-0.50 mm. longa, 0.36 mm. lata; rete denso, pellucido, ad ⅔ folii chlorophylloso; cellulis basilaribus marginalibus quadratis, diam. 10 μ, juxtacostalibus rectangularibus vel rhombeis, mediis et superioribus hexagonis, parietibus valde incrassatis, 30-40 μ longis, 10 μ latis; costa ultra medium evanida, basi 30 μ crassa.

This species is essentially characterized by its slender, julaceous, arcuate branches, its small leaves, almost as wide as long and strictly imbricate even when moist, and its compact areolation, the median and upper cells with walls much thickened, especially at the angles. It

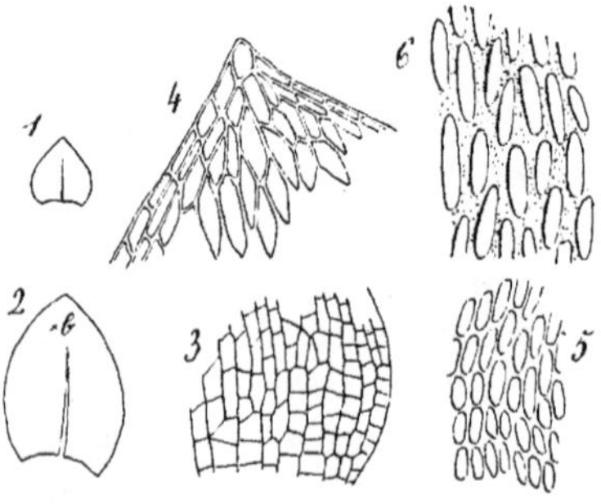

Fig. 11.—*Bryum cinereum* Thér. 1, leaf, × 17; 2, another leaf, × 30; 3, basal areolation, × 130; 4, apical cells, × 130; 5, upper cells at point *b*, × 130; 6, the same, × 200.

may be compared to *B. amblyolepis* Card., but that has soft, laxly foliolate innovations, larger leaves with laxer areolation, and the upper leaf cells not incrassate.

BRYUM (Argyrobryum) ARSENEI Thér., sp. nov.

(Fig. 12)

Puebla (724).

Habitu *Acaulon mutico* (Schreb.) sat simile. Caules et innovationes valde breves, vix 1 mm. longi, gemmiformes. Folia sicca et humida dense imbricata, concava, cordata, suborbicularia, obtusa vel subacuta, marginibus planis, 0.4-0.5 mm. longa, 0.4-0.5 mm. lata; rete pellucido, vix ubique chlorophylloso, cellulis basilaribus quadratis, ceteris breviter rectangularibus, mediis et superioribus breviter hexagonis, parietibus parum crassioribus, viridibus, 25-30 μ longis, 10-12 μ latis, costa apicem attingente, basi 30-36 μ lata. Pedicellus rubellus, 7 mm. altus, capsula pendula, oblonga, collo brevi, ruguloso, attenuato, 1.5 mm. longa, operculum convexo-conicum, haud mamillatum, annulus latus (2-3 cell.), peristomii dentes aurantiaco-purpurei, 0.30 mm. longi, trabeculis prominentibus, membrana ad ½ dentium producta, processus lanceolati, perforati, cilia inaequalia, nunc evoluta appendiculataque, nunc rudimentaria, sporae laeves, 8-10 μ crassae.

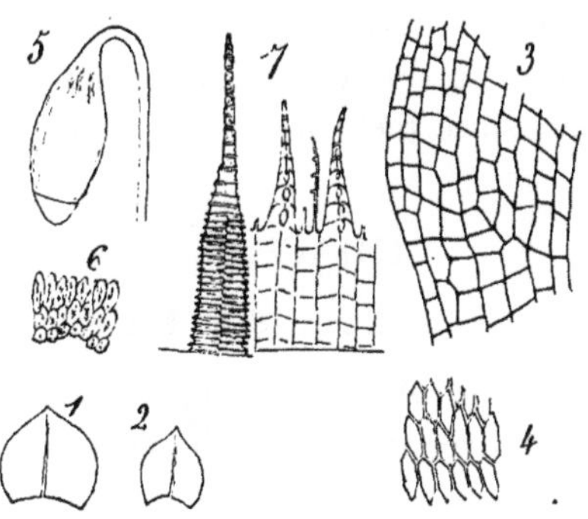

FIG. 12.—*Bryum Arsenei* Thér. 1, 2, leaves, × 17; 3, basal areolation, × 130; 4, upper cells, × 130; 5, moist capsule, × 12; 6, annulus, × 90; 7, portion of peristome, × 90.

BRYUM ARGENTEUM L. Sp. Pl. 1120. 1753

Puebla, Boca del Monte (4675); Mayorazgo (5974); Morelia (7604, 7884, 7964).

BRYUM ARGENTEUM L., forma

Morelia, Loma del Zapote (4640).

A form which fluctuates between the variety *majus* Schwaegr. and *B. amblyolepis* Card.

BRYUM ARGENTEUM L. var. COSTARICENSE R. & C. Bull. Soc. Bot. Belg. 31[1]: 167. 1893

Puebla: Cerro Guadalupe, Cerro Tepoxúchitl, Rancho Guadalupe (689, 4522, 4606, 4816, 4864). Michoacán: Loma Santa Maria (4914, 7856); Cerro Azul (4931); Loma del Zapote (7605); Jesús del Monte (7616); Campanario (7944). Distr. Federal: Mixcoac (9429).

It will be readily understood that all these specimens are neither identical among themselves nor an exact match for the var. *costaricense*. Also I regard this variety as one which ought to be included among the forms of the var. *lanatum* (P. B.) Bry. Eur., as it is scarcely distinguishable save by the paler color of the tufts and by having the leaves chlorophyllose to the middle or a little beyond.

BRYUM MINUTULUM Sch.; Besch. Mém. Soc. Sci. Nat. Cherbourg **16**: 197. 1872

Puebla, Camino de Cholula (4861).

BRYUM INSOLITUM Card. Rev. Bryol. **36**: 112. 1909

Puebla (908); Rancho Posadas (4805).

The latter plant corresponds well with the description; I feel less certain of the other, since the fruit is young and the capsules unformed.

MNIACEAE

MNIUM ROSTRATUM Schrad. in L. Syst. Nat. ed. 13. **2**^{2}: 1330. 1791

Morelia (7903); Cerro Azul (5054); Campanario (7709).

The last specimen is a robust form with much larger leaves.

BARTRAMIACEAE

ANACOLIA INTERTEXTA (Sch.) Jaeg. Adumb. **2**: 699. 1877-78

Cerro Azul (4783, 4976); Cerro San Miguel (5100, 7501); Campanario (7563).

This species is absolutely different from *Anacolia setifolia* (Hook.) Jaeg., and one can hardly understand how Mitten could have confused it with that. Reading the two descriptions is enough to satisfy one. C. Müller (Syn. **1**: 574. 1848) thus describes the areolation of *A. intertexta:* "Folia . . . e cellulis ubique aequaliter minutissimis, densis firmis rotundatis areolata." And Mitten (Journ. Linn. Soc. **12**: 268. 1869) says of *A. setifolia.* "Folia . . . basi cellulis angustis pellucidis areolati . . . cellulis angustis obscuris." Accordingly, the characters given by Brotherus in Engler & Prantl (Nat. Pflanzenfam. **1**^{3}: 634. 1904) as referring to *A. setifolia* should be understood as belonging to *A. intertexta* (cf. l. c. 635. *fig. 478*).

ANACOLIA INTERTEXTA (Sch.) Jaeg. var. **ARISTIFOLIA** Thér., var. nov.

Cerro San Miguel (4873, 5076).

The variety differs from the type in that the leaves are narrower at the base, more gradually and more finely acuminate, less strongly revolute and for a shorter distance, and less sulcate; also in that the costa is excurrent into a long slender awn, and that the median cells are narrower (6-7 μ) with less thickened walls.

ANACOLIA SUBSESSILIS (Tayl.) Broth. in E. & P. Nat. Pflanzenfam. 1[3]: 634. 1904

Esperanza (4727); Hacienda Batán (4939); Morelia (7908); Jesús del Monte (7963). Several forms.

This species has the lamina of the leaves of two layers of cells and the areolation obscure, while *A. intertexta* has the lamina of a single layer and the areolation pellucid.

PHILONOTIS VIRIDANS Card. Rev. Bryol. 38: 36. 1911

Loma Santa Maria (7646). Dioicous!

This species is surely close to *P. graminicola* (C. M.) Jaeg. I am unable to detect any difference in habit, size, form, and measurements of the leaves, areolation, and length of pedicel. (It is true that Brotherus in E. & P. Nat. Pflanzenfam. 1[3]: 646. 1904, says that *P. graminicola* has a pedicel 3 cm. long, but in a specimen I have from C. Müller the seta is not over 12 mm. long.) The two species differ, then, solely in the inflorescence: autoicous in *P. graminicola,* dioicous in *P. viridans*. Is this enough? I should admit readily that *P. viridans* is a dioicous form of *P. graminicola.*

PHILONOTIS CURVATA (Hampe) Jaeg. Adumb. 1: 545. 1873-74

Puebla, Camino de Cholula (4860). Autoicous!

This species differs from *P. radicalis* (P. B.) Brid. in its smaller size, floral verticels, shorter seta, and smaller capsule. The species is new to Mexico and to North America.

PHILONOTIS ELEGANTULA (Tayl.) Jaeg. Adumb. 1: 543. 1873-74

Andameo (4826); Loma Santa Maria (4898); Campanario (7632, 7634b, 7637, 7639). Identifications confirmed by Dismier.

A species new to the Mexican flora; the greater portion of the specimens differ from the type in their more robust size. *P. amblyoblasta* C. M. is distinguished from *P. elegantula* by the oval-cordate, proportionately wider leaves, and by the much stronger costa (50-60 μ, as against 30-36 μ), which retains its width nearly to the apex of the leaf and is exactly percurrent. However, it is undoubtedly a closely related species.

PHILONOTIS MARCHICA (Willd.) Brid. Bryol. Univ. 2: 23. 1827

Campanario (7538, 7709a).

This species has not been known previously from Mexico.

PHILONOTIS JAMAICENSIS (Mitt.) Card. Rev. Bryol. 38: 102. 1911

Morelia, Loma Santa Maria (4911, 5098).

BREUTELIA TOMENTOSA (Sw.) Sch.; Paris, Ind. Bryol. 155. 1894

Puebla (4950). A form with secund leaves.

BREUTELIA INTERMEDIA (Hampe) Besch. Prodr. Bryol. Mex. 60. 1871

Cerro San Miguel (5072); Campanario (7447, 7540, 7925).

Since all the material of this species and the preceding is sterile, the determinations remain doubtful.

POLYTRICHACEAE

ATRICHUM MUELLERI Sch. var. **CONTERMINUM** (Card.) Thér., comb. nov.

Atrichum conterminum Card. Rev. Bryol. 37: 5. 1910.

Cerro Azul (5095); Campanario (7562, 7641, 7753); Cerro San Miguel (5039, 5056, 5094).

Cardot recognized (cf. Rev. Bryol. 38: 37. 1911) that the majority of the characters, save the height of the lamellae, by which he separated his *A. conterminum* from *A. Muelleri* were not constant. The material from Cerro San Miguel, however, weakens this latter character. Thus, no. 5056 shows 4 lamellae of 4 or 5 cells, and nos. 5039 and 5094, collected at the same locality on another date, show lamellae of 6 to 8 cells. The height of the lamellae, therefore, is no more constant here than in the case of our *A. undulatum* P. B. Since the plants from Cerro San Miguel link *A. conterminum* closely to *A. Muelleri,* it becomes necessary to reduce Cardot's species to a variety.

POGONATUM CYLINDRICUM Sch.; Besch. Mém. Soc. Sci. Nat. Cherbourg 16: 111. 1872

Cerro Azul (5052); Cerro San Miguel (5060, 7544), Campanario (7458); Carríndapaz (7543).

Naming these mosses has been laborious. They are certainly related to *P. ericaefolium* Besch. and *P. Lozanoi* Card., but they differ in their more robust habit, longer leaves with plane or involute (not revolute) margins, larger teeth, higher lamellae (of 4 or 5 cells), and longer capsule. It occurred to me to compare this material with *P. comosum* Sch. and *P. cylindricum* Sch., species belonging to the group of *P. ericaefolium,* according to Brotherus. The Director of the Royal Botanic Garden at Kew was kind enough to lend me the

Schimper types, preserved in that collection. To my surprise I found that the moss labeled *P. comosum* by Schimper is not a *Pogonatum* but is identical with *Polytrichum alpiniforme* Card. The tables are turned, however, since so far as I can see, *Pogonatum cylindricum* Sch., which Brotherus (E. & P. Nat. Pflanzenfam. 1[3]: 691. 1905) makes a synonym of *P. comosum*, differs only in unimportant characters from the specimens collected by Brother Arsène.

POGONATUM BARNESI Card. Rev. Bryol. 38: 38. 1911

Cerro San Miguel (5057, 5068); ruisseau de Santa Maria (5096); Campanario (7577, 7578, 7943).

A rather variable species.

POGONATUM TOLUCENSE (Hampe) Besch. Mém. Soc. Sci. Nat. Cherbourg 16: 107. 1872

Cerro San Miguel (5088, 5099); Campanario (7461); Loma Santa Maria (7861); Andameo (4838).

POGONATUM TOLUCENSE (Hampe) Besch. var. CHIAPENSE (Broth.) Thér., comb. nov.

Pogonatum chiapense Broth. in Card. Rev. Bryol. **37**: 5. 1910.

Andameo (4840); Jesús del Monte (7605).

Judging from specimens collected at Las Chiapas (ex herb. Levier), *P. chiapense* Broth. differs so little from *P. tolucense* that I consider it preferable to subordinate the first species to the second as a variety. It is distinguished by the less dentate leaves with the sheathing portion oblong, scarcely dilated, and very gradually narrowed into the lamina. No attention is paid to the differences in the length of the seta. for anyone knows the value of this character in the various species of *Pogonatum!* The plant I received from Pringle under the name of *P. chiapense* (no. 10698) does not differ from *P. tolucense.*

POLYTRICHUM ANTILLARUM Rich.; Brid. Bryol. Univ. 2: 138, 747. 1827

Puebla (4954); Xúchitl, 2800 meters (7983).

POLYTRICHUM JUNIPERINUM Willd. Fl. Berol. Prodr. 305. 1787

Campanario (7443, 7572).

HEDWIGIACEAE

HEDWIGIA ALBICANS (Web.) Lindb. Musc. Scand. 40. 1879

Puebla: Cerro Tepoxúchitl (4852); Malinche (6001); Cerro San Miguel (4878). Often accompanied by *Braunia Andrieuxii.*

BRAUNIA LIEBMANNIANA Sch.; Besch. Mém. Soc. Sci. Nat. Cherbourg 16: 185. 1872

Cerro Azul (4536, 4539, 4550, 4776, 4780); Cascade de Coincho (4714).

This imperfectly understood species differs from *B. squarrulosa* (Hampe) Broth. by the less robust growth, the stems and branches being only about half as thick, by the smaller leaves, which when moist are spread out nearly at right angles, and by the shorter cells. The inflorescence is autoicous!

For a long time I hesitated to refer these collections of Brother Arsène to *B. Liebmanniana* on account of inability to verify in this material two characters mentioned by C. Müller, viz: *Leaves imbricate, inflorescence hermaphrodite.* The difficulty was solved only by comparison with Schimper's type (collected by Liebmann), which the authorities at Kew, whom I thank most appreciatively, were so kind as to lend me. This showed the leaves sometimes imbricate, sometimes a little patent; hence the leaf arrangement is not a character of much value. I dissected also a perichaetium without being able to discover a single antheridium; but below the female flower I found a thick bud, full of antheridia and paraphyses and surrounded by short leaves of different shape from the perichaetial ones. This bud is a true male flower, and the species is autoicous, not hermaphrodite as C. Müller declares.

I believe that the following explains the source of the error: When the male flower was examined, among the ripe inflated antheridia were some narrow withered organs which I suppose to be aborted antheridia, but which C. Müller doubtless took for archegonia. This hypothesis seems to be confirmed by the phrase which follows the description: "*Hermaphrodita antheridiis et paraphysibus copiosissimis archegoniis paucis.*" The male flowers of Brother Arsène's plants are exactly like those just described, and present the same peculiarity. To my knowledge, *B. Liebmanniana* has not been collected in Mexico since Liebmann's time. It is true that in the great herbaria, especially that at Paris, there are specimens collected by Bourgeau and by Hahn which are so named by Bescherelle. But the error of determination is clear; these plants, which really belong to *B. secunda,* have pedicels at least 1 cm. long; they are also paroicous, which character is doubtless the source of Bescherelle's error. Finally, this author did not know the type of *B. Liebmanniana,* as is shown by his placing the closely related to *B. Liebmanniana* and *B. squarrulosa* in two different genera.

BRAUNIA PLICATA (Mitt.) Jaeg. Adumb. 2: 87. 1869-70

Campanario (7933).

This number represents a form with the leaf cells less incrassate than those of the Quito type.

Cardot founded a variety *canescens* (Rev. Bryol. 38: 38. 1911) on Pringle's no. 10627 from Cuyamaloya. The plant I received from Pringle under this number is not the same as that which Cardot saw, and does not differ from *B. secunda*. It is useful to note this error, since it may exist in other sets of the exsiccatae.

BRAUNIA SECUNDA (Hook.) Sch.
and
BRAUNIA ANDRIEUXII Lor.

These two species are very closely related, and the distinctive characters are few. *B. secunda* may be recognized by the slender branches, whose leaves are very patent when moist, and by the slightly plicate stem leaves, which are strongly revolute from base to apex. *B. Andrieuxii* has short, thick branches with erect-patent leaves, and the stem leaves are plane or very narrowly revolute toward the base only, as well as being broader and more abruptly acuminate. These characters, however, are not always associated on the same plant. Some forms have the cauline leaves plane-margined but less broadly oval than those of *B. Andrieuxii;* others have the stem leaves of *B. Andrieuxii* and the branch leaves very patent; still others (no. 4851) have leaves plane-margined but only very slightly plicate, etc. It is necessary to recognize that the characters given for each species lack constancy, and that *B. Andrieuxii* is only an extreme form of *B. secunda*. For these reasons I have reduced it to varietal rank.

While studying the numerous specimens from Brother Arsène and those in the Pringle collection, I was astonished to find that all of them that could be referred to *B. secunda* were paroicous, while those with the characters of *B. Andrieuxii* were autoicous. This would be a specific difference, if *B. secunda* were always paroicous, but such is not the case. According to the description, and to the precise statements of Bridel, Schwaegrichen, and Schimper, the type of this species is autoicous. Furthermore, I have recently seen a specimen (Holzinger, Musc. Ac. Bor. Am. no. 480) of *B. secunda* which certainly has an autoicous inflorescence. The Mexican plants collected by Brother Arsène and by Pringle prove that the inflorescence of *B. secunda* is variable; I have therefore grouped these plants as forma *paroica*.

BRAUNIA SECUNDA (Hook.) Sch. f. PAROICA

Michoacán: Santa Clara (leg. Abb. Treviño, 4884); Morelia, Çerro Azul (4556); Campanario (7442, 7571).

BRAUNIA SECUNDA (Hook.) Sch. var. ANDRIEUXII (Lor.) Thér., comb. nov.

Braunia Andrieuxii Lor. Moosst. 164. 1864.

Morelia (7904); Carríndapaz (7542); Dos Tetillas (7658); Andameo (4832, 4847); Cerro San Miguel (4869, 4872, 5045, 5077, 5104, 7541); Punguato (4880, 4882, 5047, 5050, 5092); Cerro Tepoxúchitl (4523, 4851).

BRAUNIA SECUNDA (Hook.) Sch. var. CRASSIRETIS Thér., var. nov.

Morelia, Campanario (7448).

This variety resembles *B. secunda* in the paroicous inflorescence and the position and shape of the leaves; but it differs clearly in the more deeply plicate cauline leaves, which are less widely revolute and terminated by a short, hyaline acumen, also in the rhacomitrioid areolation like that of *B. cirrifolia* (cells elongate with much thickened, sinuate walls), and, finally, in the longer (17 to 20 mm.) pedicel. It is perhaps a new species.

LEUCODONTACEAE

LEUCODON CURVIROSTRIS Hpe. Icon. Musc. *pl. 16.* 1844

Jalapa (Veracruz), 1400 m. (7970, 7992); Hacienda Batán (4963, 4971); Esperanza (4660, 4670, 4734, 4803, 7979); Xúchitl (8005); Cerro Azul (4528, 4537).

PRIONODONTACEAE

PRIONODON MEXICANUS Thér., sp. nov.

(Fig. 13)

Jalapa (8000); Cerro Azul (4980); Campanario (7569).

Habitu *P. Piradae* Par. sat similis, differt foliis brevioribus latioribusque, magis dentatis, magis abrupte in acumen duplo breviorem contractis, valde fragilis.

This species may be distinguished from *P. laeviusculus* Mitt. by its leaves, which are stiff, imbricate when dry, very clearly plicate longitudinally, and with a somewhat longer and more slender acumen. From *P. densus* (Sw.) C. M. it differs in its shorter yet more robust stems and in having the erect-imbricate leaves nearly twice as wide and abruptly contracted to a much shorter acumen which is less sharply dentate, etc.

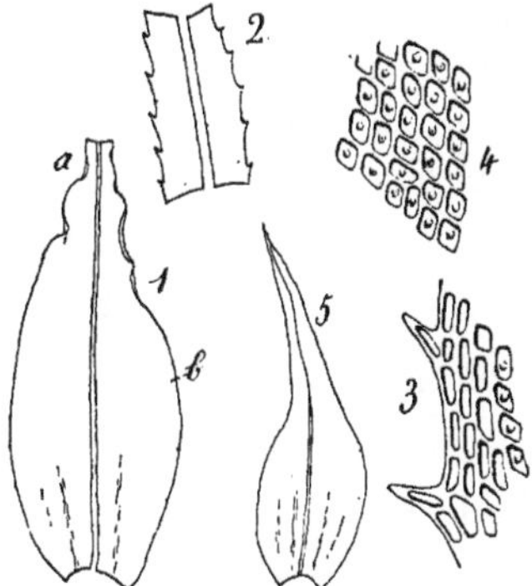

FIG. 13.—*Prionodon mexicanus* Thér. 1, mature, broken leaf, × 12; 2, portion of the acumen near *a*, × 30; 3, leaf-margin near *b*, × 200; 4, median cells, × 200; 5, young leaf, × 12.

PRIONODON ARSENEI Thér., sp. nov.

(FIG. 14)

Esperanza (4747).

Caulis robustus, 4-8 cm. altus, ramosus, ramis brevibus, crassis, dense foliosis. Folia erecto-appressa, plicata, fragilia, e basi oblonga

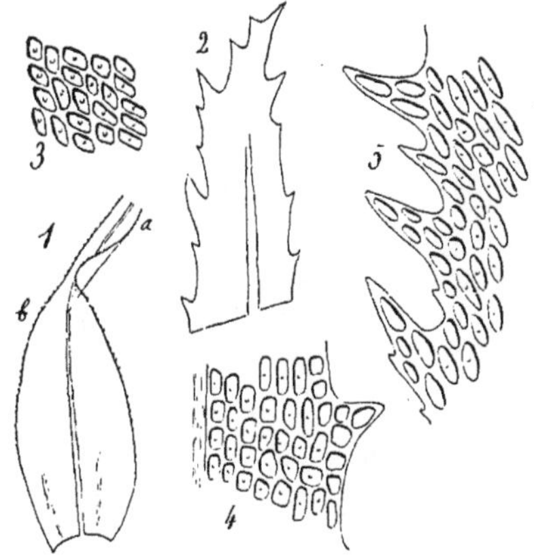

FIG. 14.—*Prionodon Arsenei* Thér. 1, leaf, × 12; 2, acumen of a young leaf, × 90; 3, median cells, × 200; 4, margin of leaf near *a*, × 200; 5, margin of leaf near *b*, × 200.

abrupte in acumen loriformum contracta, fere e basi runcinato-dentatis, cellulis ovatis, papillosis, diam 15 μ, basilaribus juxtacostalibus linearibus, porosis, marginalibus parvis, quadratis, valde incrassatis; costa angusta infra summum evanida.

The leaves of this species are notable for the large teeth which are often again dentate, like those of *P. ciliatus* Besch. from Reunion. *P. lycopodium* (C. M.) Jaeg., from New Grenada, is easily distinguishable from our species by its habit, its elongate, often flagelliform branches, its flexuose, undulate, crisped leaves, etc.

ENTODONTACEAE

ENTODON ERYTHROPUS Mitt. var. **MEXICANUS** Card. Rev. Bryol.. 37: 11. 1910

Loma Santa Maria (5091); Carríndapaz (7579a); Teocalli de Cholula (4859); Hacienda Alamos (4630); Rancho Guadalupe (4603, 4611); Tlaxcala, Santa Ana Chiautempan (4854 p. p.).

ENTODON ERYTHROPUS Mitt. var. **INTERMEDIUS** Thér., var. nov.

Campanario (7512, 7518, 7522, 7580).

The leaves are small and very concave as in the var. *Muenchii*, but the plant has the size, areolation, peristome, and annulus of var. *mexicanus*. It is possible to look upon this variety as forming a transition between var. *mexicanus* and var. *Muenchii*, and it justifies Cardot's action in joining *E. Muenchii* Broth. to the polymorphous *E. erythropus*.

ENTODON ERYTHROPUS Mitt. var. **MUENCHII** (Broth.) Card. Rev. Bryol. 37: 11. 1910

Fort Guadalupe (4619, 4623); Cerro Guadalupe (654, 656, 684, 686); Hacienda Alamos (4693, 4695, 4762, 4765, 4767, 4782): Rincón (4566, 4568); Jesús del Monte (7615). Distrito Federal, Mixcoac (9459).

The variety appears to me to differ from var. *mexicanus* by the rather less robust size, the less complanate branches, and the smaller, very concave leaves, which are more laxly areolate and have fewer quadrate basal cells and a larger costa.

ENTODON ERYTHROPUS Mitt. var. **BREVISETUS** Card. Rev. Bryol. 40: 39. 1913

Morelia: Bosque San Pedro (4572).

ENTODON ABBREVIATUS (Br. Eur.) Jaeg. Adumb. 2: 356. 1875-76

Cerro Azul (4779).

ERYTHRODONTIUM CYLINDRICAULE (C. M.) C. M. Bull. Herb. Boiss. 5: 208. 1897

Xúchitl, at foot of Orizaba, 2800 m. (7990). Morelia: Cascade de Coincho (4715); Cerro Azul (5081); Campanario (7466, 7527, 7530, 7633a).

No. 7466 is a form with shorter and more concave leaves; no. 7633a has shorter setae (6 to 7 mm.).

ERYTHRODONTIUM DENSUM (Hook.) Par. Ind. Bryol. ed. 2. 2: 158. 1904

Puebla: Hacienda Santa Bárbara (4516). Mexico: El Oro (leg, *Rangel,* 4881). Morelia: Cerro Azul (4534, 4542, 4553, 4787); Campanario (7450, 7525); Andameo (4846 p. p.).

The specimens numbered 4534, 4542, 4787, 4846 p. p., and 7450 have the cauline leaves shorter and more quickly contracted to a short acumen; they doubtless represent Cardot's var. *brevifolium* (cf. Rev. Bryol. 37: 12. 1910). No. 7525 is a more slender form with filiform branches; the solitary capsule seen has a short, conic operculum.

ERYTHRODONTIUM PRINGLEI Card. Rev. Bryol. 37: 11. 1910

Morelia: Cascade de Coincho (4710); Campanario (7519).

PLATYGYRIELLA HELICODONTOIDES Card. Rev. Bryol. 37: 9. 1910

Cerro Azul (4546, 4551, 4778, 4795, 4796), Jesús del Monte (7614).

PLATYGYRIELLA IMBRICATIFOLIA (R. S. Will.) Thér., comb. nov.

Erythrodontium imbricatifolium R. S. Will. in Card. Rev. Bryol. 37: 12. 1910.

Morelia: Jesús del Monte (7602, 7603, 7606, 7618, 7624); Cerro San Miguel (5101); Bosque San Pedro (4584); Carríndapaz (7953)

Since several of the plants are fertile, it is possible now to determine that the systematic position of this species, mentioned by Cardot in Rev. Bryol. 37: 12. 1910, is not in the genus *Erythrodontium* but in *Platygyriella*. Actually, the capsule is annulate, the exostome has the structure of that of *P. helicodontoides,* and, finally, the endostome is composed of a short exserted membrane bearing slender fragile cilia that nearly equal the teeth in length.

PYLAISIA SCHIMPERI Card. Bull. Herb. Boiss. 7: 373. 1899

Morelia: Campanario (7457).

A species not previously known from Mexico.

PYLAISIA SUBFALCATA Sch. Bry. Eur. (46-47:) Pylaisaea 3. 1851 (nom.); Besch. Mém. Soc. Sci. Nat. Cherbourg 16: 147. 1872

Puebla: Hacienda Batán (4938, 4970 p. p.).

PYLAISIA FALCATA Sch. var. **INTERMEDIA** Thér., var. nov.

Puebla: Esperanza (4663, 4667, 4678, 4688a, 4735).

The leaves are as long as those of *P. falcata*, the acumen long and slender, the alar cells small and obscure, the capsule short as in the species, but in the direction of the leaves and in their median areolation the moss recalls *P. subfalcata*. It is probably an intermediate form, unless it be a new species. That the length of the seta is greater than in either of the two related species would seem to confirm the latter view. Unfortunately the lack of capsules in good condition renders examination of the peristome impossible.

ROZEA BOURGEANA Besch. Mém. Soc. Sci. Nat. Cherbourg 16: 142. 1872

Morelia: Campanario (7565).

FABRONIACEAE

FABRONIA PERIMBRICATA C. M. Flora 83: 334. 1897

Morelia: Loma Santa Maria (7866).

A species not previously collected in Mexico.

FABRONIA FLAVINERVIS C. M. Linnaea 38: 645. 1874

Morelia: Calzada de Mexico (5066). Veracruz: Córdoba (s. n.).

FABRONIA PATENTIFOLIA Card. Rev. Bryol. 37: 49. 1910

Puebla (4512, 4814).

In the Revue Bryologique 38: 41. 1911, Cardot expresses the view that this species is only a variety of the preceding; I consider it, on the contrary, sufficiently distinct from *F. flavinervis*. Besides the characters mentioned by Cardot, it has a capsule attenuate at base into a rather long neck, a conical obtuse operculum, a peristome made up of 8 pellucid, striate teeth, and small (13 μ) spores. *F. flavinervis* has an oblong capsule, abruptly contracted into a short neck, a convex, rostrate operculum, a peristome made up of 16 opaque, papillose teeth, and spores 15 to 18 μ in diameter.

RHACOPILACEAE

RHACOPILUM TOMENTOSUM (Sw.) Brid. Bryol. Univ. 2: 719. 1827

Puebla: Hacienda Alamos (4698, 4719). Morelia (7739, 7895); Campanario (7453, 7521); Cerro San Miguel (7441); Loma Santa Maria (4918, 4928, 7867, 7868 p. p.); Santa Clara (4887, leg, *Abb. Treviño*). Tlaxcala: Santa Ana Chiautempan (4854).

It is possible to separate these specimens into two groups, one composed of plants with large stipules that are almost as wide and as long as the leaves, which latter are crisped when dry and not complanate-distichous. These forms I judge to represent *R. latistipulaceum* Card. (Rev. Bryol **38**: 41. 1911). The second group contains the plants with leaves erect-distichous when dry, little or not at all crisped, and the stipules not more than half the size. This group represents *R. tomentosum.*

But there are some specimens that do not readily fall into these groups; that is, the characters mentioned above are neither constant nor always concurrent in the same plant. The stipules, especially, display much variation in size. Further, it is the opinion of Mrs. E. G. Britton that the species should be united.

SMITHSONIAN MISCELLANEOUS COLLECTIONS
VOLUME 78, NUMBER 3

E. H. Harriman Fund

THE CLASSIFICATION AND DISTRIBUTION OF THE PIT RIVER INDIAN TRIBES OF CALIFORNIA

BY
C. HART MERRIAM, M. D

(Publication 2874)

CITY OF WASHINGTON
PUBLISHED BY THE SMITHSONIAN INSTITUTION
DECEMBER 31, 1926

The Lord Baltimore Press
BALTIMORE, MD., U. S. A.

THE CLASSIFICATION AND DISTRIBUTION OF THE PIT RIVER INDIAN TRIBES OF CALIFORNIA

By C. HART MERRIAM

CONTENTS

INTRODUCTION

The present paper embodies some of the results of field work among the Pit River tribes begun in 1907 and continued at intervals for 20 years. From 1887 to 1910 the work with Indians was incidental to zoological and botanical studies conducted while mapping the Life Zones for the U. S. Biological Survey, but since 1910 it has

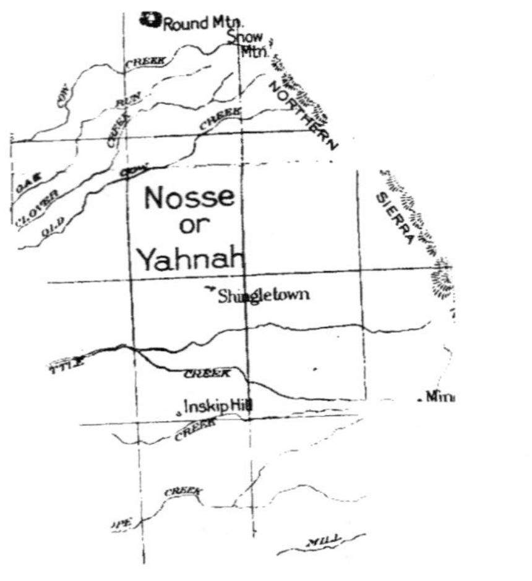

MAP OF THE PIT RIVER TRIBES—ACHOMAWAN STOCK

By C. Hart Merriam

1926

ENGRAVED AND PRINTED BY THE U.S. GEOLOGICAL SURVEY

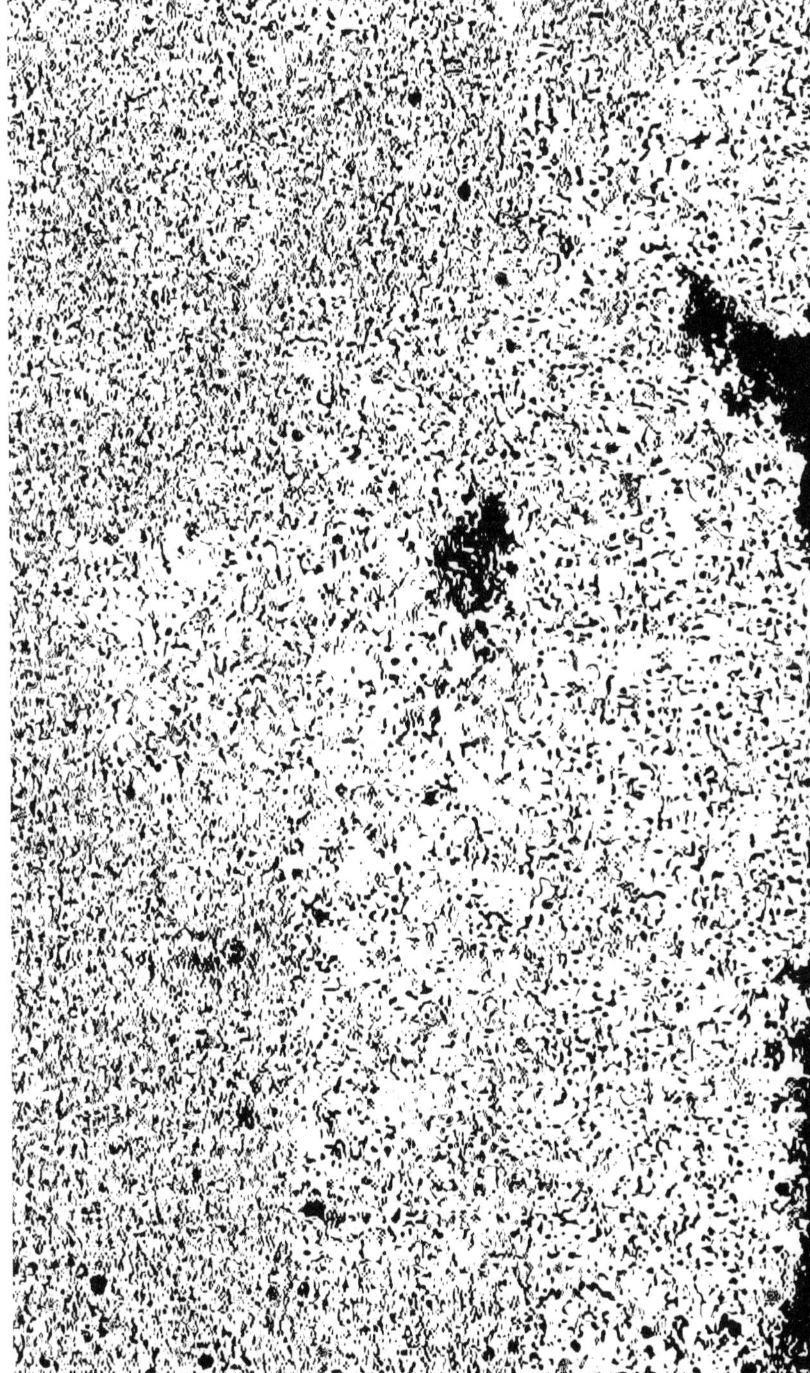

been continued under a special fund contributed by Mrs. E. H. Harriman for my investigations as research associate under the Smithsonian Institution.

The lamentable dying off of the old people, in many cases the last survivors of tribes on the verge of extinction, is a matter of grave concern and has seemed to demand the stressing of this work in advance of everything else—for the gathering of vanishing data is a duty we must all recognize.

The material here published relates to the classification and distribution of the Pit River tribes. The results in general anthropology, mythology, ethnozoology and ethnobotany are reserved for later communications.

My work in anthropology has been done *not* for the ultra specialist in linguistics, but for the average educated American who wants to learn about our aboriginal inhabitants. The alphabet employed therefore gives the usual sounds of the letters in the English alphabet. The only addition is the super [ch] to denote the sound of *ch* in the German *buch*—a sound lacking in English but common in many Indian languages.

Disquieting special and abnormal combinations and usages—such as *c* for *sh* [spelling shut, *cut*] ; *c* for *th* [spelling that, *cat*] ; *tc* for *ch* [spelling church, *tchurtch*] ; *s* for *sh* [spelling sham, *sam*] ; *ts* for *s* [spelling sad, *tsad*] ; *dj* for *j* [spelling bluejay, *bludjay*] ; *au* for *ow* [spelling how, *hau*] ; *x* for aspirated *h* or *k*, and so on—have not been adopted.

In writing Indian words many anthropologists are opposed to the use of the hyphen. Thus, no less an authority than Roland Dixon, in reviewing a scholarly contribution to the ethnology of California, states: "The advisability of such extensive hyphenation as is here used is open to question, and it is to be hoped that in further publications the forms will be given without this unnatural separation, convenient though it may be in some ways." (*Am. Anthropologist*, Vol. 6, p. 715, 1904.)

I am of the opposite view, believing that the liberal employment of the hyphen in the separation of syllables is most helpful both to the transcriber and the student.

Many of the names used by the several tribes for neighboring tribes are the place names of the localities where the principal or 'ruling' villages were situated—and in more than one case the village bore the same name. They are here recorded as spoken, although it should be kept in mind that when properly used in the tribal sense, the terminal *le, me, se, te, we, ye,* or *che,* or at least *e,* should be added.

In the lists of names of the tribes it often happens that several spellings are given for the same name. This means that different members of the tribe have given variations or different pronunciations.

The information here presented has been obtained at different times from a number of Pit River Indians of different tribes. Several of the informants are no longer living. Those who have helped most are Istet Woiche (known to the whites as William Hulsey), leader of the *Mo-des'-se;* Samuel and Robin Spring, *As-tah-ke-wī'-che;* Robert Rivis and Harry Wilson, *Ap-woo'-ro-kā'e;* Jack Williams (chief), Sam Steel and old Pete, *Ham-mah'-we;* Harry George (chief) and old Billy Quinn and wife, *At-wum'-we;* Davis Mike and wife, *Å-choo'-mah'-we.* Much of the *Å-choo'-mah'-we* material was obtained 20 years ago.

The accompanying map (frontispiece) shows the distribution of both groups of Pit River tribes (*Achomaw'-an* and *Atsook'-an* Families) and also the contiguous parts of the ranges of all the surrounding tribes.

Failure to receive proof of the map necessitated a correction of the position of the tribal name *Hā-we-si'-doo* after the map had been printed, and prevented changing the color of the *Yah'nah* area, which too closely resembles that of the adjoining *At-soo-kā'-e* tribe.

My daughter, Zenaida Merriam, who for years has accompanied me in my fieldwork among California Indians, has rendered efficient help in the preparation and proof reading of the present paper.

C. HART MERRIAM.

October, 1926.

THE PIT RIVER TRIBES OF CALIFORNIA—ACHOMAWAN STOCK

GENERAL DISTRIBUTION

The Pit River tribes dwell in the northeastern corner of California from the Big Bend of Pit River and Montgomery Creek easterly to Goose Lake and the Warner Range, where they reach within 10 or 15 miles of the Nevada line, and within 6 or 7 miles of the Oregon line.

On the west they occupy both slopes of the heavily forested northern extension of the Sierra Nevada, spreading thence easterly over the broken lava country and open deserts of the western part of the Great Basin area—though by the accident of the present course of Pit River, which cuts deep canyons through the mountains, some of their lands now drain into the interior of California.

The earliest name I have seen for the Pit River Indians is "*Palaihnih* or *Palaiks,*" published without enlightening information by Horatio Hale in 1846 (*Ethnography of the Wilkes Exploring Expedition,* pp. 218, 569, 1846).[1] The term was adopted by Gallatin (1848), Latham (1850, 1854, and 1860), Berghaus (1852), Ludewig (1858), and others, and has been recently revived by Kroeber (*Handbook Inds. Calif.*, p. 282, 1925).

The term *Achomawe,* now in general use, seems to have been introduced by Stephen Powers in 1874 (*Overland Monthly,* Vol. 12, pp. 412-415, May, 1874) and quickly superseded "Palaihnih or Palaiks." With variations in spelling it was again used by Powers in 1877, and accepted by Yarrow (1881), Mallery (1881), Curtin MS (1889), Powell (1891), *Handbook Am. Inds.* (1907), Dixon (1908), Chamberlin (1910), Kroeber (1925) and other present day writers.

While for a number of years the tribes under consideration were commonly referred to collectively as constituting the *Achomawan Stock,* and later as the *Shasta-Achomawi* [still later reduced to Shastan], it should be borne in mind that the word Achomawe is *not* in use among themselves except in the form *Ah-choo′-mah′-we* (or *Ă-ju′-mah-we*), the specific name of one of the tribes—the one in Fall River Valley. They have no collective or general name for themselves.

It should be borne in mind also that the Pit River group is by no means homogeneous, the component tribes speaking two very distinct languages, as pointed out by Dixon in 1907,[2] and emphasized by my much more extensive vocabularies. I regard the differences as of Family rank.

CLASSIFICATION

The term *Family,* according to my ideas of classification, is of distinctly lower rank than *Stock;* a Stock may include several very distinct families. Twenty years ago, when discussing the use of these terms, I said:

> Ethnologists use the terms "stock" and "family" interchangeably, regarding them as synonymous, and drop at once from stock to tribe, giving no heed to divisions of intermediate rank. And if evidence of relationship, however

[1] Apparently from the Klamath word *p'laikni,* meaning from above, or upper country, applied to members of their own tribe on Sprague River.—Gatschet, *Klamath Dictionary,* p. 269, 1890.

[2] The Shasta-Achomawi: A new linguistic stock, with four new dialects.—Amer. Anthrop., Vol. 7, No. 2, pp. 215-216, April-June, 1907.

remote, is detected between two or more stocks the practice is to merge such stocks under a common name and pool the contained tribes—as if the aims of science were served by the abolition of group names and by mixing together in a common jumble a rabble of tribes of diverse relationships! (*Am. Anthropologist,* vol. 9, No. 2, p. 340, April–June, 1907.)

In the case of the Indians now under consideration, I deem it a mistake to stretch the *Shastan Stock*—as has been done by Dixon and Kroeber—by referring to it such strikingly unlike tribes.

The Pit River tribes, as already stated, fall naturally into two widely different groups—a northern and a southern. Of these tribes, nine speak slightly variant dialects of the northern group—which may be called *Ăchoomah'-an;* while only two speak similarly variant dialects of the southern group—which may be called *Atsookā'-an.* The names of the nine *Ăchoomah'-an* tribes (from west to east) are: Mo-des'-se, To-mal-lin'-che-moi', Il-mah'-we, Ă-choo'-mah'-we, At-wum'-we, As-tah-ke-wi'-che, Hā'-we-si'-doo, Ko'-se-al-lek'-tĕ, and Ham-mah'-we. The names of the two *Atsookā'-an* tribes are At-soo-kā'-e and Ap-woo'-ro-kā'e.

The classification here suggested may be readily seen from the following table:

CLASSIFICATION OF THE ACHOMAWAN STOCK

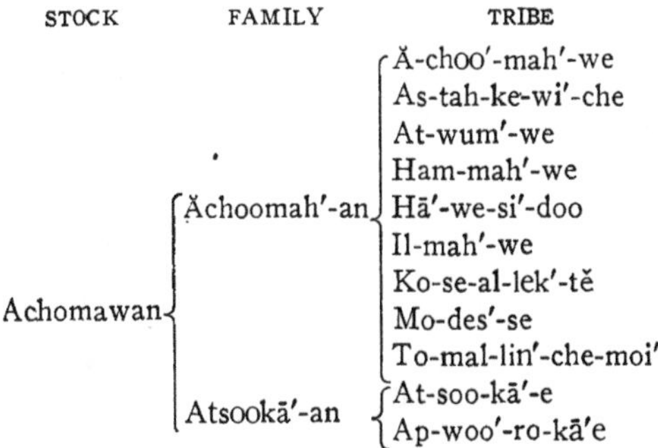

STOCK	FAMILY	TRIBE
Achomawan	Ăchoomah'-an	Ă-choo'-mah'-we
		As-tah-ke-wi'-che
		At-wum'-we
		Ham-mah'-we
		Hā'-we-si'-doo
		Il-mah'-we
		Ko-se-al-lek'-tĕ
		Mo-des'-se
		To-mal-lin'-che-moi'
	Atsookā'-an	At-soo-kā'-e
		Ap-woo'-ro-kā'e

COMPARATIVE VOCABULARIES OF THE ACHOOMAH'-AN AND ATSOOKĀ'-AN LANGUAGES

An examination of the accompanying vocabularies, arranged to exhibit in parallel columns the differences and similarities of the languages of the two sets of tribes, will suffice to show the grounds on which I have elevated the *At-soo-kā'-e* and *Ap-woo'-ro-kāe* tribes to family rank, under the name *Atsookā'-an* Family, thus separating them from the Pit River tribes proper—the *Achoomah'-an* Family.

Among the tribes of each of the two families, many words differ dialectically. Such words are here omitted, the purpose of the accompanying table being to bring out the striking resemblances and differences between the languages of the two groups.

In the case of birds, a number of species are named in imitation of their note or call. Such onomatopoeic names, being essentially the same in all the tribes, are not included in the vocabularies. Omitted names of this kind are:

Sparrow Hawk	Lik'-le'-kah
Raven	Kahk'
Crested Jay	Kahs-kah'-sah
California and Woodhouse Jays	Tsi'-tsi'-ah
Poorwill	Po-luk'
Valley Quail	Sah-kah'-kah

COMPARISON OF WORDS IN ACHOOMAH'-AN AND ATSOOKĀ'-AN LANGUAGES

	Achoomah'-an Tribes	*Atsookā'-an Tribes*
1	Ham-mis	Choo' (or Jew')
2	Hahk'	Ho-ke'
3	Chas'-te	Kis'-ke
4	Hah-tah'-mah	Hahk-kow'
5	Lah'-too	Her-rah'-pah-kin'-nah
6	Mus-oo'-itch	Che-roo-poot chuk'-ke
10	Mah-lo'-se	Chew'-wik-she
People	Is' (or Ês)	Ow'-tĕ' (How-tĕ')
Man	Yal'-le-yu	Kahs-we-wah'-ho'
Woman	Ah'-me-tā-o-chan	Min-nĕ-re'-char'-rah
Old woman	We-ah-chah'-loo	Pā'-chah hā-char-rah
Father	-nā-chah	-tah'-tah
I	It (or Êt)	Ahk (or Ok)
You	Me'	Me'-e
Head	Lah	Nah-hah'
Forehead	Ool'	Aw[ch]-ah-de'
Eye	Ah'-sah	O'-ye
Nose	Yah-me	Yu'-de
Ear	E-saht'	Ahs-mahk'
Throat	Hal-lōk	E-tsis'
Mouth	Ahp'	Ah-poo'
Lips	Poo'-lip	Poo'-lip
Tongue	Ip-le'	Ah-pă-le'
Neck	Wahp'-te	Oop-ke (or Ōp-ke')
Knee	Kaw-hwi	Bul-lōts
Arm	Lah-pow'	Rah-pow'
Hand	Êl	E[ch]-de'
Nails	Al-le-tsah'	Me-choo'-rah

Belly	Pin-nōk′	O′-me′
Chest	Ko-kōs	Too-koo′-rah
Breasts	E-chit	Ah-chis′-kah
Feet	Che′-koi	Chōts′-kĕ′
Skin	Ĕ′-moo	Ĕ-we′
Hair	Te′-ye	War-ri′-ne-kahs′
Teeth	E-chah′	E-tsow′
Bone	Al-lah′-te	Choot′-se′
Tendon	Pim′	E-peu′
Brain	Hem-mal′	Ches′-kah har′-rah
Heart	Hah-dah′-che	Yōp-wi′-ke
Liver	Oo-wā′	Oop′-se
Intestines	Pētch-hol	Pits-hor′-re
Blood	Ah-te	Is′-soo-re
Strong	E-pah′-che	E-pah-ke′
A dance	Tā-hā-kah-le	Yo′-pi′
A song	Dā′-se	Yā′-che′
Rabbitskin blanket	Chal-lo′-kah (or Chĕkah′-te)	Tsah-to′-ro
Moccasins	Kĕ-lah′-lah	Kā′-oo-nar′-rah
Buckskin	Tek-wah′-le	Tips-we′-re
Snowshoes	Tah-pah′-ge	Wir′-re
Shell necklace	Tah-se′-ge	Yo-kĕ-nas′-we′
House	Te-loo′-che	Yow′-te
Ceremonial house	Ahs-choo′-e	Tsum-mah′-hah′
Fire	Mah-lis′	Ah′-hah-we′
Fireplace	To′-ah-ēt	Pah-ye′-tse
Coals	Howk	Chōk′-che′
Smoke	Mah-kets	Rāk′-poo
Fire drill	Tow-wah′-che	Wŏ-te′-kĕr-ras′
Bow	Loo-poo-e′-se	Too-mir-re-ye (To-mid′-ye)
Arrow	Les-tah-ke	Kahp′-ste′
Stone arrow point	Tah-pus′-to-me	Mik′-ke-dĕ′
Quiver	Too-ni′-e	Wits′-tahs′
Sling	Chă-mĕ′-tĕ	Tsā-te′-mah
Digging stick	Kahs′-wah	Hahs′-wah′
Cooking basket	Pah-poo′-gah	Taw[ch]-o-de′
Baby basket	Tōts-chah′-me	Yah[ch]-per-re′
Mortar basket	Kal-lo′-wah (or Kal-loo′-wah)	Ken-no′-wah (or Ken-nah′-ho-wah)
Stone pestle	Ko′-pah	Choo′-poo′ (or Tsoo′-poo′)
Grinding stone (Metate)	Se′-lah	Chahs-koot (or Tsas-kōt′)
Handstone for rubbing	Sah′-wah (or Tsah-wah)	Tsah-wah′
Fish net	E′-kā-low′	E-ke-row′
Meat	Me′-soots′	Ah-soop′

Dried meat	We-soo′-che me′-sootch	Māk-wā ah-soop
Marrow	To-kā′-se	Yo′-kis-se′
Fish	Ahl′	Mah-tse′
Eggs	E-sah′	Ē-tsah′
Acorns	Tĕ-tahts	Tah-ke′
Salt	Tēs	Eu′-tĕ (Yu′-tĕ)
Medicine	Ted-dal′-waht-kă	Che-was′-so′-hes
Tobacco	Oop′	O-pe′
Pipe	Skōt′	Skot′
Good	Too′-se	Woo-si′
Bad	Al-lă-ho′-koi-e	Sow-kah
Lake	Loo′-poo	Poo-wow′
Water	Ahs′	Ah-che′
A spring	Sap′-te	Pah′-poo-tse
Mountain	Ah′-ko	Ah-hă-o′
Valley	At-wum′	Ahk-win′-ki
Road or trail	It-tā-oo	Ah-koo′-roo-e′
Rock (stone)	Al-lis′-te	Noo′-yah-hĕ
Sand	Tahs (or Tas)	Tis′
Mud	Ah-moo′-yah	Ah-moo′-yah
Fog	Ah-too′-mum	Wēt-să-be′mits
Rain	Ahs-che (Tas′-che)	Koi-yit′ (Ki-ye-te′)
Snow	Te′	E′-te′
Ice	Lah′-kahts	Pah-too-re′
Wind	Tā-ho′-me	We′-so′ (Ye′-soo)
Summer	Ah′-loo′-e	Ahp-noo′-e
Winter	Ahs-choo′-e	Ahs-choo′-e
Hot	Ahs-stah′-ke	Pā-o-kah-tah′-we
Cold	Ahs-tā-yu	Yes′-ko-pah′-we
Yesterday	Me′-chim	Oos′-pe′
Today	Bah′-lah	Pin′-ne-kah′-se
Tomorrow	Lo′-ko-me′	Lōk-mah′-se
Daytime	Mah-tik′-chak	Ah-se′-ye
Night	Mah-hek′-chah	Ah-pā′-nah
Now	Pahl′-mus and Wā	Kā-nē′-kah (or Ken′-ne-kah)
Not yet	Nahm′-yu-we	Ow-ko′-ŏ′ (or Ah-ko′-ŏ′)
Long ago	Ĕs-kah′-ne′-tse	Pah-lahk′-me (or Bel-lok′-mim)
Sun	Chool′	Tsun′-ne-oo′
Lightning	Tid′-dah-lum′-che	Pah-loop-lahk′
Clouds	Ah-loo′	Nah′-mis-to′-che
Shade	Wā′-lah	Mahs-ke′
Many (plenty)	Kahm (Kah′-me)	Tsah-ko′
Big	Wah-wah′	E′-te-ke
Little	Chawk′-chă	E-win′-ke

High	Hā′-wis	Choo-pow′-al-lĕ (E′-mar-rah′)
Low (or down)	Chōk-chahk-chook-che	Ā-po-tah (Ow-oo-tā′)
Long	Wah′-wahk choo′-che	Choo-pow′-al-lĕ
Short	Chōk-chah-kah′-de	E-we-kah′-lĕ (E-win′-kah-lĕ)
Straight	Ets-poo′-e	Ē′ts-pĕh′
Crooked	E-ke′-le	Pāt-kĕ-le′
Middle	E-se′ (Is-se′)	Choo′-se
Full	Ah^ch-too-ge	O-pā′
Where	Chah′-wah	Ah′-ke′
Here	Pe′-wah	Kah′-ke
There	Hook-yĕ′ (Gook-yĕ)	Koot-skah′
Grizzly Bear	Woi′ (Woi^ch)	Be′-re-ke (Pe′-re-ke)
Black Bear	Wah-he′-mah	We′-ke-rits′
Mountain Lion	Dah-chah′-lah (Tah-tsah′-lah)	Wă-rahk′-mit-tah′
Bobcat	La-tsah′-le (Nā′-tsah-le)	O-de-ahs′-we
Coyote	Jā′-mool	Mah′-ke-tah′
Big Wolf	Tse′-moo	Mi-yah′-ke
Otter	Hah′-seu	Ser-ruk′-kă-se
Big Skunk	Hah yen′-nah	Hah yen′-nah
Little Spotted Skunk	Sen-net′-ke	Sen-not′-kah
Badger	Hahk′	Wĕ′-he-yah
Mink	Choo-pah′	Yut-poo′-re-yu
Weasel	Yahs′	Nim′-che-ge (Noo′-che-ge)
Mole	Wahl′-lakh-moo′-sah	Pis-tik′
Elk	Bow	We-did′-e-ke (Wă-re′-rik-ke)
Deer	To′-se (Do′-se)	Mah′-koo
Antelope	Chā-kah-kow′-we	Wahs′-te
Bighorn (Mt. Sheep)	Lah-mah′-te	Wim′-mĕ-ke
Beaver	Pōm (Poom)	Hah-yah′-dah
Ground Hog (Marmot)	Ah-mahl′	Po-ye′-ke
Gray Ground Squirrel	Ah-chaht′	Het′-war′-ruk
Gray Tree Squirrel	To-wah′-te	Wahs′-war′-re
Chipmunk	To-kis′	Tsoot-wi′-ah
Pocket Gopher	Is-taht′	O-yā′-he (Yam-yā′-tsah)
Kangaroo Rat	Te-noon′-dah	Tis-wah-dā′-de
White-footed Mouse	Yā-ne′-nah	Tsoom′-se
Wood Rat (round tail)	Tah′-me-yoos′	Tōk-pā-dah
Cottontail	Hah-waht′	Dā′-buk-ke
White-tail Jackrabbit	Waht′-waht	Pah-koo′-rah
Tail	E-pe′	Tip′-poo-wer′-re (Tēp-hwe′-re)

Horns or Antlers	Poo′-wah	We′-pah-ke
Excrement	Ahp′-se	Wĕ′-ke
Hole or Burrow	Woos′-che	O′-ye
Golden Eagle	Low-we′-chah	Pit′-we (Ēt′-we)
Bald Eagle	Hem-mah′-lah	Pot-pe-did′-de
Red-tail Hawk	Mah-tit′-ke	Tis′-so-tah
Fish Hawk	Poo′-tis	Tōk-tōk′-is-se
California Condor	Ahm′-pin-ne	Oom′-pin-ne
Turkey Buzzard	Noots′-ke	Chah′-te (Tsah′-te)
Great Horned Owl	Chum-mah-hah′-loo	Soo′-kah-how
Crow	Ah-we′-chah	Ah-i′-e-se
Blue Grouse	Tam-moom	Wā′-mah-hi′-wah
Mountain Quail	To-ahk′	Toi-toi′-se
Meadowlark	Ham′-me-choon′-tah	Tēts′-ko-lah′
Nighthawk	Pĕ-sek′	Se′-yu-tahk
Kingfisher	Cho′-lo-ah′-mah	Kil′-lĕ-sis′-se′ (Ker′-re-sis′-se)
Pileated Woodpecker	Mahk-mah′-kah	Waht-tah-di-wah
Lewis Woodpecker	Dă (Tĕ)	Chew′-che-wah′
Hummingbird	Chā-pah′-sah	Pă-tsoos′
Pelican	Mah-mil′	Mah-me′-dah
Great Blue Heron	Pel-lah-kes′	Pel-lah-kās′
Mudhen	Al-loo-tsōk	Al-loo′-chuk
Birds (all kinds)	Che′-kah	Che′-kah
Eggs	E′-sah′	Ē′-chah
Feathers	Taw-hā	Ē-top′
Rattlesnake	How′-tah (Hoo′-tah)	Ahs-cho′-me
Gartersnake	Ho-mā′	Hel′-lo-ki′-wah
Scaly Lizard (Sceloporus)	Chā′-hā-nahts	Sik′-tan-nah′
Turtle	Hah′-pits	Ah′-pits
Fish	Ahl′	Mahk′-tse
Salmon	Ahl-lis′	Ah′-nē
Trout	Sal′-le-pi′	Naht′-tă-ye
Sucker	Laht-hā	Bik′-ye
Dragon Fly	Hal-loo′-mah (Hal-le-yoom′)	Chets′-koo-mah
Butterfly	Wahl-wahl′-lah	Top′-lah-lah (Hah′-lah-lah′)
Mosquito	Al-le-hah	Po-ri′-wah
Flea	Ah-che-tsah′	Che-kah-pĕ′-re
Yellowjacket	Chā′-yu	Mo-mo-mis′-e-se
Spider	Chah-hah′	Wek-mah′
Douglas Spruce	Lah-so′	Mah′-tis ōp
Juniper	Ko′-se-mel′-lo	Mah′-hah ōb
Incense Cedar	Lah-too′	Nah-tōb
Yew	Mus-swe′-lo	Paht-soo′-ye

Black Oak	Tat'-tah-cho'	Tah'-ke ōb
Mountain Mahogany	Choos-leō	Pah'-he-bah ōb
Elder	Hal'-le	Po'-raw-ko'-bah ōb
Manzanita	Pat-choi'-yo	Wā-yar ōb
Wild Plum	Paht-te'-lo	Paht-kōb
Chokecherry	Poo-le-lo' (Bul-le-lo')	Choo'-e-wup ōb
Serviceberry	Pe'-tah-lo	Pāk-ne-kōb
Sagebrush	Pah'-tah	Ko-pōp ōb
Big Round Tule	Aht'	Wahm' ōb
Flat Tule (Cattail)	Poo-sahk'	Keu-să-rōb
Grass	Has'-te	Kah-se'-re
Flower	Ah-mahl'	New'-ye-ne
Leaf	Ah-tah'-pah	Lah-tahb'
Root	Wah-too'	Wah-too'
Tree	Ah-soo'	Ahts'-we'

THE HIGH AND LOW LANGUAGES

Attention should be called to the interesting fact that in addition to the language of the common people, there is a *"high language,"* spoken by the chiefs, medicine men, and other leaders. I have found *"high languages"* also among the *Me'-wuk* and a few other tribes; they were not understood by the common people. Such languages are worthy of special investigation as likely to throw light on ancient relationships.

In the case of the Pit River Indians, thirty-four words of the "high language" of the *Mo-des'-se* were given me by *Istit Woiche,* venerable leader of that tribe. No special effort was made to obtain them—they were given incidentally and always mentioned as *"high words"* to distinguish them from the common words, called by him "low words." These are brought together in the following table. Doubtless many more might be secured.

Fragmentary List of "High" and "Low" Words of the Mo-des'-se Language

	LOW WORD	HIGH WORD
Father	No-nă'-chah	To-wah-e'
Mother	Ne-noo'-chah	Tah-te'-e
Old man	Tah-he'-yu-chah	We'-ah-che' yah'-le-yu
My wife	E-too ah-me-tā'-o-chan	Ham-mis
Orphan	Mah-hă'-noo-chum	Sōm-tah'-le
Lover	Das'-poo-i'-me	Kah'-choo dol'-le lók'-te
Puberty	De'-mah-cho'-ke	Del'-lā-hah'-che
Pregnant	Sow'-too-we	E'-se'
Proud	E'-sim to'-je	Lah-sats to'-je
Ashes	Mah'-soo-ke	To'-ko

Smoke	Mah-kāts′	Bahk′-mŏ
Smoke hole	Mah-kets′ ah-pit′	Mak-kets′ oo′-tad-de ki′-mit
Poker	Se′-kah	Te-nah′-lā yēs-tă
Light from fire	Tŭ-mā-ki′-e	Tip-le′-che
An earthquake	De-go-wi′-ye	Tĕ-poo′-mĕ
Clear	Tah-saht-we	Tā-ho-chow choo-we
Rain	Tahs′-che	Too-tsim′-che
Lightning	Tid′-dah-lum′-che	Til-loo-plok′-che
Earthquake	Te-poo′-me	Tig-goo-wi′-e
Sunset	Cho′-lo-to ches′-che	To-lōp-chij′-je
First	It-tahm′	Pā′-tum
Last	Kes′-tum	Ke′-tahk-chum
Soon	Too-wil-chahn	Hoo′-tah
Common	Tah-kow′-chan	Kah-de′-kah dok′-me
Plenty	E-kah′-che	Kum′
Many (or much)	E-kah′-che	Kum′
Behind	Tin-ne-kah too′-je	Te-ah′-too
Trade	Tah-we′u-che	Tan′-nah mut′-se
This	Pe′	Pe-kah
The same	Mah′-mo chi′-e	Ah-mits-kah ahn′-che
Mt. Shasta	Êt-ah′-ko′	Ā-te chan′-nah
Doctor	Tse-kēo′	Hi-ă-too and Wah-sahk′-chan′-ne
Woodpecker	Chis-to′-tol	An′-nă-wah se′-lah
Hand me (pass to me)	Ste-lah	Stĕ-goo′-yahk

CONTRASTING ENVIRONMENTS

The *Ăchoomah′-an* tribes are exposed to an unusual range of physical and climatic conditions and, as would be expected, are profoundly influenced by the local environment. Thus the *Mo-des′-se,* the most westerly tribe, inhabit the dense humid forest of the west slope of the Northern Sierra—a habitat comparable with that of some of the northern Coast tribes; while the *At-wum′-we, As-tah-ke-wi′-che, Hă′-we-si′-doo, Ko′-se-al-lek″tĕ* and *Ham-mah′-we* inhabit open or broken country, in the main arid and treeless—a habitat comparable with the sagebrush plains and open deserts of northern Nevada and southern Idaho—the home of the Northern Piute, Bannok, and Shoshone tribes.

The necessity for adapting themselves to such diverse environments is a matter of life and death, for the exposures to sunlight and wind and storm vary from dark shady forests to barren open deserts; the animals and plants available for food, clothing, and medicine are strikingly different, requiring diverse methods of capture and treatment; the ceaseless struggle for existence differs vastly in intensity, and even the method of guarding against enemies implies unlike forms of vigilance.

It is inevitable that the continued pressure of such widely different environments for a long period of time—measured by thousands of years—should produce notable differences in the lives, habits and beliefs of the people, and if continued long enough might be expected to affect the language and even the physiognomy.

DISTINCTIVE ANIMALS AND PLANTS OF THE HUMID AND ARID AREAS

Persons familiar with the climatic requirements of our western animals and plants will be interested in the accompanying tabular statement of distinctive species of the contrasting areas:

DISTINCTIVE ANIMALS AND PLANTS OF THE MO-DES'-SE AREA (HUMID FOREST, SEE PLS. 1 & 2) COMPARED WITH THOSE OF THE AT-WUM'-WE AREA (ARID DESERT, SEE PLS. 9 & 12)

MO-DES'-SE AREA Big Bend to Montgomery Creek	AT-WUM'-WE AREA Big Valley, Lassen County
MAMMALS	
	Buffalo (extinct)
	Antelope (*Antilocapra americana*)
	Badger (*Taxidea taxus neglecta*)
Fisher (*Martes pennanti pacifica*)	Groundhog (*Marmota flaviventer*)
Marten (*Martes caurina*)	
California Little Spotted Skunk (*Spilogale phenax latifrons*)	Great Basin Little Spotted Skunk (*Spilogale gracilis saxatilis*)
Ring-tail Civet (*Bassariscus raptor*)	
Aplodontia (*Aplodontia major*)	Whitetail Jackrabbit (*Lepus campestris sierrae*)
California Blacktail Jackrabbit (*Lepus californicus*)	Oregon Blacktail Jackrabbit (*Lepus calif. wallawalla*)
California Cottontail (*Sylvilagus auduboni*)	Desert Cottontail (*Sylvilagus nuttalli*)
Snowshoe Rabbit (*Lepus americanus klamathensis*)	Idaho Pigmy Rabbit (*Brachylagus idahoensis*)[1]
Gray Squirrel (*Sciurus fossor*)	
Pine Squirrel (*Sciurus douglasi albolimbatus*)	
Flying Squirrel (*Glaucomys sabrinus flaviventris*)	
Chipmunk (*Eutamias amoenus*)	Great Basin Chipmunk (*Eutamias pictus*)

[1] Not yet collected in Big Valley but obtained a few miles farther east.

Gray long-tail Groundsquirrel (*Citellus douglasi*)

Roundtail Woodrat (*Neotoma fuscipes*)

Oregon short-tail Groundsquirrel (*Citellus oregonus*)

Kangaroo Rat (*Dipodomys californicus*)

Bushytail Woodrat (*Neotoma cinerea occidentalis*)

BIRDS

Blue Grouse
Mountain Quail
Crested Bluejay
Acadian Owl
Pileated Woodpecker
California Woodpecker
Blackheaded Grosbeak
Oregon Robin

Sagehen
Columbian Sharptail Grouse

Woodhouse Jay
Burrowing Owl
Magpie (Rocky Mt. species)
Horned Lark
Mountain Mocker (Oroscoptes)

TREES AND SHRUBS

Ponderosa Pine (*Pinus ponderosa*)
Sugar Pine (*Pinus lambertiana*)
White Fir (*Abies concolor*)
Douglas Spruce (*Pseudotsuga*)
Incense Cedar (*Libocedrus*)
Yew (*Taxus brevifolia*)
Valley Oak (*Quercus lobata*)
Mt. Live Oak (*Quercus chrysolepis*)
Tanoak (*Quercus densiflora echinoides*)
Tree Maple (*Acer macrophyllum*)
Vine Maple (*Acer circinatum*)
Madrone (*Arbutus menziesi*)
Chinquapin (*Castanopsis sempervirens*)
Tree Dogwood (*Cornus nuttalli*)

Juniper (*Juniperus occidentalis*)
Sagebrush (*Artemisia tridentata*)
Rabbitbrush (*Chrysothamnus*)
Gutierrezia sp. ?
Peraphyllum ramosissimum
Purshia tridentata

Klamath River Mountain Mahogany (*Cercocarpus macrourus*)
Redbud (*Cercis occidentalis*)
Oso (*Osmaronia cerasiformis*)
Manzanita (*Arctostaphylos patula*)
Thornapple (*Crataegus rivularis*)
Huckleberry (*Vaccinium ovatum*)
Coffeeberry (*Rhamnus californica*)
Hazel (*Corylus californica*)
Azalea (*Azalea occidentalis*)
Smilax vine (*Smilax californica*)
Mountain lilac (*Ceanothus integerrimus*)

Great Basin Mountain Mahogany (*Cerocarpus ledifolius*)

Buckbrush (*Ceanothus cordulatus*)
Mahala mats (*Ceanothus prostratus*)
Kitkadizza (*Chamaebatia foliolosa*)
Wild Grape (*Vitis californica*)

THE ACHOOMAH'-AN FAMILY

Boundaries.—The northern boundary of the nine *Achoomah'-an* tribes (collectively) is a nearly straight line about 80 miles in length running obliquely from a point about 16 miles northnortheast of Bartles (approximately 33 miles NNE of Big Bend) to the east side of Goose Lake. This line if continued westerly would strike Mount Shasta—as I have been told by old people of several of the tribes.[1]

The *eastern boundary,* starting from Willow Ranch[2] or the nearby mouth of Fandango Creek on the east side of Goose Lake, follows this stream in a southeasterly direction to Fandango Valley, and continues thence southerly along the crest of Warner Mountains—the time-honored boundary between the Pit River Indians and their hereditary enemies, the Northern Piutes—and over the summit of Warren Peak (Buck Mt.) and on to Buckhorn Mt. and the vicinity of Red Rock where, turning southwesterly, it reaches the neighborhood of Termo and continues thence southerly to the divide south of Madeline Plains.

The *southern boundary,* starting from the west side of Pit River west of the mouth of Montgomery Creek, follows Montgomery Creek to its upper waters, where it turns north to the southwest corner of Goose Valley,[3] which it skirts; thence north to Lookout Mountain on the ridge between Goose and Burney Valleys, whence its course is nearly due east to the pass immediately north of Bald Mountain (crossing Hat Creek between Carbon and Cassel); then northeast to Muck Valley, and east to Hayden Hill, where it turns abruptly to the southeast and south, deflected by the long tongue-like southward extension of the *Ham-mah'-we,* to Madeline Plains.

The *western boundary,* starting on the north at a point about 16 miles north (or northnortheast) of Bartles, passes southerly, skirting the western edge of Big Bear Flat to the high ridge 5 or 6 miles south of Bartles, which it follows westerly and southwesterly around the

[1] Old Indians have told me that their northern boundary abuts against an air line running from Mt. Shasta northeast-by-east to a point on Goose Lake.

[2] Some Indians put the northern limit at Sugar Hill, 5 or 6 miles south of Willow Ranch.

[3] Goose Valley (an unfortunate name) must not be confused with Goose Lake Valley about 90 miles to the northeast.

valley of Kosk Creek and the Big Bend of Pit River to a point on the west side of Pit River a couple of miles southwest of the mouth of Montgomery Creek.

Surrounding tribes.—The *Achoomah'-an* tribes collectively are in contact with seven tribes, five of which belong to different linguistic stocks: On the west the *Northern Wintoon* and *O-kwahn'-noo-tsoo;* on the north the *Modok;* on the northeast and east the *Northern Piute;* on the south the *At'-soo-kā'-e* and *Ap-woo'-ro-kā'e;* on the southwest for a short distance, the *Nos'-se* or *Yah'-nah.*

THE MO-DES'-SE TRIBE

PLATES 1-4

The *Mo-des'-se* are the westernmost of the Pit River tribes.[1] From Big Bend of Pit River in the center of their territory they spread northerly to include the basin of Kosk and Nelson Creeks, southerly to Montgomery Creek, westerly to the nearest mountains on the west side of Pit River (from opposite the mouth of Montgomery Creek northward); easterly to Chalk Mountain and the divide separating the waters of Kosk and Nelson Creeks from those of streams further east.

South of the main body of the tribe, whose headquarters are at Big Bend, are two other bands—by some Indians regarded as distinct tribes—the *Ah-me'-che* at 'The Cove.' about midway between Big Bend and Montgomery Creek; and the *E'-poo'-de* whose principal rancheria was on Montgomery Creek two or three miles above its mouth.

[1] In August, 1855, Lieut. Henry L. Abbot (now the venerable General, aged 95) was engaged in surveying the Pit River country with Lieut. Williamson. In his report he states:

"We passed many pits about six feet deep and lightly covered with twigs and grass. The river derives its name from these pits, which are dug by the Indians to entrap game."—Pacific R. R. Reports, Vol. VI, p. 64, 1857.

The Mo-des'-se Indians tell me that these pits, called by them *Ah-pahm',* were dug lengthwise in the trails. They were 5 or 6 feet deep and about 6 feet long by 3 wide "to fit the body of the deer." They were made narrow to prevent the deer from jumping out, and the sides were narrowest at the top, which was usually reinforced by a hewn plank projecting a little way over the edge. The opening was covered with slender sticks and leaves. A large piece of maple bark was hung above the trail, a little to one side, to warn the Indians so they would not fall in. Once a little girl was caught. Then the chiefs came together and put a stop to pit trapping.

Mo-des'-se Villages at and Near Big Bend

Mo-dess'.—The ruling village, situated just east of the mouth of Kosk Creek (*An-noo-che-che*) on the north bank of Big Bend Pit River. (Place and rancheria same name.)

Lah'-lah-pis'-mah.—At Hot Springs on the south side of Big Bend, opposite *Mo-dess'*.

Oo-le'-moo-me.—On south side of Big Bend just east of the actual bend. (Place and rancheria same name.)

Tah'-sah.—On a sandy gravel flat on south side of river half a mile east of Hot Springs and near south end of Rope ferry. (Place and rancheria same name.)

Al-loo-satch-hā.—Small group of houses on south side of river only about 400 yards above *Lah'-lah-pis'-mah* and hardly entitled to rank as a distinct village.

Adjacent tribes.—The *Mo-des'-se* were in contact with five tribes: the Shastan *O-kwahn'-noo-tsoo'* [now extinct as a tribe] on the northwest and north, the related *To-mal-lin'-che-moi'* on the east; the *At-soo-kā'-e* on the southeast; the *Yah'-nah* or *Nos'-se* on the south; the *Northern Wintoon* on the southwest.

Names used by Mo-des'-se for Themselves and Other Tribes and Bands

Related tribes and bands:

A-choo-mah'-we (*Ah'-choo-mow'-we; Hā'-wis-aht'-wum-we*).—Tribe occupying Fall River Valley, including Pit River from Big Valley Mountains down stream to falls about 3 miles below mouth of Fall River .... [*A-choo'-mah'-we*]

Ah-doo-wah'-no-che-kah-te (*At-too-an'-noo-che; At-wah-noo'-che*, slurred *Tah-wahn'-noo'-che*).—Hat Creek tribe.................. [*At-soo-kā'-e*]

Ah-me'-che (*Ah'-mits*).—*Mo-des'-se* band at 'The Cove' on Pit River, midway between Big Bend & Montgomery Creek (5 or 6 miles north of Montgomery Creek)................................ [*Ah-me'-che*]

Ah-mit'-at-wum, Ah-mits'-che.—Dixie Valley and Eagle Lake tribe...... [*Ap-woo'-ro-kā'e*]

As-tah-ke'-wah (*As-tah-pe'-wah* and *Han-too'-che*).—Tribe in Hot Springs (Canby) Valley .... [*As-tah-ke-wi'-che*]

At-too-an'-noo-che. Hat Creek tribe. See also *Ah-doo-wah'-no-che*.... [*At'-soo-kā'-e*]

At-wah-noo'-che.—Tribe in Hat Creek and Burney Valleys. See also *At-too-an'-noo-che* .... [*At'-soo-kā'-e*]

At-wum-chan'-ne.—Likely Valley tribe...................... [*Ham-mah'-we*]

Chah-wahs'-te' chan-ni'-che.—Goose Valley band of *To'-mal-lin'-che-moi*. [Another name for *E-tsah'-tah*].

Choo-kā'-che.—Band of *Ap-woo'-ro-kā'e* on Beaver Creek (south of Snell Ranch).

Dah-bō^{ch}-e.—Band of *At-wum'-we* in Ash Creek Valley, southeast of Big Valley............. [Pronounced *Tŏ-bŭ^{ch}-we* by the *Ham-mah'-we*]

E'-poo'-de (*E'-pud-de*).—*Mo-des'-se* band on lower Montgomery Creek and adjacent part of Pit River. Former village on site of present Montgomery Creek postoffice .... [*E'-poo'-de*]

E-tsah'-tah.—Goose Valley band of *To-mal-lin'-che-moi.* Also called *Chah-wahs'-te chan-ni'-che* .... [*E-chah'-tah-we*]

Ham-mah'-we.—Tribe on South Fork Pit River from Likely Valley east to Jess Valley and south to include Madeline Plains.. [*Ham-mah'-we*]

Ham'-mă-o-ket'-tal-le.—Another name for Big Valley tribe. Same as *To-tā'-o-me* .... [*At-wum'-we*]

Han-too'-che.—Tribe in Stone Coal and Hot Springs or Canby Valley. See also *Ah-stah-ke-wah* .... [*As-tah-ke-wi'-che*]

Hā'-wis-aht'-wum-we.—Another name for Fall River Valley tribe....... [*Ā-choo'-mah'-we*]

Il-mah'-we.—Pit River tribe (from falls 3 or 4 miles west of Fall River Mills down to Pecks Bridge); includes Cayton Valley and reaches upper Bear Creek. The ruling village, *Il-mah'*, was at site of Pecks Bridge and occupied both sides of the river .... [*Il-mah'-we*]

Ko'-se-al-lik'-tah.—Tribe at forks of upper Pit River, on Alturas Plain.. [*Ko'-se-al-lek'tĕ*]

Mŏ-des'-se.—Tribe with headquarters at Big Bend Pit River, reaching upstream to Deep Creek. Their name for their own tribe.... [*Mo-des'-se*]

Tah-wahn'-noo'-che.—Hat Creek tribe. See also *At-too-an'-noo-che*...... [*At'-soo-kā'-e*]

To-mal-lin'-che-moi'.—Tribe between *Il-mah'-we* and *Mo-des'-se;* headquarters at junction of Screwdriver Creek with Pit River (on Lindsay's place, 6 or 7 miles in air line below Pecks Bridge)......... [*To-mal-lin'-che-moi'*]

Too-hat'-mah.—Cayton Valley band of *Il-mah'-we* .... [*Too-hat'-mah*]

Too-tā'-o-me (*Too-tā'-o-mal'-le*).—Still another name for *At-wum'-we,* the Big Valley tribe; also called *Ham'-mă-o-ket'-tal-le*.... [*At-wum'-we*]

Wah-num-che'-wah.—Burney Valley band of *At-soo-kā'-e*.. [*Oo'-kah-soo'-we*]

Unrelated tribes and bands:

Ah-lah'-me. Klamath Lakes tribe [*Klamath*]. Given by Gatschet as *Ala'mmimakt ish* in Klamath language.

Ah-poo'-e.—*Northern Piute,* of Fort Bidwell and Alkali Lakes in Surprise Valley.

Ah-tah'-me-kah'-me.—Tribe "toward Redding." Said to talk same as *Te'-si-che* [*Nos'-se* or *Yah'-nah*].

Ā-te' ("Squaw Creek tribe").—Tribe south and east of Mount Shasta, on upper McCloud River, Squaw Creek, and in Fox Mountain region [*O-kwahn'-noo-tsoo*].

Bă-kah'-mah'-le.—See *Pah'-kĕ-mah'-le.*

Che-yu'-wit.—Tribe on Squaw Creek. [May be band of *O-kwahn'-noo-tsoo.*]

Ek-pe'-me, E^ch-pe'-me.—*Northern Wintoon* tribe; McCloud River *Wintoon.*

Hā'-wis-se-kahs'-te.—Trinity River *Wintoon.*

Loo'-too-ah'-me.—*Modok* tribe.

O-se'-low-wit.—*Wi-dal'-pom* band of *Northern Wintoon* [usually written *Ydalpom*].

Pah'-kĕ-mah'-le (*Pah-rah'-mah-le; Bă-kah'-mah'-le*).—Northeastern *Midoo* of Big Meadows [*No-to-koi'-yo*].

Pas-sā'-put-che.—*Yah'-nah* of Round Mountain region. Same as *Te'-si-che.*

Shas'-te'-che (*Sas'-te'-che*).—*Shaste* of Yreka and Shasta Valleys. [They call themselves *Ge'-katch* or *Ke'-katch.*]

Te'-si-che. *Nōs'-se* of Round Mountain, Millville, Red Mountain and North Cow Creek [*Nos'-se* or *Yah'-nah*]. Same as *Pas-sā'-put-che.*

Names used for Mo-des'-se by Other Tribes

NAME	TRIBE USING NAME
E-tah'-me [Nickname]	*A-choo'-mah'-we*
Mah'-dā'-se	*A-choo'-mah'-we*
Mo-des'-se	*At-wum'-we* and *As'-tah-ke-wi'-che*
Po'-mah-de'-he; Po'-mah-re'-ye	*At-soo-kā'-e* and *Ap-woo'-ro-kā'e*
Poo'-e-choos'; Poo'-e-soos'	*Wintoon*

Geographic Names used by Mo-des'-se

Alturas: Town (County seat of Modoc Co.) near junction of South Fork with main Pit River.................................Ko'-se-al-lek'-tah

Bagley Mt.: On west side Pit River southwest of Big Bend............ Now'-wahs-nem'-chah

Bald Mt.: Solitary butte 7 miles south of Fall River Mills. (Indian race ground) ....E-pah'-kŏ-mah

Bald Peak, or 'Bally': Bare rock dome at head of Nelson Creek on east side of Kosk Creek Basin.....................Mah-how'-mah-dā ah'-ko

Also apparently ....Pah'-mah-ye'-mah

Bear Creek: Upper part of Fall River.........................Soolt'-mah'

Bear Mt.: Six or 7 miles northeast of Bartles....Sōl'-mah oo-te-choo-kah'-te

Beaver Creek: Entering Pit River from south just west of Pittsville. Choo-kā'

Bee Knoll: Hill on southwest side Pit River near Hayes, about 4 miles southeast of Big Bend............................Ah' soo-tah ke'-wah

Big Bend Pit River: The ruling village of same name was on north side ....Mŏ-dess'

Big Bend: South side.....................................Oo-le'-moo-me

Big Basin of Kosk Creek: North of Big Bend, enclosed on north and east by Horseshoe Ridge............................Tal-lo'-we tam'-me

Big Valley: A large flat desert valley about 20 miles wide (east-west), reaching from Modoc Lava Beds south to Hayden Hill......To-tā'-ah-me

Big Valley Mts.: Timbered range on west side of valley........Tā-wahl'-me

Bluejay Mt.: On west side Pit River about 2 miles northwest of mouth of Montgomery Creek..........Chim-chim'-min-nuk (Chin'-che-min-nok)

Brock Butte: Between Squaw Creek and Pit River......Ah'-lis-te te'-we-che

Buck Mt.: On west side Pit River north of Bluejay Mt.........Sal'-le ah'-ko

Buckeye Place: On Pit River 2 miles below Big Bend........Pah'-sil-lo'-mah

Burney Butte: A lofty volcanic peak between Burney Creek and Hat Creek ....Ah-po'-hah

Burney Creek: A long stream flowing north to Pit River, which it enters a mile or two west of Pecks Bridge..................Wah′-num che′-wah

Burney Valley: Small valley on Burney Creek northeast of village of same name; belongs to At-soo-kā′-e tribe..................Wah′-num at-wum

Butte: Small butte in Fall River Valley, 2 miles east of Fall River...... Yah-pah′-mah

Cayton Valley: On north side Pit River between Pecks Bridge and Soldier or Fort Mt..Lah-hat′-mah

Chalk Mt.: On north side Pit River 5 or 6 miles east or southeast of Big Bend .....Too-le-pah′-ah-te ah′-ko and Chool′-oo-te-mah-wah-mah-gā′-oi

'The Cove': On east side of Pit River, midway between Big Bend and Montgomery Creek ....Ah′-mitz′

Crater Peak: (McGee and Manzanita Mts.)...................Che-wah′-ko

Dixie Valley: A springy valley south of the ridge which separates the At-wum′-we from the Ap-woo′-ro-kāe. An important stronghold of the Ap-woo′-ro-kāe ....Ah-mēt′

Eagle Lake: A large and beautiful body of water northnorthwest of Susanville; in territory of Ap-woo′-ro-kā′e.....................Ah-sit′

Fall River 'City' or 'Mills': At junction of Fall and Pit Rivers........ In′-choo-te-ĕ-o

Flat: Between Soldier and Saddle Mts.........................Tā-lahs′-te

Goose Valley: Northwest of Burney; home of the E-chah′-tah-iss′ band of To-mal-lin′-che-moi′. Not to be confused with Goose Lake Valley, 90 miles to the northeast.....................E-chah′-tah; E-tsah′-kah

Goose Valley Mountains: Long ridge just west of Goose Valley..Che-yow′-wit

Its highest peak, at north end.............Tam′-mahk-de-nā-lah-teu′-choi

Grasshopper Valley: North of Eagle Lake; belongs to Ap-woo′-ro-kā′e.. Choo′-e at′-wum; At-to-maw′-wah

Grizzly Peak: Eight miles north of Big Bend.............Ah′-ko la-chah′-ke

Hatchet Creek: Next creek north of Montgomery Creek, flowing west to Pit River; belongs to Mo-des′-se.....................Ah-pil′-choo-me

Hat Creek: Rising on Mt. Lassen and flowing northerly to Pit River; formerly called Canoe Cr. Headquarters of the At-soo-kā′-e tribe.... Had-de-we′-oo; Had-dik-wē′u

Horseshoe Ridge: Lofty ridge north and east of Big Bend (running easterly from Grizzly Peak; then south around heads of Kosk and Nelson Creeks to Chalk Mt.)........................Ah′-ko Wil′-lah-le

Hot Springs: On south bank of Big Bend; for ages a famous health resort; much used by Indians during my visit in 1907...Lah′-lah pis′-mah

Hulsey Ridge: Low ridge occupied by road to Wm. Hulsey's house, about 2 miles south of Big Bend..........................How′-we-nim′-che

Il-mah′: Stretch of Pit River from falls 3 miles below (west of) mouth of Fall River to a point a little below Pecks Bridge, 10 miles farther downstream; territory of the Il-mah′-we. The ruling village (Il-mah′) was on both sides of the river at Pecks Bridge.....................Il-mah′

Kettle Mt.: On west side Pit River between Big Bend and 'The Cove'.. Ah′-ko Ĕs-chahk-chā′-hā

Kosk Creek: Entering Big Bend from north. Mo-dess′ village was on east side of its junction with Pit River......................An-noo-che′-che

Lassen: Lassen Butte or Mt. Lassen........................Yet-te′ chan′-nah

Little Cow Creek: About 7 miles south of Montgomery Creek; in territory of Yah'-nah ....Te'-sah-dā'-e ju'-me

Lookout Mt.: Culminating point on ridge between Goose and Burney Valleys, on boundary between At-soo-kā'-e and Mo-des'-se tribes.... Soo'-kaht ah'-ko

McArthur: Small white settlement in southeastern part of Fall River Valley ....Hā'-wis at'-wum

McCloud River: Large river tributary to lower Pit River from north. One of the principal populated areas of the *Northern Wintoon*..It-tă-choo'-mah

Medicine Lake: A strange place, partly sheltered on north and northeast by Mt. Hoffman and Glass Mt.; on or near the Achoomahwe-Modok Boundary; noted battleground of *Modok* and Pit River Indians and well known to all tribes from the *Shaste* and *Wintoon* on the west to the *Northern Piute* on the east; coveted for its inexhaustable store of obsidian for arrow and spear points....Saht'-tit (from *Saht'*, obsidian)

Montgomery Creek: Stream flowing west from northern Sierra to Pit River; boundary between Mo-des'-se and Yah'-nah............E'-pur'-re

Nelson Creek: Tributary to Big Bend from north, just east of Kosk Creek; in territory of Mo-des'-se.....................Ah-lis' choo'-chah

Pecks Bridge: A well known locality on Pit River just east of mouth of Burney Creek and southwest of Cayton Valley. Close to boundary between *Il-mah'-we* and *To'-mal-lin'-che-moi'*. Former village *Il-mah'* occupied both sides of river here. At present, site of one of the great hydroelectric plants of the Pacific Gas & Electric Co..........Boo-lo'-wit

Pit River: A long stream; in times of highwater flowing out of Goose Lake; cutting deep canyons through the northern part of the Sierra Nevada, and emptying into Sacramento River a dozen miles north of Redding. Present source of much of the hydroelectric power of Pacific Gas & Electric Co. ....Ah-choo'-mah

Pit River above Fall River..............................Hem-mā'-ju-me

Pit River below Fall River...............................El-mah'-ju-me

Pit River Ridge: A high mountainous ridge rising from west side Pit River, and parallel to it, from Big Bend south to Mongtomery Creek. Scene of many battles between Mo-des'-se and Wintoon............ Tah'-mahk-te-nā'-lah teo'-oo-too or Tah'-mahk te-nā'-lah teu'-choi-oo too'-loo-ke

Ridge: On north side Pit River extending westerly from Chalk Mt. toward mouth of Nelson Creek...............................Oo'-le-ke'-wah

Roaring Creek: Tributary to Pit River from east, north of Hatchet Creek ....Huk'-de mahs-hā; Huk'-te-mah-hā

Rock Creek: Tributary to Pit River from north, in territory of To'-mal-lin'-che-moi' ....Choo-lah'-mi choo'-me

Round Mt.: Conspicuous hill 3 miles southwest of Montgomery Creek settlement, in territory of *Nos'-se* or *Yah'-nah*...............Bus-sā'-put

Saddle Mt.: Elongated two-humped mountain at southwest corner of Fall River Valley; immediately south of Fort Mt. and northwest of Fall River City.

The north hump..........Tah-mah-meu'-chah (slurred, Tah-mah'-mootch)

The south hump...Ah-pah-tạ̄'-me

Sandy Flat: On south side Pit River at Big Bend Bridge, ½ mile east of Hot Springs ....Tah'-sah

Shasta, Mt.: Highest peak in Northern California...................... E-tah'-ko or Itah'-ko (low word); Ā'-te chan'-nah (high word)

Soldier Mt. (also called Fort Mt.): Isolated volcanic mountain on west side of southern part of Fall River Valley, immediately north of northwest end of Saddle Mt. Old Fort Crook was at its northeast base ....Sim'-lōk

South Fork Pit River: Stream flowing west and then north (through a long narrow marshy valley) to join North Fork near Alturas, the two uniting to form the main Pit River..................Ham'-mah de'-kah

Squaw Creek: Tributary from north to McCloud River, west of Pit River ....Che-yu'-wit

Wengler: Old sawmill settlement; abandoned postoffice site.............. Woi hoo' e-tĕ-o (from *Woi*, Grizzly Bear, and *e-tĕ-o*, trail)

Wengler: Sawmill site......Che'-tah-tahk'-che-we (from *Che'-tah-tahk*, bracken fern, and *che'-we*, flat)

THE TO-MAL-LIN'-CHE-MOI' TRIBE

The *To-mal-lin'-che-moi'* lie between the *Mo-des'-se* and *Il-mah'-we*, occupying the deep canyon of Pit River from the middle of the loop south of Chalk Mountain, upstream to near the mouth of Burney Creek, just west of Pecks Bridge. The ruling village (of same name) was on the north side of Pit River at the junction of Screwdriver Creek, 6 or 7 miles below Pecks Bridge. On the north they apparently reached the mountains 4 or 5 miles southeast of Bartles, but the exact northern limit is unknown. On the south they occupied the whole of Goose Valley, reaching within a mile or two of the small settlement of Burney. The Goose Valley band (*E-tsah'-tah-iss*, also called *Chah-wahs'-te-chan-ni'-che* by the *Mo-des'-se*) is hemmed in both east and west by lofty ridges, while on the north the deep canyon of Pit River renders communication with the main body of the tribe difficult.

They were in contact with 4 tribes: on the west, the *Mo-des'-se;* on the northwest for a short distance, the *O-kwah'-noo-tsoo;* on the north and east, the *Il-mah'-we;* on the south, the Burney Valley branch of *At'-soo-kā'-e*. Their territory is in the main forested.

Names used for To-mal-lin'-che-moi' by Other Tribes

NAME	TRIBE USING NAME
E-chat'-tah-we.—Band in Goose Valley	*A-choo'-mah'-we*
E-tsah'-tah-iss and *Chah-wahs-te' chan-ni'-che.*—Band in Goose Valley	*Mo-des'-se*
Oo'-we-che'-nah.—Band in Goose Valley	*At-soo-kā'-e*
To-mal-lin'-che-moi'	*Mo-des'-se*

THE IL-MAH'-WE[1] TRIBE

PLATES 5-7

The *Il-mah'-we* occupied a rather narrow strip between the *To-mal-lin'-che-moi'* on the west and the *Ă-choo'-mah'-we* on the east. On Pit River their territory extended only from near the mouth of Burney Creek just west of Pecks Bridge, upstream to the falls on Pit River, about 3 miles below Fall River Mills. On the north it included Cayton Valley and is said to have reached Bear Mountain; on the south it embraced about 4 miles of the lower course of Hat Creek, stopping abruptly on an east-west line crossing Hat Creek about midway between Carbon and Cassel postoffice, where it abutted against the northern boundary of the *At'-soo-kā'-e*.

Their country is called *Il-mah';* their people, *Il-mah'-we;* their language, *Il-mā'-wah.*

The ruling village, *Il-mah',* long since destroyed, occupied both sides of the river at the site of Pecks Bridge. Another village, *Wen'-ne-hah'-le,* was in the canyon at the falls three miles below the mouth of Fall River; and still another, *Mah-pe'-dah-dā',* at the junction of Salmon Creek with Pit River.

The *Mo-des'-se* tell me that the *Il-mah'-we* were formerly the dominant tribe on Pit River.

Their territory, like that of the adjoining *To-mal-lin'-che-moi'* was mainly forested.

The *Il-mah'-we* were in contact with four tribes: the *To-mal-lin'-che-moi'* on the west; the *O-kwahn'-noo-tsoo* on the northwest; the *Ă-choo'-mah'-we* on the north and east; the *At'-soo-kā'-e* on the south.

NAMES USED FOR IL-MAH'-WE BY OTHER TRIBES

NAME	TRIBE USING NAME
Il-mah'-we	*Mo-des'-se, At-wum'-we* and *Ă-choo'-mah'-we*
Po-mah-re'-ye	*At'-soo-kā'-e* and *Ap-woo'-ro-kā'e*
Too-hat'-mah (slurred *Hat'-mah*).—Band in Cayton Valley	*Mo-des'-se* and *Ă-choo'-mah'-we*

[1] The name was written *Illmawces* by Stephen Powers in 1874 (*Overland Monthly,* Vol. 12, pp. 412-415, May, 1874); spelling changed in his *Tribes of California,* 1877, to *Il-mâ-wi* (p. 267, 1877), and adopted by Mooney in *14th Ann. Rept. Bur. Eth.* (for 1892-1893), p. 1052, 1896. [Mention only.]

THE Ă-CHOO'-MAH'-WE TRIBE

Plate 8.

The *Ă-choo'-mah'-we* occupy a considerable extent of country east of the *Il-mah'-we,* their territory comprising the drainage basins of White Horse Valley, Fall River Valley, and Pit River from the south end of Big Valley Mountains westerly to Pit River falls, 3 miles below Fall River Mills. On the north they claim part of the Lava Beds country and the southern parts of Medicine Lake and Glass Mt., in which region many battles were had with their enemy the Modok.

Their western boundary crosses Pit River directly south of Saddle Mt., from which it runs northwesterly to Fort Mt. (known also as Soldier Mt.) and thence northerly and northwesterly to Bear Mt. and the territory of the Modok.

The eastern boundary runs from Glass Mt. southerly to Buck Butte and Round Mt.; thence southeasterly along the low divide between White Horse Valley and Egg Lake marsh, south of which it follows the summit of Big Valley Mountains to the western edge of Muck Valley.

The southern boundary extends from Conrad Lake to a point a mile or a mile and a half north of Bald Mt. and thence northeasterly to the west side of Muck Valley.

The larger part of their domain is a broad open flat-bottom valley interrupted by low volcanic buttes and ridges, well watered and grass covered on the south but continuing northward into the arid region of the Lava Beds.

North of Pit River both eastern and western borders are forested with Ponderosa pines and other trees; south of Pit River the land is partly forested and partly open.

The largest town in their territory is Fall River Mills (or 'City'); smaller settlements are Pittsville, McArthur, Glenburn, and Dana.

The *Ă-choo'-mah'-we* are in contact with six tribes: the *Il-mah'-we* and *O-kwahn'-noo-tsoo* on the west; the *Modok* on the north; the *At-wum'-we* on the east; the *Ap-woo'-ro-kā'e* and *At-soo-kā'-e* on the south.

Former Ă-choo'-mah'-we Villages

The rancheries of the Ă-choo'mah'-we tribe were located along the courses of Pit River and Fall River; most of them were of small size.

Bi'-yu-mah.—At site of McArthur.

Dā-lahs'-te.—On lower part of Fall River at site of first house north of Intake.

Daw[ch]*-tah'-pit.*—On south side Pit River about 2 miles below Fall River mouth.

Dē'-el mach'-mah (or *Êl-mach-mah*).—At Intake of Pacific Gas & Electric Co. on Fall River near Fall River Mills.
Dis-naht'-chah.—On both sides Pit River between Fall River and McArthur.
In'-ju-tā-eu'.—At present Fall River Mills, at junction of Fall River with Pit River. Old house holes may still be seen.
Joo'-wahl-oo'.—On Pit River at site of Pittville.
Lo'-kah-le.—On south side Pit River opposite McArthur.
Nahn-chah-me'-nah.—[Not located.]
Paht-to'-mah.—On Fall River above *Dā-lahs'-te* (toward Dana).
Sok'-too-wah'-dā.—On Fall River above *Paht-to'-mah.*
Soo-kă'-mah.—[Not located.]

NAMES USED BY Ă-CHOO'-MAH'-WE FOR THEMSELVES AND OTHER TRIBES AND BANDS

Related tribes and bands:

Ă-choo'-mah'-we, Ă-ju'-mah'-we or *Ah-choo-mah'-we'-che.*—Their name for their own tribe.................................... [*Ă-choo'-mah'-we*]
Ah'-mit'-che and *Kaw-le-wah.*—Dixie Valley tribe......... [*Ap-woo'-ro-kāe'*]
As'-stah-ke'-wah. Tribe in Hot Springs or Canby Valley.. [*As-tah-ke-wi'-che*]
At'-to-maw'-wah.—Band in Grasshopper Valley just north of Eagle Lake .... [band of *Ap-woo'-ro-kā'e*]
At-wum'-chan'-ne and *E-tah'-me.*—Big Valley tribe........... [*At-wum'-we*]
At-wum'-noo-che, Had'-de-we'-we and *To-ah-no'-che.*—Hat Creek tribe... [*At-soo-kā'-e*]
E-chat'-tah-we.—Band in Goose Valley just north of Burney........... [band of *To-mal-lin'-che-moi'*]
E-tah'-me.—Tribe in Big Valley, Lassen County.............. [*At-wum'-we*]
Also, nickname for *Mo-des -se*
Had'-de-we'-we, At-wum'-noo-che and *To-ah-no'-che.*—Hat Creek tribe.. [*At-soo-kā'-e*]
Hā'-we-si'-too.—Tribe between Alturas and Goose Lake...... [*Hā-we-si'-doo*]
Il-mah'-we.—Pit River tribe next below the Ă-ju'-mah'-we...... [*Il-mah'-we*]
Kaw-le-wah and *Ah-mit'-che.*—Dixie Valley tribe.......... [*Ap-woo'-ro-kāe*]
Ko'sal-lek'-te.—Band at Alturas .... [*Ko'-se-al-lek'-tĕ*]
Mah-dā'-se.—Tribe at Big Bend Pit River...................... [*Mo-des'-se*]
To-ah-no'-che.—Hat Creek tribe........................... [*At-soo-kā'-e*]
Too-hat'-mah or *Hat'-mah.*—Cayton Valley band of *Il-mah'-we.*

Unrelated tribes and bands:

Ah-poo'-e.—Tribe in Surprise Valley...................... [*Northern Piute.*]
Al-lah'-me.—Klamath Lake tribe.................. [" In *Yan-nox* country "]
Bug-gah'-mah-le and *Puk-kah'-mah.*—Tribe in Big Meadows............ [*No'-to-koi'-yo Midoo*]
Ek-pe'-me.—McCloud River Wintoon................. [*Northern Wintoon*]
Loo'-too-am'-me.—Tribe of Tule Lake and northern part of Goose Lake.. [*Modok*]
Puk-kah'-mah and *Bug-gah'-mah-le.*—Tribe in Big Meadows............ [*No'-to-koi'-yo Midoo*]
Sas'-te'-che.—Name for both *Sas'-te* (*Shas'-te*) and *O-kwahn'-noo-tsoo'.*
Te'-si-che.—Tribe from Round Mt. south............. [*Yah'-nah* or *Nōs'-se*]

Names used for Ă-choo'-mah'-we by Other Tribes

NAME	TRIBE USING NAME
Ă-choo'-mah'-we, Ah'-choo-mow'-we, and *Hā'-wis-aht'-wum-we*	*Mo-des'-se* and *Ham-mah'-we*
Ă-ju'-mah'-we	*As'-tah-ke-wi'-che* and *At-wum'-we*
Ă-ju'-mah-we and *Ham-mow-ēs'*	*At-wum'-we*
Too'-e-chow'-we; Too'-e-tsow'-we; Te-tsow'-we	*Ap-woo'-ro-kā'e* and *At'-soo-kā'-e*

Geographic Names used by the Ă-choo'-mah-we

Bald Mt. seven miles south of Fall River Mills	Pak'-kă-mah' [E-pak-kă-mah'?]
Big Valley	To-tā'-me
Big Valley Mts. between Fall River Valley and Big Valley	Tā'-wahl'-me
Burney Butte	Ah-po'-hah
Cayton Valley west of Fall River Valley	Too-hat'-mah
Crater Peak (may include McGee Peak), between Mt. Lassen and Burney Butte	Yā-te'-chah-lah
Dixie Valley	Am-mē't
Fall River	Ah-choo'-mah'
Fall River Valley, the broad middle part	Ah-cho-mo-de-kah-te
Rancheria site in lower northwest part of Fall River City	Te'-kah-te
Fort Crook at northeast base of Fort (or Soldier) Mt.	Soolt'-mah
Goose Valley just north of Burney	Chă'-tah
Grasshopper Valley north of Eagle Lake	At'-too-maw'-wah'
Hat Creek	Hat'-te we'-oo'
Hoffman, Mount	Tahs-se'-wah
Klamath Lakes country	Yan'-nox
Lassen Peak	Yet'-te'-chă-nah
McCloud River	E-chah'-tah
Medicine Lake	Saht
Modok country, Tule Lake to Goose Lake	Loo'-too-am
Mountain (small isolated mountain or high hill) south of Saddle Mt. on south side Pit River southwest of Fall River City	Chah'-chek'-kel
Pine Creek	Am-mook'-mah
Pit River	Ă-ju'-mah ā-chim and E-le-mah'
Poison Lake	As-sit'
Saddle Mt. just south of Soldier Mt. and northwest of Fall River City (the north hump)	Tah'-mah'-mootch'
The southern hump	Ah-pah-tā'-me
Salmon Creek	Mah-pe'-dah-dā'
Shasta, Mount	Yet'
Soldier Mt. ('Fort Mt.') just west of Fort Crook	Sim'-lōk
South Fork Pit River	Ham'-mah-de'-kah

THE AT-WUM′-WE TRIBE

PLATES 9-12

The *At-wum′-we* occupy an extensive desert area known as Big Valley, on the east side of the northern Sierra Nevada, between the related *Ă-choo′-mah′-we* on the west and the *As-tah-ke-wi′-che* and *Ham-mah′-we* on the east. Their territory is larger than that of any of the other Pit River tribes, extending from Medicine Lake and Glass Mountain southerly to Hayden Hill and the headwaters of Ash Creek, and in a west-east direction from Big Valley Mountains and the low divide east of White Horse Reservoir easterly to Scheffer Mountain, Cottonwood Creek, and Upper Ash Creek, including Round Valley and the drainage area of Rush Creek northeast of Adin. This area except along the streams is an arid or semi-arid desert plain comprising Big Valley and a part of the Lava Beds, bordered by strips of forest in the mountains. [See also pp. 12-14.]

The western boundary—the line separating them from the *Ă-choo′-mah′-we*—begins at Glass Mountain on the north, runs southerly and southeasterly to Big Valley Mountains, which range it follows south to Muck Valley in the great southern loop of Pit River. The southern boundary is almost a straight line running east from Muck Valley to Hayden Hill, where it turns abruptly to the southeast to enclose the headwaters of Ash Creek. The eastern boundary starts from a point just north of Mowitz Butte and runs directly south for about 22 miles to Stone Coal Valley on Pit River, where it bends to the southeast to enclose the upper waters of Ash Creek.

In Big Valley proper the *At-wum′-we* had only one permanent winter village. Its name was *Ah-pe′-dah-dā* and it was located on Ash Creek between Adin and Lookout.

The *At-wum′-we* band in Ash Creek Valley is called *Ko-sel-lat′-to-mah* by themselves, and *Dah-bo′ᶜᵏ-e* by the *Ham-mah′-we* and *Mo-des′-se.* It is separated from the *Ham-mah′-we* by the ridge or divide between Ash Creek and South Fork Pit River.

The *At-wum′-we* band in Round Valley, northeast of Adin, is called *Se-te′-wah.*

The white settlements within their area are Bieber, Adin, and Lookout.

The *At-wum′-we* are in contact with five tribes: the *Ă-choo′-mah′-we* on the west; the *Modok* on the north; the *As-tah-ke-wi′-che* and *Ham-mah′-we* on the east; the *Ap-woo′-ro-kā′e* on the south.

NAMES USED BY AT-WUM'-WE FOR THEMSELVES AND OTHER TRIBES

Related tribes:

Ah-pe'-dah-dā' (*Yah-pe'-dah-dā*).—Principal village and land of *At-wum'-we*, located on Ash Creek in Big Valley, 7 miles west of Adin.

A-ju'-mah'-we and *Ham-mow-ēs'*.—Fall River tribe........*A-choo'-mah'-we*

A-mits'-che (*Ah-mitch'-e*).—Dixie Valley tribe.............*Ap-woo'-ro-kā'e*

Ah-lahm'-se-ge.—Northeastern band of *Hā'-we-sī'-doo* (east of southern part of Goose Lake)

As-tah'-re'-wi'-ge.—Hot Springs or Canby Valley tribe......*As-tah-ke-wi'-che*

At-wum'-we'.—Their name for their own tribe.................*At-wum'-we*

Chah'-lahk-se.—*Hā'-we-sī'-doo* band at west base of Cedar Mt., 11 miles ENE of Alturas....................................*Chah'-lahk-se*

Dal-mo'-mi-che (*Dal-mo'-mah, Tal-mo'-mah*).—*Ko'-se-al-lek'-te* band at Essex Hot Spring 10 miles west of Alturas.............*Dal-mo'-mi-che*

Doo'-me-lit' (*Too'-me-lit'*).—*Ko'-se-al-lek'-te* village about 3 miles southwest of Alturas...*Doo'-me-lit'*

Had'-de-we'-we and *Hat'-te-we'-we es*.—Hat Creek tribe..........*At-soo-kā'-e*

Ham-mah'-we.—Likely Valley tribe..........................*Ham-mah'-we*

Ham-mow-ēs' and *A-ju'-mah'-we*.—Fall River tribe..........*A-choo'-mah'-we*

Ho-mā'-wet.—*Ko'-se-al-lek'-te* village and band about 6 miles NNE of Alturas ....*Ho-mā'-wet*

Il-mah'-we.—Pit River tribe below Fall River Mills.............*Il-mah'-we*

Ko'-sah-lek'-tah.—Tribe at Alturas, reaching north about 8 miles, and south 6 miles.......................................*Ko'-se-al-lek'-tē*

Ko'-sel-lat'-to-mah.—*At-wum'-we* band in Ash Creek Valley.......... *Ko'-sel-lat'-to-mah*

Mo-des'-se.—Tribe at Big Bend Pit River.......................*Mo-des'-se*

Se-te'-wah.—*At-wum'-we* band in Round Valley (northeast of Adin)... *Se-te'-wah*

Unrelated tribes:

Ah-poo'-e (*A-poo'-e*).—*Northern Piute* of Fort Bidwell region.

Ek-pe'-me.—Wintoon of west side Pit River opposite mouth of Montgomery Creek ....*Northern Wintoon*

Kă-bă'-mah-le and Pă-kah'-mah'-le.—Big Meadows tribe..*No-to-koi'-yo Midoo*

Loo'-too-ah'-me (*Loot-wah'-me*).—Modok-Klamath (collectively).

Te'-si-che.—Tribe in Round Mt. region................*Yah'-nah* or *Nōs'-se*

NAMES USED FOR AT-WUM'-WE BY OTHER TRIBES

NAME	TRIBE USING NAME
Ahk-koo'-e	*At-soo-kā'-e*
Ah-ko-we'-e'; Ahk-we"	*Ap-woo'-ro-kā'e*
At-wum'-chan'-ne	*A-choo'-mah'-we*
At-wum'-zan'-ne	*Ham-mah'-we*
At-wum'-we	*As-tah-ke-wi'-che*
*Dah-bō*ch*-e*.—Band in Ash Creek Valley	*Mo-des'-se*

E-tah'-me .... *Ă-choo'-mah'-we* and *Ham-mah'-we*
Ham'-mă-o-ket'-tal-le; To-tā'-o-me; Too-tā'-o-mal'-le .. *Mo-des'-se*
Tō-bŭ^ch^we.—Band in Ash Creek Valley............... *Ham-mah'-we*

Geographic Names used by the At-wum'-we

Adin .... *Til'-le-nah-ge'*
Bald Mt. .... *E-pah'-goo-mah*
Big Valley ("our valley") .... *At-wum'*
Big Valley Mountains.......................... *Til-lok'-lok e-doo'-lin* and *Too-woos'-tah-tse*
Dixie Valley .... *Ah-mit'*
Grasshopper Valley .... *At-o-maw'-wah*
Hat Creek .... *Hat'-te we'-we*
Hayden Hill .... *Tsah-mā'-heu* (old original name)
Hayden Hill .... *De-tse-lag-gā doo-we-ēt* (new name)
Hoffman, Mount .... *Tahs'-se'-wah*
Lassen Volcano .... *E-te'-chan'-nah*
Madeline Plains .... *Sel'-lat-um'*
Medicine Lake .... *Saht'* [obsidian]
Muck Valley pond.............................. *Ah'-te-ke*
Pit River .... *Ă-ju'-mah*
Round Valley (just northeast of Adin)............ *Sow'-we sop'-te*
Shasta, Mount .... *Yet*
South Fork Pit River........................... *Ham-mow'-tā-kaht'*
Willow Creek (east side Hayden Hill to Big Val.).. *Tsah'-he-se*

THE AS-TAH-KE-WI'-CHE TRIBE

Plates 13-15

The *As-tah-ke-wi'-che* territory lies between that of the related *At-wum'-we* on the west and *Hā-we-si'-doo* and *Ko'-se-al-lek'-te* on the east, extending from a little north of Mowitz Butte and Timbered Mt. southerly to Stone Coal Valley, Scheffer Mt., and the ridge running from Scheffer Mt. easterly to the valley of South Fork Pit River, which it approaches at a point nearly opposite Signal Butte.

The center of population was in Hot Springs or Canby Valley and its eastern extension to old Centerville, thus enclosing O-pah'-wah Butte (formerly called Centerville Butte and Rattlesnake Butte). The As-tah-ke-wi'-che were helpful to our troops at the time of the Modok war and were promised (by General Crook) an area 20 or 25 miles square around O-pah'-wah Butte—which of course was never given them.

Villages of the As'-tah-ke-wi'-che.—There were two permanent villages: *As'-tah-re'-wah,* on the knoll at Hot Spring 3 miles east of Canby; and *Han-too'* (or *Han-teu'*) in Stone Coal Valley.

The *As-tah-ke-wi'-che* were in contact with five tribes: the *At-wum'-we* on the west; the *Modok* on the north; the *Hā'-we-sī'-doo* and *Ko'-se-al-lek-tĕ* on the east; the *Ham-mah'-we* on the south.

Names used by As-tah-ke-wi'-che for Themselves and Other Tribes

Related tribes:

Ă-ju'-mah'-we.—Fall River tribe.............................[*Ă-choo'-mah'-we*]
Ă-mitch'-e.—Dixie valley tribe.............................[*Ap-woo'-ro-kā'e*]
As-tah-re-wī'-se (*As-tah-ke-wi'-se*).—Their name for their own tribe.... [*As-tah-ke-wi'-che*]
At-wum'-jen'-ne.—Likely Valley tribe, on South Fork Pit River, reaching east to mountains and south to Madeline Plains. Talk same as *As-tah-ke-wi'-che* ....[*Ham-mah'-we*]
At-wum'-we.—Big Valley tribe...............................[*At-wum'-we*]
De-baw'-kĕ-e.—Ash Creek Valley band of At-wum'-we.......[*De-baw'-kĕ-e*]
Ham-mah'-we.—Likely Valley tribe.........................[*Ham-mah'-we*]
Han-too'.—As'-tah-ke-wi'-che band in Stone Coal Valley on Pit River.... [*Han-too'*]
Hat'-te-we'-we (*Had'-de-wi'-we*).—Hat Creek tribe...........[*At-soo-kā'-e*]
Hā-we-sā'-doo (*Hā-we-sī'-doo*).—Next tribe on the east, between our tribe and the Piute, reaching north on east side Goose Lake to Willow Ranch near mouth of Fandango Creek.................[*Hā-we-sī'-doo*]
Ko'-se-al-lek'-te (*Ko-sal-lek'-tah*).—Tribe on Alturas Plain.[*Ko'-se-al-lek'-te*]
Mă-too-tsā'.—Band on southwest side of Goose Lake between our tribe and *Modok* ....[Band of *Hā-we-sī'-doo*]
Mo-des'-se.—Tribe at Big Bend Pit River.....................[*Mo-des'-se*]

Unrelated tribes:

Al-lah'-me.—Klamath Lakes tribe.
Ă-poo'-e.—*Piute* tribe.
Kah-bah'-mah'-le.—*No-to-koi'-yo* tribe of Big Meadows (now Lake Almanor).
Loo'-too-ah'-me.—*Modok* tribe.

Names used for As-tah-ke-wi'-che by Other Tribes

NAME	TRIBE USING NAME
Ah-stah'-ke'-wi'-che	*Ă-choo'-mah'-we*
As-tah-ke'-wah, As-tah-pe'-wah, and *Han-too'-che*	*Mo-des'-se*
As'-tah-ke'-watch and *As'-tah-kah-we-zo'*	*Ham-mah'-we*
As'-stah-ke-wah	*Ă-choo'-mah'-we* and *Ap-woo'-ro-kā'e*
As-tah'-re'-wi-ge	*At-wum'-we*

GEOGRAPHIC NAMES USED BY AS-TAH-KE-WĪ'-CHE TRIBE

Alturas Plain .... Ko'-se-al-lek'-tah
Ash Creek Valley (belongs to At-wum'-we) .... De-baw'-kĕ'-ĕ
Big Valley .... At'-wum'
Centerville or Rattlesnake Butte, about 8 miles east of Canby [name now changed by Govt. Geog. Board to Opahwah] .... O-pah'-wah
Forested mountains between Canby and Big Valley........ Hal-le'-wit
Hot Springs or Warm Springs Valley (Canby Valley).... Ah'-stah-re'-wah; As'-tah-ke'-wah
Little Hat Creek...................................... Tal'-le-mo'-mah
Mountains south of Canby Valley....................... Del-ah'-ko
Pit River .... Ă-ju'-mah
Round Valley, immediately northeast of Adin (belongs to *At-wum'-we* of Big Valley)......................... Se-te'-wah
South Fork Pit River.................................. Tol-lok'-ko'-be
Warner Mountains, between Alturas and Surprise Valley (especially Eagle Peak)............................ Wah'-dahk-cho'-se

THE HA-WE-SĪ'-DOO TRIBE

The *Hă'-we-sī'-doo* held the country on both sides of the southern part of Goose Lake and thence southward to the northern part of Alturas Plain, about 8 miles north of Alturas; and from Ingalls Swamp and Mill Spring on the west to the summit of Warner Mountains on the east.[1] From Goose Lake easterly the intertribal line between themselves and the Piute started from or near Willow Ranch[2] on the shore of the lake and followed up Fandango Creek 5 or 6 miles, continuing easterly to Fandango Valley, where several battles with the Piute were fought.

From Fandango Valley southward to Cedar Peak the eastern boundary follows the crest of Warner Range. The southern boundary is a nearly straight line from Big Sage Reservoir to Cedar Peak.

Their territory is mainly an open desert plain dotted in the hills with junipers and mountain mahogany. Their principal rancheria was on the mesa 8 or 10 miles north of Alturas.

VILLAGES OF THE HĂ'-WE-SI'-DOO

Ah-lahm'-se-ge.—On east side of southern part of Goose Lake between Sugar Hill and Davis Creek.

Chah'-lahk-se.—At or near west base of Cedar Mountain, about 11 miles northeast of Alturas.

---

[1] The name of the band on the east side of Goose Lake is *Ah-lahm'-se-ge;* that of the band on the west side is *Mă-too-tsā'*.

[2] Some members of the tribe say that the line was at Sugar Hill, 5 or 6 miles south of Willow Ranch.

Hā-we-si'-doo.—On the mesa about 10 miles north of Alturas. The ruling village, from which the tribe took its name.

Mă-too-tsā'.—On west side of southern part of Goose Lake, about 14 miles north of Alturas.

Adjoining tribes.—The *Hā'-we-sī'-doo* were in contact with four tribes: the *As'-tah-ke-wi'-che* on the west; the *Modok* on the north; the *Northern Piute* on the northeast and east; the *Ko'-se-al-lek'tĕ* on the south.

THE KO'-SE-AL-LEK'-TE TRIBE

PLATES 16 AND 17

The *Ko'-se-al-lek'-tĕ* are a small tribe centering on the Alturas Plain and reaching easterly to the crest of the Warner Mountains. Their territory, nearly square in outline, is sandwiched in between that of the *Hā-we-si'-doo* on the north and the *Ham-mah'-we* on the south. Their northern boundary extends easterly from Big Sage Reservoir to Cedar Mountain (or perhaps the adjacent Bald Mt., called Cedar Mt. on some maps); the eastern boundary follows the summit of Warner Range from Cedar Mt. to Warren Peak (called Buck Mt. on some maps); the southern boundary is a straight line from Warren Peak to Signal Butte on South Fork Pit River (4 miles north of the mouth of Fitzhugh Creek) and continues westerly for 10 or 12 miles; the western boundary, apparently, is a north-south line from Big Sage Reservoir southward, passing a little west of Essex Hot Spring and continuing to intersect the latitude of Signal Butte. It thus includes the treeless desert region of Rattlesnake Creek, valuable stretches of upper Pit River and the lower part of South Fork Pit, together with the several streams that flow westerly from Warner Range to South Fork Pit River.[1]

KO'-SE-AL-LEK'-TĔ VILLAGES AND BANDS

Del'-mo-mi'-che or *Tal'-mo-mi'-che* band (village *Del-mo'-mah*).—At Essex Hot Springs 10 miles west of Alturas; westernmost band of the Ko'-se-al-lek'-tĕ.

Ham-mah'-le-lah'-pe.—Village and band on Pine Creek near mouth of canyon where creek comes out from Warren Peak of Warner Range. The territory of this band adjoins on the south that of the Ham-mah'-we.

Ho-mā'-wet.—Village and band at foot of Warner Range about 6 miles NNE of Alturas.

Ko'-se-al-lek'-tah.—Ruling band; village on site now occupied by Alturas.

Too'-me-lit or *Doo'-me-lit.*—Village and band about 3 miles SW of Alturas.

[1] I have been reluctant to accord tribal rank to such small divisions as those of the *Hā'-we-si'-doo* and *Ko'-se-al-lek'-te,* but since the adjacent related tribes call them 'tribes' and refuse to regard them as a part of themselves, and since each comprised a number of villages, no other course seems open.

THE HAM-MAH'-WE TRIBE

PLATES 18-21

The *Ham-mah'-we, Ko'-se-al-lek'-te* and *Hā'-we-si'-doo* are the easternmost of the Pit River tribes. They form a north-south series, one above another, from Madeline Plains to Goose Lake, and on the east all three abut against the Northern Piute. All three reach the crest of the Warner Mountains but owing to the slightly easterly trend of the southern part of the range, the *Ham-mah'-we* attain a longitude a trifle nearer the rising sun than do either of their more northern neighbors.

The territory of the *Ham-mah'-we* lies between that of the *At-wum'-we* and *Ap-woo'-ro-kāe* on the west and the *Northern Piute* on the east. It is broadest at the north, narrowing rapidly to the south.

The northern boundary is a remarkably straight line 35 miles in length, running nearly due east from Scheffer Mountain to Warren Peak (known also as Buck Mountain, altitude 9700 ft.), crossing South Fork Pit River at Signal Butte about 4 miles north of the junction of Fitzhugh Creek with South Fork Pit River, midway between Likely and Alturas. The western boundary runs southeasterly from Scheffer Mountain (or perhaps from Stone Coal Valley west of Scheffer Mountain) following the divide east of Cottonwood and Ash Creeks, and thence south to the east side of Grasshopper Valley.

The eastern boundary follows the crest of Warner Mountains from Warren Peak southward to the end of the range, beyond which it curves to the southwest between Cedar Creek and Red Rock Valley, continuing apparently to McDonald Peak and thence southwesterly to the end of Fredonyer Mts. south of Madeline Plains.

The main part of the territory of the *Ham-mah'-we* is a broad open plain bisected by the marshy valley of South Fork Pit River and known as Likely Valley. It includes Madeline Plains on the south, and Jess and West Creek valleys on the east. It is limited on the west by the hills west of Madeline Plains and on the east by the divide between Cedar Creek and Red Rock Creek—Red Rock Valley belonging to the *Piute*.

The *Ham-mah'-we* are in contact with four tribes: the closely related *As-tah-ke-wī'-che* and *Ko'-se-al-lek'-te* on the north; the *At-wum'-we* and *Ap-woo'-ro-kā'e* on the west and southwest; the *Northern* Piute on the east and southeast.

Former Permanent (Winter) Villages of the Ham-mah'-we

Bo'-yah.—In Jess Valley.
Sah-lah'-wit.—In West Creek Valley.
Tat'-nah-hōm'-zah.—On Fitzhugh Creek near its junction with South Fork Pit River.
Til-luk-ko'-be (*Tul-luk-ko'-be*).—In Likely Valley near present settlement of Likely.

Names used by the Ham-mah'-we for Themselves and Other Tribes and Bands

Related tribes and bands:

A-choo'-mah'-we.—Fall River Valley tribe......... [*A-choo'-mah'-we*]
Ah-mitch'-e (*Ah-mit'-se" Ah'-mit*).—Dixie Valley and Eagle Lake tribe........................ [*Ap-woo'-ro-kā'e*]
As'-tah-ke'-watch, As-tah-re'-wah and *As-tah-kah-we-zo'.*—Hot Springs or Canby Valley tribe.... [*As-tah-ke-wi'-che*]
At-wum'-zan'-ne.—Big Valley tribe.............. [*At-wum'-we*]
Chah'-lahk'-se.—Hā-we-si'-doo Band 11 miles NNE of Alturas .... [*Chah'-lahk-se*]
Del-mo'-mah.—Band between Alturas and Canby Valley .... [Band of *Ko'-se-al-lek'-tĕ*]
E-tah'-me.—Another name for Big Valley tribe.... [*At-wum'-we*]
Hah-dik'-yu'-we.—Hat Creek tribe.............. [*At-soo-kā'-e*]
Ham-mah'-le-lah'-pe.—Band of Ko-se-al-lek'-tĕ on Pine Creek .... [*Ham-mah'-le-lah'-pe*]
Ham-mah'-we.—Their name for their own tribe; used also by related tribes. Also called *Tul-lok-ko'-be* or *Do-lu-ko'-be,* the name of So. Fork Pit River............................ [*Ham-mah'-we*]
Hā-we-si'-doo.—Tribe extending from Goose Lake south to about 8 miles north of Alturas........ [*Hā-we-si'-doo*]
Ko'-să-lek'-tah.—Tribe on Alturas Plain........... [*Ko'-se-al-lek'-tĕ*]
Pat'-yu-lo'-mit.—Ham-mah'-we bands in West Creek and Jess Valleys..................... [*Pat-yu-lo'-mit*]
Tŏ-bŭ'ch-we, Dah-boch-e or *Bah-boch-e.*—Ash Creek band of *At-wum'-we* .... [*Dah-boch-e*]

Unrelated tribes:

Ah-poo'-e.—Tribe in and east of Surprise Valley .... [*Northern Piute*]
Loo'-too-ah'-me (*Loot-wah'-me*).—Tribes north of Pit River tribes...... [*Modok* and *Klamath,* collectively]
Pah-kah'-mah-le'.—Big Meadows Midoo... [*No-to-koi'-yo Mi'-doo*]

Names used for Ham-mah'-we by Other Tribes

NAME	TRIBE USING NAME
As-pe-se' and *Ah-pis'-se-ye*	*Ap-woo'-ro-kā'e*
At-wum-chan'-ne and *Ham-mah'-we*	*Mo-des'-se*
At-wum'-jen'-ne and *Ham-mah'-we*	*As'-tah-ke-wī'-che*
Ham-mah'-we	*At-wum'-we* and *As-tah-ke-wī'-che*

GEOGRAPHIC NAMES USED BY THE HAM-MAH'-WE

Crooks Canyon, west of Likely................Soo'-dah pe-dā'-ge
Blue Lake three miles south of Jess Valley and five east of forks of West Creek...........Tes-ahp'-te
Eagle Lake ....Aṣ-soo-soo' kah'-te
Eagle Peak of Warner Range..................Wah'-dok-tsoo'-ge
Fitzhugh Creek nine miles north of Likely......Tan'-nó-hum'-jah
Small hill south of Fitzhugh Creek..........Yah[ch]-poo'-mah
Grasshopper Valley ....At'-too-um-wah'
Horse Lake ....Tahs'-te at'-wum (slurred, Tahs-taht'-wum)
Hot Spring on South Fork Pit River east of Likely ....Te'-mah kahtch'-hit
Indian Reservation about four miles southeast of Likely ....Choo-loo'-ko'-pe
Lava hills (sagebrush and juniper) south of Indian Reservation ....Ko-pah'-ko
Pointed peak south of Reservation..........Ten'-ne-heu'-it
Likely Valley ....Doo-loo'-ko'-be
Madeline Plains ....Sel-lat'-too-um
Sagebrush Hole or Basin.....................Pah'-tah lahts ah'-mit
Signal Butte two miles north Fitzhugh Creek..San-kow'-jā
Snowstorm Mt. on Piute-Hammahwe boundary (probably McDonald Peak)...............Choos'-leu ah'-ko
South Fork Pit River.........................Dō-lŭ'-ko'-be; Doo-loo'-ko'-be; or Tŭ-lo-ko'-be
Warren Peak, locally called Buck Mt..........Tahk'-tah-gā'-wah
Small butte at east base Warren Peak........Itch'-it

THE ATSOOKĀ'-AN FAMILY

The *Atsookā'-an* family comprises only two tribes: the *At-soo-kā'-e,* known as the Hat Creek tribe, and the *Ap-woo'-ro-kā'e,* known as the Dixie Valley tribe. Heretofore they have been classed among the Pit River tribes, but their language, as pointed out by Dixon and confirmed by my much fuller vocabularies, is widely different.

The area they occupy lies immediately south of that of the Pit River tribes (*Achomah'-an* Family), extending from a point about 8 miles east of the settlement of Montgomery Creek, easterly to Hayden Hill, the Fredonyer Mountains (between Eagle and Horse Lakes), and the divide east of Willow Creek Valley. In a north-south direction they reach from about a mile north of Cassel on Hat Creek south to Lassen Volcano, and farther east from Hayden Hill to the divide south of Eagle Lake. Their territory in an east-west direction measures about 65 miles; in a north-south direction, in the widest part, about 35 miles.

The boundary between the two tribes has been located for me by old Indians of both tribes in exactly the same way.

THE AT-SOO-KĀ'-E TRIBE

PLATES 22 AND 23

The *At-soo-kā'-e,* or Hat Creek tribe, inhabit the country north of Mt. Lassen from Lassen Peak to a mile or a mile and a half north of Cassel postoffice.

The northern boundary, beginning at the southwest corner of Goose Valley, runs easterly for about 4 miles (skirting the south edge of this valley a little more than a mile north of the latitude of Burney) and then turns north to Lookout Mountain (a peak in the ridge between Goose and Burney Valleys 4¼ miles north of Burney postoffice) where it bends abruptly east, following apparently a straight line for 16 miles to the pass between Bald Mountain and the hill immediately north of it, whence it turns abruptly southeast, becoming the eastern boundary and continuing in the same direction in a straight line for 19 or 20 miles to the east side of Grass Valley, where it changes to southwest, passing Poison Lake and reaching the lava mountains about 12 miles east of Lassen Peak. The western boundary, from the headwaters of Montgomery Creek, runs southerly along the crest of the northern Sierra Nevada for about 10 miles to Snow Mountain, where it turns southeast and continues for about 24 miles, by way of the west side of Bunch Grass Valley and Noble Pass, to Lassen Peak.

The territory of the *At-soo-kā'-e* thus includes the greater part of Burney Valley, the whole of Burney Butte, Tamarack, McGee and Crater Peaks, Bald Mountain on the northeast, Bunch Grass Valley on the west, Grass Valley on the east, the entire drainage basins of Lost Creek and of Burney and Hat Creeks except a few miles of their lower courses near Pit River, a multitude of small lakes, and the tremendous lava ridges that spread northerly from Mt. Lassen.

It is an exceedingly mountainous country strewn with lava and beset with lofty extinct volcanoes. The greater part is forested with coniferous trees intermixed with oaks, and, strange to say, with an isolated area of Digger pines (*Pinus sabiniana*) which extends from about 3 miles west of Cassel on Hat Creek easterly for 16 miles to a point on the west side of Beaver Creek Valley 3 miles east of the east base of Bald Mt. (which is partly clothed with Ponderosa pines). East of the narrow valley of Beaver Creek the Digger pines continue to the rim of the deep canyon of the Great South Loop of Pit River, thus penetrating well into the territory of the *Ap-woo-ro-kāe.* On the north they cross Pit River immediately west of Fall River Mills and reach their northern limit 4 or 4½ miles beyond, there entering the territory of the *Ă-choo'-mah'-we*. This forest is of unusual interest, consisting not only of Digger and Ponderosa pines, but also of an abundance of large junipers, and of mountain mahogany of both desert and California species, and including among the shrubs such subarid species as redbud, aromatic sumac, servisberry, mountain manzanita, and in places sagebrush.

Contrasted with this rock strewn forest area is the Valley of Hat Creek, a deep swiftly flowing stream bordered by grassy meadows and marshes fed by numerous springs, some of large size.

The *At'-soo-kā'-e* are in contact with seven tribes: the *Nos'-se* or *Yah'-nah* on the west and southwest, the *Mo-des'-se* on the northwest, the *To-mal-lin'-che-moi'*, *Il-mah'-we* and *Ă-choo'-mah'-we* on the north, the related *Ap-woo'-ro-kā'e* on the east, the *Mi'duan No-to-koi'-yo* on the south.

NAMES USED BY THE AT'-SOO-KĀ'-E FOR THEMSELVES AND OTHER TRIBES

Related tribes:

Ahk-koo'-e.—Tribe in Big Valley............ [*At-wum'-we*]
At'-soo-kā'-e (*Ah'-tsoo-kā'-e*).—Their name for their own tribe........................ [*At'-soo-kā'-e*]
Oo'-kah-soo'-e.—'Tribe' in Burney Valley.... [Band of *At'-soo-kā'e*]
Oo'-we-che'-nah.—'Tribe' in Goose Valley.... [Band of *To-mal-lin'-che-moi'*]
Po'-mah-de'-he.—Tribe at Big Bend of Pit River .... [*Mŏ-des'-se*]
Too-e-chow'-we.—Tribe in Fall River Valley.. [*Ă-choo'-mah-we*]
Wah'-doo-kā-e.—Tribe in Dixie and Eagle Lake Valleys .... [*Ap-woo'-ro-kā'e*]

Unrelated tribes:

Ah-tso-hen'-ne-ye'.—Modok tribe .... [*Loo'-too-ah'-me*]
Hen'-nah.—Tribe in Surprise Valley.......... [*Northern Piute*]
Ok'-pis-se'.—Tribe south of Honey Lake..... [*Wash'-shoo*]
Pe'-kah-soo'-e.—Tribe in Big Meadows....... [*No-to-koi'-yo Midoo*]
Te'-mow-we.—Tribe at Round Mountain...... [*Nŏs'-se* or *Yah'-nah*]

NAMES USED FOR AT'-SOO-KĀ'-E BY OTHER TRIBES

NAME	TRIBE USING NAME
Ah-doo-wah'-no-che-kah-te, At-too-an'-noo-che, At-wah-noo'-che, and *Tah-wahn'-noo'-che*	*Mo-des'-se*
Ah'-tsoo-kā'-e	*Ap-woo'-ro-kā'e*
Hah-dik'-yu'-we	*Ham-mah'-we*
Hat'-te-we'-we and *Hat'-te-we'-we es*	*At-wum'-we* and *As-tah-ke-wī'-che*
At-wum'-noo-che, To-ah-no'-che and *Had'-de-we'-we*	*A-choo'-mah'-we*
Wah-num-che'-wah.—Band in Burney Valley	*Mo-des'-se*

Geographic Names used by the At'-soo-kă'-e

Burney Butte	Ah-po'-hah
Burney Valley	Oo'-kah-soo'-ĕ'
Dixie Valley	Op'-wah-de'-wah
Fall River Mills	Te-chow-e'-wah
Hat Creek	E-dits-te'-e
Lassen, Mt.	Per-roo'-e-ke'-nah
McGee and Crater Peaks	Bop'-ske
Pit River	Po'-mah-rah
Shasta, Mount	Wi'-ke

THE AP-WOO'-RO-KĂ'E TRIBE [1]

Plates 24-27

The *Ap-woo'-ro-kă'e* or Dixie Valley tribe owned a considerable area immediately east of their near relative, the *At'-soo-kă'-e,* extending from the west side of Beaver Creek Valley, a mile or two east of Bald Mountain, easterly to Hayden Hill, and southeasterly to include Grasshopper Valley, Eagle Lake, and Willow Creek Valley or Basin, just over the ridge north of Susanville. The northern boundary, beginning on the west about a mile northeast of Bald Mountain, runs northeasterly, crossing Beaver Creek and the west arm of the south loop of Pit River to Muck Valley, whence it immediately recrosses Pit River (the eastern arm of the loop) and continues easterly in a nearly straight line for 15 miles to Hayden Hill. The east boundary, beginning at Hayden Hill, follows a southeasterly course along the low divide between Dry Valley and Grasshopper Valley and continues in the same general direction to the Fredonyer Mountains between Eagle Lake and Horse Lake, turning thence southward and following the divide east of Willow Creek Valley. The southern boundary follows the divide between the Susanville plain on the south and the valleys of Eagle Lake and Willow Creek on the north, continuing westerly along the same divide to the upper part of Pine Creek. The western boundary is the intertribal line already described as the eastern boundary of the *At'-soo-kă'-e.*

The territory of the *Ap-woo'-ro-kă'e* thus includes a very short stretch of Pit River (the southern part of the South Loop south of Muck Valley), the greater part of the valley of Beaver Creek, the whole of Horse Creek Valley, Dixie Valley, Grasshopper Valley, the

[1] The pronunciation of the last syllable of the name *Ap-woo'-ro-kă'e* is not perfectly clear. The terminal *e* is not distinctly uttered—as it is in the case of the *At-soo-kă'-e.* The word may be written *Ap-woo'-ro-kă''* (prolonging the *ă* sound) or *Ap-woo'-ro-kă'e*—as here adopted.

valleys of Eagle Lake and Willow Creek, and numerous volcanoes, craters and buttes. The area of the tribe therefore is considerably greater than that of the related *At'-soo-kā'-e.*

The country of the *Ap-woo'-ro-kā'e* lacks the continuous coniferous forests of the At'-soo-kā'-e. At the same time some of the higher ridges are forested with yellow pine [Ponderosa pine], but the greater part of the area is open rocky sagebrush country dotted with juniper and mountain mahogany, in places of sufficient size and abundance to form low open forests—always attractive and welcome to the traveler—here and there alternating with moist grassy valleys and little lakes. Besides these, is the large and beautiful body of water known as Eagle Lake, bordered on the east by a juniper-covered lava ridge, and on the west, at a little distance, by continuous coniferous forests that push north from the valuable timber lands of the *No-to-koi'-yo Midoo.*

In marked contrast, and beginning only 4 or 5 miles north of Eagle Lake, is the bare alkali flat known as Grasshopper Valley which, though only 10 miles in length by half that in width, is a pointed reminder of the extensive desert flats not far away.

The *Ap-woo'-ro-kā'e* are in contact with six tribes, namely: the related *At'-soo-kā'-e* on the west, the Fall River *Ä-choo'-mah'-we* on the northwest, the Big Valley *At-wum'-we* on the north, the *Ham-mah'-we* of Madeline Plains on the northeast, the *Northern Piute* on the east and southeast, the *Mi'doo No'-to-koi'-yo* on the south.

Former Villages of the Ap-woo'-ro-kā'e[1]

Ap'-wah-re'-wah.—In Dixie Valley; said to have been a large town. [Can it be another name for *Rats-ow'-we-ke?*]

Pits-ă-roo'-hoo (or *Pit-să-roo'-hoo*).—In Willow Creek Valley.

Ras-să-re'-wah.—On Beaver Creek.

Rats-ow'-we-ke.—On Indian Creek at east end of Dixie Valley, at foot of mountain ridge of same name. Principal village.

Sik-kah'-kek.—Another village on Beaver Creek.

---

[1] Harry Wilson, one of the head men of the Ap-woo'-ro-kāe, gave me the names of two villages on the Susanville side of Antelope Ridge, namely, *Nor'-ro-witch' hăch,* on or near present site of Susanville; and *Nor'-rah-chă'-e-che,* at south base of Antelope Ridge 5 or 6 miles northeast of Susanville. Another member of the tribe tells me that these villages belonged, not to his tribe, but to the Big Meadows tribe (*No-to-koi'-yo Midoo*) which probably is correct. I assume that Harry Wilson misunderstood my question.

Names used by the Ap-woo'-ro-kā'e for Themselves and Other Tribes

Related tribes:

Ahk-we", Ah-koo-we'-e'.—Big Valley tribe........... [*At-wum'-we*]
Ah-pe-se', Ah-pis'-se-ye. Likely Valley tribe on upper South Fork Pit River............................ [*Ham-mah'-we*]
Ah'-tsoo-kā'-e.—Hat Creek tribe...................... [*At'-soo-kā'-e*]
Ap-woo'-roo-kā'e, Ap-poo'-ro-kā" (*Ă-poo'-ro-kā'e*) *Ap-wah'-roo-kā", Ap-wur'-roo-kā'e.*—Their name for their own tribe............................ [*Ap-woo'-ro-kā'e*]
As-stah-ke'-wah.—Hot Springs Valley tribe.......... [*As-tah-ke-wi'-che*]
Po-mah-re'-ye.—Pit River tribe, next below Fall River *Mills* .... [*Il-mah'-we*]
Too'-e-tsow'-we (slurred *Te-tsow'-we*).—Fall River tribe .... [*Ă-choo'-mah'-we*]

Unrelated tribes:

Ah'-tsoo-hah-ne'-ye; Ah'-tsoo-hun'-ne .... [*Modok*]
Hen'-nah .... [*Piute*]
Mah'-nah-tse'-e; Mah'-nah'-tse .... [*Washoo*]
Pe'-kah-soo'-e; Te'-ke-soo-we.—Big Meadows tribe.. [*No-to-koi'-yo Mi-doo*]
Te'-mow-we .... [*Nōs'-se* or *Yah'-nah*]

Names used for the Ap-woo'-ro-kā'e by Other Tribes

NAME	TRIBE USING NAME
Ah-mit'-at'-wum and Ah-mits'-che.—Band in Dixie Valley	*Mo-des'-se*
Ah-mitch'-e and Ah-mit'-se	*At-wum'-we, As-tah-ke-wi'-che,* and *Ham-mah'-we*
At'-to-maw'-wah.—Band in Grasshopper Valley	*Ă-choo'-mah'-we*
Choo-kā'-che.—Band on Beaver Creek	*Mo-des'-se*
Kaw-le'-wah and Ah-mit'-che	*Ă-choo'-mah'-we*
Wah'-doo-kā-e	*At'-soo-kā'-e*

Geographic Names used by the Ap-woo'-roo-kā'e

Bald Mt. .... We-puk'-kah-me
Beaver Creek .... Ko-too'-roo-si'
Davis Creek, Big................................ Ti'-yu mur-ră si'-te-ke
Davis Creek, Little............................. Wah-mo-pōwk'-noo-e
Big Valley .... Ahk-woo'; Ah-ko-wah'
Crater Butte Lake, eight miles east-southeast of Poison Lake .... Chā'-heu poo-wow'
Dixie Valley .... Ap-wah-re'-wah
Dry Grass Valley, about 3 miles north of Poison Lake .... Po-sĕ'-te
Dry Valley .... Too'-ser-ritch'-e
Eagle Lake .... At-sip sook-ă-de'; Ah-tsip soo'-kă-de'

Grasshopper Valley .... Ak-we yu-bow
Hat Creek .... At-soo-kā'-yah
Hayden Hill .... Tsah-mā'-heu
Lassen, Mt. .... Wi'-ko hin'-ki
Likely Valley, on Upper South Fork Pit River.. Ah-pe-se'
Mountains between Dixie Valley and Big Valley. Ahk-wi-heu; Ahk-we'-ă-heu'
Muck Valley .... Ah'-te-ke
Pit River (the main stream).................. Po'-mah-rah
Pit River, Great South Loop between Muck Valley and Horse Creek.................. Jak-chah-se'
Poison Lake .... Her-rup'-mah
Reservoir Valley .... Too-loo-we'-too-ke
South Fork Pit River........................ Ham'-mah-de'-he
Willow Creek (east side Hayden Hill to Big Valley) .... Tsah'-he-se
Willow Cr., east of Eagle Lake............... Pe-cher'-ro-oo'
Willow Creek Mountain....................... Pe-cher'-rŏ ah-koo

ALPHABETIC LIST OF PIT RIVER TRIBES AND BANDS

NAME	LOCATION AND AUTHORITY
Ă-choo'-mah'-we (Ah-choo-mah'-we; Ah-choo-mah-we'-che; Ă-ju'-mah'-we)	Fall River tribe; their name for themselves; also used by Mo-des'-se and related tribes. Pronounced *Ạ-ju'-mah-we* by the At-wum'-we and Ȧs-tah-ke-wi'-che. [*Adzuma'wi* Handbook, 1907, from Curtin MS., 1889; DeAngulo. 1926.] See also *Hā-wis-aht'-wum-we.*
Ah-doo-wah'-no-che	Modesse name for At'-soo-kā'-e. See At-too-wah'-noo-che.
Ah-koo-we'-e' (Ahk-koo'-we; Ahk-we")	Ap-woo'-ro-kā'e and At-soo-kā'-e name for Big Valley At-wum'-we.
Ah-lahm'-se-ge	Northeastern band of Hā'-we-si'-doo, east of southern part of Goose Lake. Their name for themselves; used also by the At-wum'-we, As-tah-ke-wi'-che, Ko'-se-al-lek'-tĕ and Ham-mah'-we.
Ah-me'-che (Ah'-mits)	Mo-des'-se name for closely related band at 'The Cove' on Pit River.
Ah-mit'-che (Ah-mit'-se, Ah-mits'-che, Ah-mitch'-e, Ah-mit, and Ah'-mit at'-wum)	Variants of name applied to Dixie Valley Ap-woo'-ro-kā'e by all the Pit River tribes proper (Achoomah'-an Family).
Ah-pe'-dah-dā' (Yah-pe'-dah-dā')	Large At-wum'-we village on Ash Creek in the open flat of Big Valley 7 miles west of Adin. Ruling band.

Ah-pis-se'-ye (Ah-pe-se') .... Ap-woo'-ro-kā'e name for Ham-mah'-we.

Ah-stah'-ke-wi'-che .... See As-tah-ke-wi'-che.

Ă-ju'-mah-we .... As-tah-ke-wi'-che and At-wum'-we pronunciation of Ă-choo'-mah'-we. Used also by members of other tribes.

Ak[h]owigi .... Given by Dixon (1908) as At-soo-kā'-e name for Beaver Creek band [of Ap-woo'-ro-kā'e].

Ap'amadji .... Given by Dixon (1908) as Ă-choo'-mah'-we name for Burney Valley band of At-soo-kā'-e. See *Wah'-num-che-wah.*

Ap-woo'-ro-kā'e .... Dixie Valley tribe, extending to and including Eagle Lake; their name for themselves. [Name written *Apwaraki* by Kelsey MS., 1906; *Apwarukē'i* by Dixon, 1908.]

As'-tah-kah-we-zo' .... Ham-mah'-we name for As-tah-ke-wi'-che.

As-tah-ke-wi'-che[1] (As-tah-re-wi'-se). Tribe in Canby (Hot Springs) Valley; their name for themselves (pronounced As-tah-re'-wi'-che). Called As-tah-ke'-watch, As'-tah-re'-wah and As'-tah-kah-we-zo' by the Ham-mah'-we; As-tah'-re-wi'-ge by the At-wum'-we; Ah-stah-ke-wi'-che by the Ă-choo'-mah'-we and slurred As-tah-ke'-wah (name of Hot Springs Val.) by neighboring tribes. *Astakiwi* of Kroeber, 1925.

A-tco-mâ-wi (Mallery, 1881)........ See Ă-choo'-mah'-we.

At-soo-kā'-e (Ah'-tsoo-kā-e) .... Hat Creek tribe; their name for themselves; used also by Ap-woo'-ro-kā'e. [Written *Atsugē'wi* by Dixon, 1905 and 1908; *Atsuge* by Mason, 1904, and by DeAngulo, 1926].

At'-to-maw'-wah .... Ă-choo'-mah'-we name for Grasshopper Valley band of Ap-woo'-ro-kā'e.

At-too-wah'-noo-che (At-too-an'-noo-che; slurred Ah'-wah-noo'-che) ... Mo-des'-se name for Hat Creek tribe (At-soo-kā'-e).

Atuami, A-tu-a-mih .... See At-wum'-we.

At'-wum-chan'-ne (At-wam-chun'-ne) .... Mo-des'-se name for Ham-mah'-we.

At-wum'-jen-ne .... As-tah-ke-wi'-che name for Ham-mah'-we.

At-wum'-noo-che .... Ă-choo'-mah'-we name for At-soo-kā'-e.

---

[1] Name written *Astakaywas* and *Astakywich* by Stephen Powers in 1874: *Es-ta-ke'-wach* by Powers in 1877; *Astaqke'wa* in Handbook Am. Inds. (after Curtin MS., 1889), 1907; *Astaghiwawi* by DeAngulo in 1926.

At-wum'-we .... Big Valley tribe; their name for themselves; used also by As-tah-ke-wi'-che. [Written *A-tu-a'-mih* by Powers, 1877; *Atuami* by Kroeber, 1925.]

At-wum-zan'-ne and E-tah'-me .... Ham-mah'-we names for At-wum'-we. [Written *Atwamzini* by DeAngulo, 1926.]

Bah'-bo^{ch}-e (Dah-bo^{ch}-e, or Tŭ-bŭ^{ch}-we) .... Ham-mah'-we name for Ash Creek band of At-wum'-we. See Ko-sel-lat'-to-mah.

Big Valley Indians (Gatschet, 1890). See At-wum'-we.

Chah'-lahk-se .... Hā-we-si'-doo village and band at west base of Warner Mts. about 11 miles northeast of Alturas.

Chah-wahs'-te' chan-ni-che .... One of the Mo-des'-se names for E-tsah'-tah, the Goose Valley band of To-mal-lin'-che-moi'.

Chenoya (Chenoyana, Chunoi'yana, Teunoiyana) .... Yah'-nah name for At-soo-kā'-e. Handbook 1907 from Curtin MS. 1885.

Choo-kā'-che .... Mo-des'-se name for Beaver Creek band of Ap-woo'-ro-kā'e. See Ras-să-re'-wah and Sik-kah'-kek.

Chu-mâ'-wa .... Given by Powers (1877) as tribe in Round Valley. The Round Valley referred to is just NE of Adin, in the NE corner of Big Valley. It was occupied by the Se-te'-wah band of At-wum'-we. The Chu-mâ-wa is probably a rendering of Ju-mah'-we, a term sometimes applied to Pit River tribes.

Dah-bo^{ch}-e (or To-bŭ^{ch}-we) .... Mo-des'-se name for Ash Creek band of At-wum'-we; called *Bah'-bo^{ch}-e* and *To-bŭ^{ch}-we* by the Ham-mah'-we; *De-baw'-kĕ-ĕ* by the As-tah-ke-wi'-che. See Ko-sel-lat'-to-mah.

Dal'-mo-mi'-che, Del'-mo-mi'-che; Tal'-mo-mi'-che (village, Del-mo'-mah, Tal-mo'-mah) .... Ko'-se-al-lek'-tĕ village and band at Essex Hot Spring, 10 miles westerly from Alturas; westernmost band of Ko'-se-al-lek'-tĕ.

Dixie Valley tribe .... See Ap-woo'-ro-kā'e.

Do-lu'-ko'-be or Tŭ-lok-ko'-be .... Name of South Fork Pit River, often applied to the Ham-mah'-we.

Doo'-me-lit .... See Too'-me-lit.

Eagle Lake Indians (Rebellion Records, 1897) .... See Ap-woo'-ro-kā'e.

E-chat'-tah-we .... A-choo'-mah'-we name for Goose Valley band of To'-mal-lin'-che-moi'.

E-poo'-de .... Mo-des'-se name for closely related band on Montgomery Creek.

Es-ta-ke'-wach (Powers, 1877)...... See As-tah-ke-wi'-che.

E-tah'-me .... One of the Ă-choo'-mah'-we and Ham-mah'-we names for Big Valley At-wum'-we; also used by Ă-choo'-mah'-we as nickname for Mo-des'-se.

E-tsah'-tah (E-tsah'-tah iss) .... Mo-des'-se name for Goose Valley band of To-mal-lin'-che-moi'. See E-chah'-tah-we.

Had'-de-we'-we .... See Hat'-te-we'-we.

Hah-dik'-yu'-we .... See Hat-te-we'-we.

Hah-te'-wah .... Given by Powers (1877) as Hot Springs Valley tribe [As-tah-ke-wi'-che]. See *Han-too'-che.*

Hamefcuttelies (Powers, 1874; Ha-mef-kut'-tel-li, Powers, 1877)..... See Ham'-mă-o-ket'-tal-le.

Ham-mah'-le-lah'-pe .... Ko'-se-al-lek'-te village and band on Pine Creek near mouth of canyon where creek emerges from west base of Warren Peak. The territory of this band adjoins that of the Ham-mah'-we on the south and extends west to South Fork Pit River. Name used also by Ham-mah'-we.

Ham-mah'-we .... Tribe from Likely Valley on South Fork Pit River to Madeline Plains; their name for themselves; used also by Mo-des'-se, At-wum'-we, As-tah-ke-wi'-che, and Ko'-se-al-lek'-tĕ. [Name written *Hu-mâ'-whi* by Stephen Powers in 1877. *Hamawi,* Kroeber, 1925.]

Ham'-mă-o-ket'-tal-le[1] .... Mo-des'-se name for At-wum'-we of Big Valley.

Ham-mow-ēs' .... One of the At-wum'-we names for Fall River Ă-choo'-mah'-we. [See also Ă-ju'-mah'-we.]

Han-too'-che (Han-too', Han-teu').. As-tah-ke-wi'-che band in Stone Coal Valley. Name used by Mo-des'-se for entire As-tah-ke-wi'-che tribe. *Hantiwi,* Kroeber, 1925.

Hat Creeks; Hat Creek tribe........ See At-soo-kā'-e.

Hat'-mah (Too-hat-mah) .... Cayton Valley band of Il-mah'-we; their own name; used also by Mo-des'-se and Ă-choo'-mah'-we.

Hat'-te-we'-we (Had'-de-we'-we, Hat'-te-we'-we es) .... As-tah-ke-wi'-che and At-wum'-we name for Hat Creek At-soo-kā'-e. Called Hah'-dik'-yu'-we by the Ham-mah'-we. [Written *Hadi'wiwi* by DeAngulo, 1926.]

[1] Called Hamefcuttelies by Stephen Powers in 1874.

Hā-we-si'-doo (Hā-we-sā'-doo, Hā-we-si'-too) .... Tribe from Goose Lake to northern end of Alturas plain; their name for themselves; used also by Ă-choo'-mah'-we, As-tah-ke-wi'-che, Ham-mah'-we and Ko'-se-al-lek'-tĕ.

Hā'-wis-aht'-wum-we .... One of the names used by Mo-des'-se for Ă-choo'-mah'-we.

Ho-mā'-wet .... Ko'-se-al-lek'-tĕ village and band at west base of Warner Range about 6 miles northnortheast of Alturas.

Hot Springs Indians (Gatschet, 1890) .... See As-tah-ke-wi'-che.

Hu-mâ-whi (Powers, 1877; Powell, 1891) .... See *Ham-mah'-we.*

Idjuigilum'idji .... Given by Dixon (1908) as Ă-choo-mah'-we name for Beaver Creek band of Dixey Valley Indians [Ap-woo'-ro-kā'e]. Called Choo-kā'-che by the Mo-des'-se.

Il-mah'-we .... Pit River tribe, from Pecks Bridge nearly to Fall River; their name for themselves; used also by the Mo-des'-se, Ă-choo'-mah'-we and At-wum'-we. Named from *Il-mah'*, their principal village.

Ko'-se-al-lek'-tĕ (Ko-sal-lek'-tah) .... Tribe occupying Alturus Plain from about 8 miles north to 6 miles south of Alturas; their name for themselves; used also by all related tribes from the *Mo-des'-se* to the *Ham-mah'-we.* They had at least 5 permanent (winter) villages. [Doubtless DeAngulo's *Qosalektawi* (1926), mentioned without locality.]

Kaw-le'-wah .... Ă-choo'-mah'-we name for Dixie Valley Ap-woo'-ro-kā'e.

Ko-sel-lat'-to-mah .... At-wum'-we band in Ash Creek Valley.

Kum' Mi'-dem .... No-to-koi'-yo Midoo name for Pit River Tribes. *Kōm'-maidŭm,* Handbook 1907 (from Dixon MS. 1904).

Lah'-lah-pis'-mah .... Band of Mo-des'-se at Hot Springs on south side of Big Bend of Pit River, opposite Mo-dess', the ruling village, which was on the north side.

Made'qsi (Handbook 1910, from Curtin MS. 1889) .... See Mo-des'-se.

Mah-pe'-dah-dā' .... Il-mah'-we band at junction of Salmon Creek with Pit River.

Mah-dā'-se .... One of the Ă-choo'-mah'-we pronunciations of Mo-des'-se, which see. Written *Madehse* by Kroeber, 1925.

Mă-too-tsā' .... As-tah-ke-wi'-che name for Hā'-we-si'-doo band on west side of southern part of Goose Lake.

Mo-aht'-was (Mo-e-twas, Palmer 1855; Mo'atwash, Mu'atwash, Gatschet 1890) .... Klamath name for Pit River Indians.

Mo-des'-se .... Tribe at Big Bend Pit River; their name for themselves. Used also by neighboring tribes. Pronounced *Mah-dā'-se* by some of the *Å-choo'-mah'-we*. [Written Made'qsi in *Handbook Am. Inds.*, 1910, from Curtin MS., 1889.] Named from Mo-dess', the ruling village.

Monctske (Russell, 1857) .... Yah'nah name for Mo-des'-se.

Mu'atwash .... See Mo-aht'-was.

Oo'-kah-soo-we (Oo'-kah-soo-we ah-di-ow'-te) .... At-soo-kā'-e name for Burney Valley band of their own tribe. Called *Wah'-num-che'-wah* by the Mo-des-se.

Oo-le'-moo-me .... Mo-des'-se name for their village at Hot Spring on south side Pit River at Big Bend.

Oo'-we-che'-nah .... At-soo-kā'-e name for Goose Valley band of To-mal-lin'-che-moi'. [See also E-tsah'-tah-iss.]

Palaihnih or Palaiks .... Name used by Horatio Hale for Pit River tribes collectively (Ethnog. Wilkes Expd. 218, 569, 1846) and adopted by Gallatin, 1848; Latham, 1850; Berghaus, 1852; Eastman (map), 1852; Ludewig, 1858; Bancroft, 1875; Gatschet, 1877; Powell, 1891.

Pikas (Hutchings, 1857) .... Given as Indian Valley name for Pit River Indians. Probably an error. The Indian Valley Indians are *No-to-koi'-yo Midoo.*

Pat'-yu-lo'-mit .... Ham-mah'-we name for their bands in West Creek and Jess Valleys.

Pit Rivers .... Common name for Pit River tribes.

Po-mah-de'-he (Po-mah-re'-he, Po-mah-re'-ye) .... At-soo-kā'-e and Ap-woo'-ro-kāe name for Mo-des'-se and Il-mah'-we. Written *Pomar'ü* by Dixon (1908).

Poo'-e-choos (Poo'-e-soos') .... Name, meaning 'easterners,' applied loosely by Wintoon to tribes east of themselves; often referring to Yah'-nah and Pit Rivers collectively. Variously written, as *Pu'-i-su, Pu'-su, Pu'-shush,* and other forms.

Qosalektawi .... Achoma'wan 'group' mentioned without locality by de Angulo, 1926. Obviously Ko'-se-al-lek'-te, which see.

Se-te'-wah .... At-wum'-we band in Round Valley, NE of Adin.

Tah'-sah .... Mo-des'-se village on south side Pit River near Rope Ferry, ½ mile east of Hot Spring at Big Bend.

Tah'-wahn-noo-che .... Slurred form of At-too-wah'-noo-che, the Mo-des'-se name for the Hat Creek At-soo-kā'-e. [See also Too'-e-tsow'-we.]

Tal'-mo-mi'-che .... See Del'-mo-mi'-che.

Tcuno'iyana .... See Chenoya.

Te-tsow'-we (Te'-chow-we) .... Ap-woo'-ro-kā'e and At-soo-kā'-e name for Fall River Ă-choo'-mah'-we. Also called Too'-e-tsow'-we by the Ap-woo'-ro-kā'e.

To-ah-no'-che .... Ă-choo'-mah'-we name for Hat Creek At'-soo-kā'-e. [See also Tah'-wahn-noo-che.]

To-bŭ^{ch}-we (Bah-bo^{ch}-e or Dah-bo^{ch}-we) .... Ham-mah'-we name for Ash Creek band of At-wum'-we. Called *Dah-bō^{oh}-e* by the Mo-des'-se.

To-mal-lin'-che-moi' .... Tribe on Pit River between Mo-des'-se and Il-mah'-we. Their own name; used also by Mo-des'-se.

Too'-e-tsow'-we (Too'-e-chow'-we, Te'-tsow-we) .... Ap-woo'-ro-kāe and At-soo-kā'-e name for Fall River Ă-choo'-mah'-we.

Too-hat'-mah (slurred, Hat'-mah) .. Cayton Valley band of Il-mah'-we; their name for themselves; used also by Mo-des'-se and Ă-choo'-mah'-we.

Too'-me-lit (Doo'-me-lit) .... Ko'-se-al-lek'-tĕ village and band about 3 miles southwest of Alturas.

To-tā'-o-me and Too-tā'-o-mal'-le ... Mo-des'-se name for At-wum'-we of Big Valley.

Tuqte'umi .... Il-mah'-we name for At-wum'-we (Handbook 1907, from Curtin MS., 1889).

Wah'-doo-kā-e .... At-soo-kā'-e name for Dixie Valley Ap-woo'-ro-kāe.

Wah'-num che-wi'-che .... Mo-des'-se name for Burney Valley band of At-soo-kā'-e. See Oo'-kah-soo-we.

Wamarī'i .... Given by Dixon (1908) as At-soo-kā'-e name for Burney Valley band. See Wah'-num-che-wi'-che.

Wen'-ne-hah'-le .... Il-mah'-we village and band at falls in canyon of Pit River 3 miles below Fall River mouth.

Yah-pe′-dah-da′ .... See Ah-pe′-dah-dā′.
Yah′-ho-re′-choo′-ish .... One of the Shaste names for Pit River Indians.
Yuca's [error for Yukes] .... Name, meaning enemies, given by Lieut. Edward Russell and Col. G. Wright in 1853 (1857) for Pit River Indians.

NAMES USED FOR PIT RIVER TRIBES (COLLECTIVELY) BY UNRELATED TRIBES

The Klamath (Lakes) tribe call the Pit River Indians.. Mo-aht′-was
The Northern Piute (Fort Bidwell) tribe call the Pit River Indians .... Ish′-she show′-we
The No-to-koi′-yo tribe call the Pit River Indians...... Kum′-mi′-dem
The Shaste tribe call the Pit River Indians............ Oo′-chah-hah-roo chah′-wish and Yah′-ho-re′ choo′ish
The Wintoon tribe call the Pit River Indians.......... Poo′-e-soos′
The Yah′-nah or Nos′-se tribe call the Pit River Indians. Chun-noi′ yah′-nah

NAMES USED BY PIT RIVER TRIBES FOR UNRELATED TRIBES OF NORTHERN CALIFORNIA

NAME USED	TRIBE USING NAME	TRIBE TO WHICH NAME IS APPLIED
Ah-lah′-me (*Al-lah′-me*).	*Mo-des′-se*, *Ä-choo′-mah′-we* and *As-tah-ke-wī′-che*.	Klamath Lakes tribe.
Ah-poo′-e (*Ä-poo′-e*). [See also *Hen′-nah.*]	*Mo-des′-se*, *Ä-choo′-mah′-we*, *At-wum′-we*, *As′-tah-ke-wī′-che*, *Ham-mah′-we*.	Northern Piute.
Ah-tah′-me-kah′-me. [See also *Pas-sā′-put-che*, *Te-sī′-che* and *Te′-mow′-we*.].	*Mo-des′-se*.	Yah′-nah or Nōs′-se.
Ah-tsoo-hen′-ne-ye′ (*Ah-tsoo-hah-ne′-ye*, *Ah-tsoo-hun′-ne*).	*At′-soo-kā′-e* and *Ap-woo′-ro-kā′e*.	Modok.
Ä-te′.	*Mo-des′-se*.	O-kwahn′-noo-tsoo′.
Bă-kah′-mah′-le (*Pah′-kĕ-mah-le; Pah-rah′-mah-le*). [See also *Pah-kah′-mah′-le*, *Bug-gah′-mah-le* and *Kă-bă′-mah′-le*].	*Mo-des′-se*.	No′-to-koi′-yo Midoo
Bug-gah′-mah-le and *Puk-kah′-mah*.	*A-choo′-mah′-we*.	No′-to-koi′-yo Midoo.
Che-yu′-wit.	*Mo-des′-se*.	Tribe on Squaw Creek; may be band of O-kwahn′-noo-tsoo′.
Ek-pe′-me (*E*ch*-pe′-me*).	*A-choo-mah′-we*, *At-wum′-we* and *Mo-des′-se*.	Northern Wintoon.

Hā-wis-se-kahs'-te.	*Mo-des'-se.*	Trinity River Wintoon.
Hen'-nah.	*At'-soo-kā'-e* and *Ap-woo'-ro-kā'e.*	Northern Piute.
Kă-bă'-mah'-le, Kah-bah'-mah'-le and *Pă-kah'-mah'-le.* [See also *Bă-kah-mah'-le.*]	*At-wum'-we* and *As-tah-ke-wī'-che.*	No'-to-koi'-yo Midoo.
Loo'-too-ah'-me (*Loot-wah'-me*), *Loo'-too-am'-me.*	*Ā-choo'-mah'-we, At-wum'-we, Mo-des'-se, As-tah-ke-wi'-che* and *Ham-mah'-we.*	Modok or Modok-Klamath collectively
Mah-nah-tse'-e.	*Ap-woo'-ro-kā'e.*	Wash'-shoo.
Ok-pis-se'.	*At-soo-kā'-e.*	Wash'-shoo.
O-se'-low-wit.	*Mo-des'-se.*	*Wi-dal'-pom* band of Northern Wintoon.
Pah-kah'-mah-le', Pah'-kĕ-mah-le, Pah-rah'-mah-le. [See also *Bă-kah'-mah'-le, Bug-gah'-mah-le, Kă-bă'-mah'-le* and *Puk-kah'-mah.*].	*Ham-mah'-we* and *Mo-des'-se.*	No'-to-koi'-yo Midoo.
Pas-sā'-put-che. [See also *Te'-si-che.*]	*Mo-des'-se.*	Yah'-nah or Nōs'-se.
Pe'-kah-soo'-e.	*At-soo-kā'-e* and *Ap-woo'-ro-kā'e.*	No'-to-koi'-yo Midoo.
Puk-kah'-mah and *Bug-gah'-mah-le.*	*Ā-choo'-mah'-we.*	No'-to-koi'-yo Midoo.
Sas'-te'-che (*Shas'-te'-che*).	*Ā-choo'-mah'-we* and *Mo-des'-se.*	Sas'-te (Shas'-te) and O-kwahn'-noo-tsoo'.
Te'-ke-soo-we.	*Ap-woo'-ro-kāe.*	No'-to-koi'-yo Midoo.
Te'-mow-we.	*At'-soo-kā'-e* and *Ap-woo'-ro-kā'e.*	Yah'-nah or Nōs'-se.
Te'-si-che.	*Ā-choo'-mah'-we, At-wum'-we* and *Mo-des'-se.*	Yah'-nah or Nōs'-se.

WORD BORROWINGS FROM WESTERN SHOSHONEAN TRIBES BY THE AT-SOO-KĀ-E AND AP-WOO'-RO-KĀE

It will be observed that the suffix '*ōb*' is employed by the *At-soo-kā'-e* and *Ap-woo'-ro-kā'e* to denote that the words to which it is attached are names of plants (see Vocabularies, pp. 10 and 11). The term is borrowed from certain western Shoshonean dialects in which, similarly, it follows the specific names of many of the trees, bushes and other forms of plant life.

And it is still more surprising to note that among the various Shoshonean tribes certain words that by chance or otherwise bear the closest likeness to those in the *Atsookā'-an* languages occur, *not* in the nearby *Northern Piute,* but in the vocabularies of the geographically remote *Monache* tribes of the western slope of the southern Sierra Nevada, notably the *Em'-tim-bitch* and *Wuk-să'-che.* This is shown by the following couplets:

	EM'-TIM-BITCH OR WUK-SĂ-CHE OR BOTH	A-SOO-KĀ-E OR AP-WOO'-RO-KĀE OR BOTH
Redbud	Kar-rah'-tah ōb	Tah-kah-kah ōb
Wild Plum	Pah-too ōb	Paht-kōb
Elder	Hoo-boo ōb	Ko-bah ōb
Acorn	Tik'-ki	Tah'-ke
Nuthatch	Kah'-dah-kah'-dă	Kah'-rah-kah-rah'
Small Rabbit (Cottontail or Brush Rabbit)	Tah-bo'-che	Dā-buk'-ke
Now	Man'-ne-kŭ	Ken'-ne-kah
Teeth	Tow'-wah	Tsow'
Dry hide	Poo-hoo	Poo'-doo

NAMES GIVEN BY JA-MUL

Names for Pit River and neighboring tribes occurring in the mythological history of the *Mo-des'-se*—names said to have been given many thousands of years ago by *Jā'-mul,* the old Coyote-man deity:

Ah-lah'-me .... Klamath tribe
Ah-poo'-e .... Northern Piute
Ah-tum'-me-kah'-me .... Tribes of Sacramento Valley
Bă-kah-mah'-le .... Tribe south of Mt. Lassen [No-to-koi'-yo]
Ek-pe'-me .... McCloud River Wintoon
Hā'-wis-se-kas'-te .... Trinity River Wintoon
Il-lo-wah'-me .... Tribes of northwestern California (collectively)
Il-mah'-we .... Pit River tribes (collectively)
It-te-pah'-tah-me e-kah ...... Yah'-nah of Round Mt. Region
Loo-too-ah'-me .... Modok tribe
Sas'-te'-che .... Shaste tribe

ILLUSTRATIONS

The photographs reproduced on Plate 6 were taken by Zenaida Merriam; all others by C. Hart Merriam.

PLATE 1. Big Bend of Pit River, Shasta County, from *Lah'-lah-pis'-mah* Hot Spring on south side of river. Looking down stream (westerly). Headquarters of the *Mo-des'-se* tribe.

PLATE 2. *Mo-des'-se* man and wife. Man, *Istet Woiche,* better known as William Hulsey. Big Bend, Pit River.

PLATE 3. *Mo-des'-se* man and wife. Man, *Istet Woiche,* better known as William Hulsey. Big Bend, Pit River. Full face front view.

PLATE 4. *Mo-des'-se* woman. Mrs. Walter Moody and daughters. The daughters are half Yah'-nah, the father, Walter Moody of Montgomery Creek, being a Yah'-nah.

PLATE 5. Burney Falls in Burney Creek Canyon, Shasta County. Territory of the *Il-mah'-we* tribe.

PLATE 6. Pit River Canyon about three miles below junction of Fall River. Territory of the *Il-mah'-we* tribe.

Fig. 1. The Falls, 3 miles below Fall River Mills.

Fig. 2. The Canyon just above the Falls.

PLATE 7. *Il-mah'-we* man. John Carmony of Cayton Valley, Shasta County.

PLATE 8. Fall River Valley, Shasta County, California. Territory of *Ă-choo'-mah'-we* tribe.

Fig. 1. Big Spring at head of Fall River. There are several of these huge springs.

Fig. 2. Fall River Valley. Looking west from near Glenburn, showing the flat open valley bordered by ponderosa pines.

PLATE 9. Big Valley, on line between Modoc and Lassen Counties. Territory of the *At-wum'-we* tribe. Showing level sagebrush plain. One of our camps in foreground.

PLATE 10. *At-wum'-we* men. Near Lookout, Big Valley, Modoc County. Chief Harry George (right); Billy Quinn (left).

PLATE 11. *At-wum'-we* women. Near former village of *Ah-pe'-dah-dā'*. Big Valley, Modoc County, California.

PLATE 12. *At-wum'-we* tribe. Frames of individual sweathouses, called *Tem'-mah-kah'-me*. Big Valley, Modoc County. The stones to be heated may be seen inside the frame.

PLATE 13. Eastern part of Hot Springs Valley between Canby and Alturas. Showing in background *O-pah'-wah Butte,* formerly known as Rattlesnake Butte (and still earlier as Centerville Butte). Eastern limit of territory of *As-tah-ke-wi'-che* tribe.

PLATE 14. *As-tah-ke-wi'-che* man. Sam Spring, at his home in Hot Springs or Canby Valley, Modoc County.

PLATE 15. *As-tah-ke-wi'-che* man. Robin Spring, at his home in Hot Springs or Canby Valley, Modoc County.

PLATE 16. West slope of Cedar Pass, Warner Range, Modoc County. Territory of Ho-mah'-wet band of *Ko'-se-al-lek'-te* tribe. The Warner Mountains are the divide between the *Achomawan* tribes on the west and the Northern Piute on the east.

PLATE 17. *Ko'-se-al-lek'-te* man. Sam Steel, Alturas, Modoc County.

PLATE 18. Typical lava flow in sagebrush desert east of Madeline Plains, Lassen County. Territory of *Ham-mah'-we* tribe.

PLATE 19. *Ham-mah'-we* man. Jack Williams, Chief. Likely Valley, South Fork Pit River, Modoc County.

PLATE 20. *Ham-mah'-we* man. Old Pete, Likely Valley, South Fork Pit River, Modoc County.

PLATE 21. *Ham-mah'-we* woman. Old Sally and her shack. Likely Valley Rancheria.

PLATE 22. Dense coniferous forest near Burney, on east side of northern Sierra Nevada, Shasta County, California. Territory of *At-soo-kā'-e* tribe.

PLATE 23. *At-soo-kä'-e* family, Hat Creek, Shasta County.

PLATE 24. Lava dike about seven miles northwest of Grasshopper Valley, Lassen County. Territory of *Ap-woo'-ro-kä'e* tribe.

PLATE 25. *Ap-woo'-ro-kä'e.* Dixie Valley tribe, Robert Rivis and wife.

PLATE 26. *Ap-woo'-ro-kä'e* man. Robert Rivis. Front and profile views. Note round head—extreme brachycephalic type.

PLATE 27. *Ap-woo'-ro-kä'e* woman. Dixie Valley tribe. Mrs. Robert Rivis. Front and side views.

Big Bend of Pit River, Shasta County, California. Photo from *Lah'-lah-pis'-mah* Hot Spring on south side of river. Headquarters of *Mo-des'-se* tribe.

Istet Woiche and wife. *Mo-des'-se* tribe. Big Bend Pit River, Shasta County, California.

Mo-des'-se man and wife. Istet Woiche, leader of tribe. Big Bend Pit River, Shasta County, California.

Mo-des'-se woman, Mrs. Walter Moody, and daughters. The girls are only half *Mo-des'-se,* their father being a *Yah'-nah*—Walter Moody, of Montgomery Creek, Shasta County, California.

Burney Falls, Burney Creek Canyon, Shasta County, California. Territory of the *Il-mah'-we* tribe.

Fig. 1.

Fig. 2.

Pit River Canyon below Fall River Mouth.

Fig. 1. Falls 3 miles below Fall River Mills.
Fig. 2. The Canyon, just above the Falls.

Il-mah'-we man, John Carmony. Cayton Valley band (*Too-hat'-mah*).

FIG. 1.

FIG. 2.

Fall River Valley, Shasta County, California. Territory of *Å-choo'-mah'-we* tribe.

Fig. 1. Big Spring, head of Fall River.

Fig. 2. The flat open valley bordered with Ponderosa pines, looking westerly from near Glenburn.

Big Valley on border where Lassen and Modoc counties meet—a broad flat sagebrush plain. One of our camps in foreground. Territory of *At-wum'-we* tribe.

At-wum'-we men. Chief, Harry George (right); Billy Quinn (left). Big Valley, Modoc County, California, near old Indian village *Ah-pe'-dah-dā'*.

Women of *At-wum'-we* tribe near former village of *Ah-pe'-dah-dā'*. Big Valley, Modoc County, California.

Fig. 1.

Fig. 2.

Frames of Individual Sweathouses (*Tem'-mah-kah'-me*) of the *At-wum'-we* tribe. Big Valley, Modoc County, California.

Fig. 1. Near Lookout.
Fig. 2. Near *Ah-pe'-dah-dā'*.

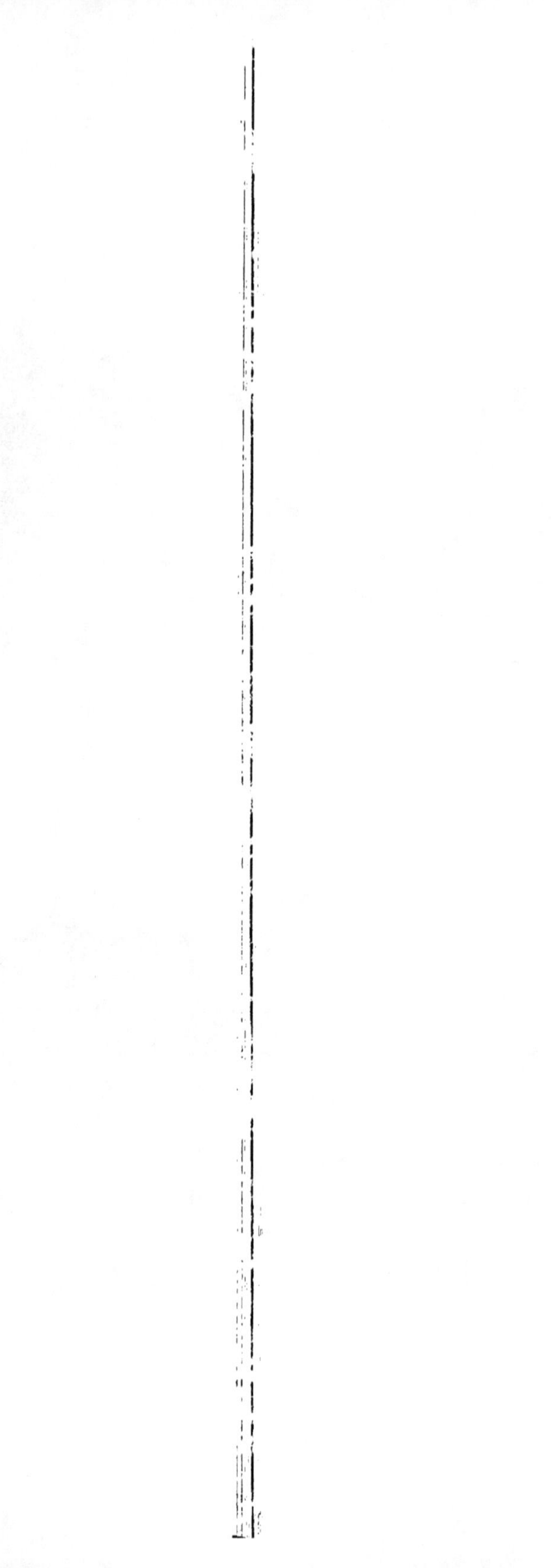

Eastern part of Hot Springs Valley. Eastern limit of territory of the *As-tah-ke-wi'-che* tribe. The conical hill in background is *O-pah'-wah* Butte, also known as Rattlesnake Butte, and in early days as Centerville Butte.

As-tah-ke-wi'-che man, Sam Spring, at his home in Hot Springs or Canby Valley, Modoc County, California.

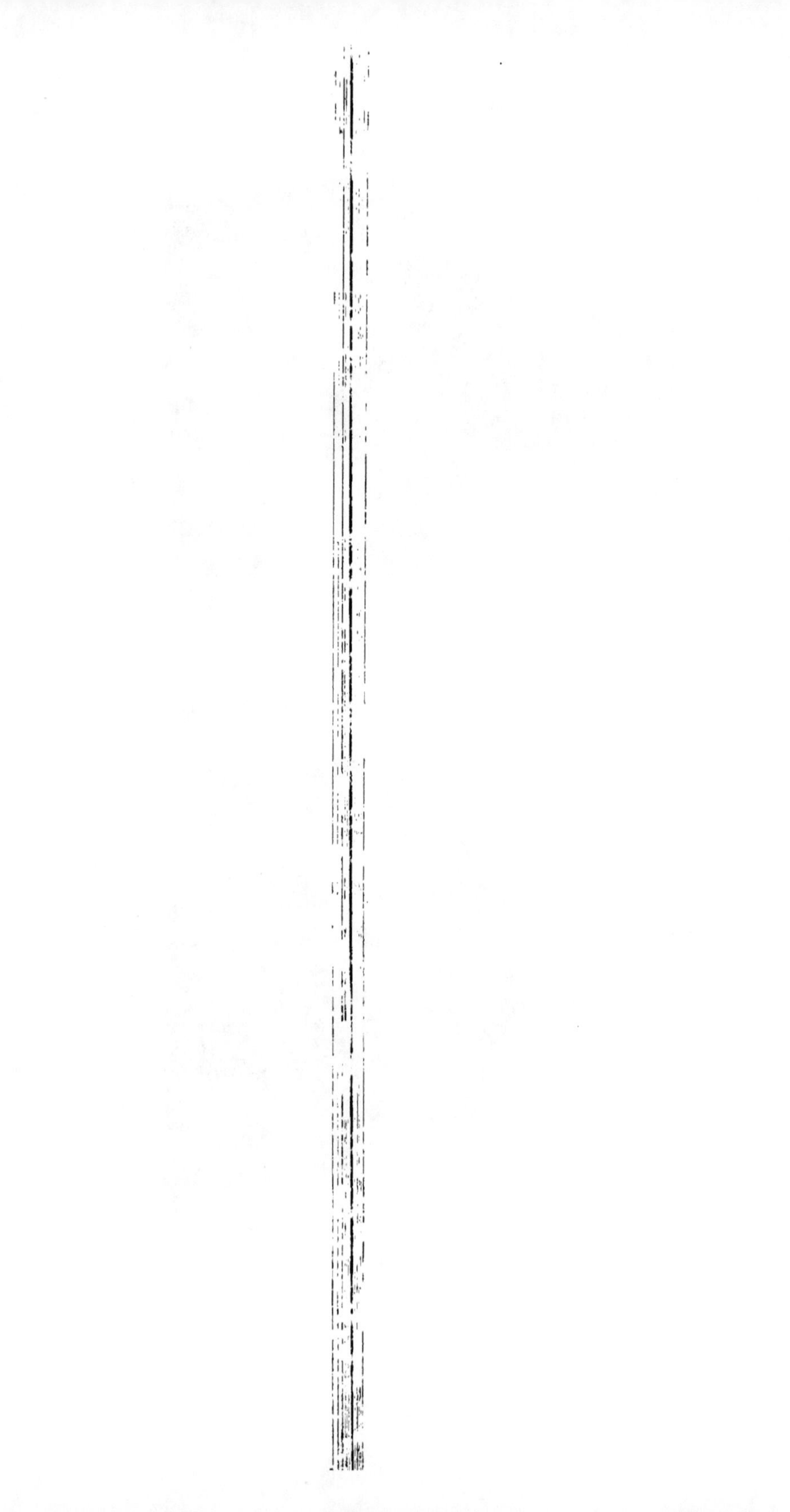

As-tah-ke-wi'-che man. Robin Spring. Hot Springs or Canby Valley, Modoc County, California.

West slope of Cedar Pass, Warner Range, Modoc County, California. Territory of *Ko'-se-al-lek'-te* tribe. This Range is the divide between the *Achomawan* tribes on the west and the Northern Piute on the east.

Ko'-se-al-lek'-te man, Sam Steel. Alturas, Modoc County, California.

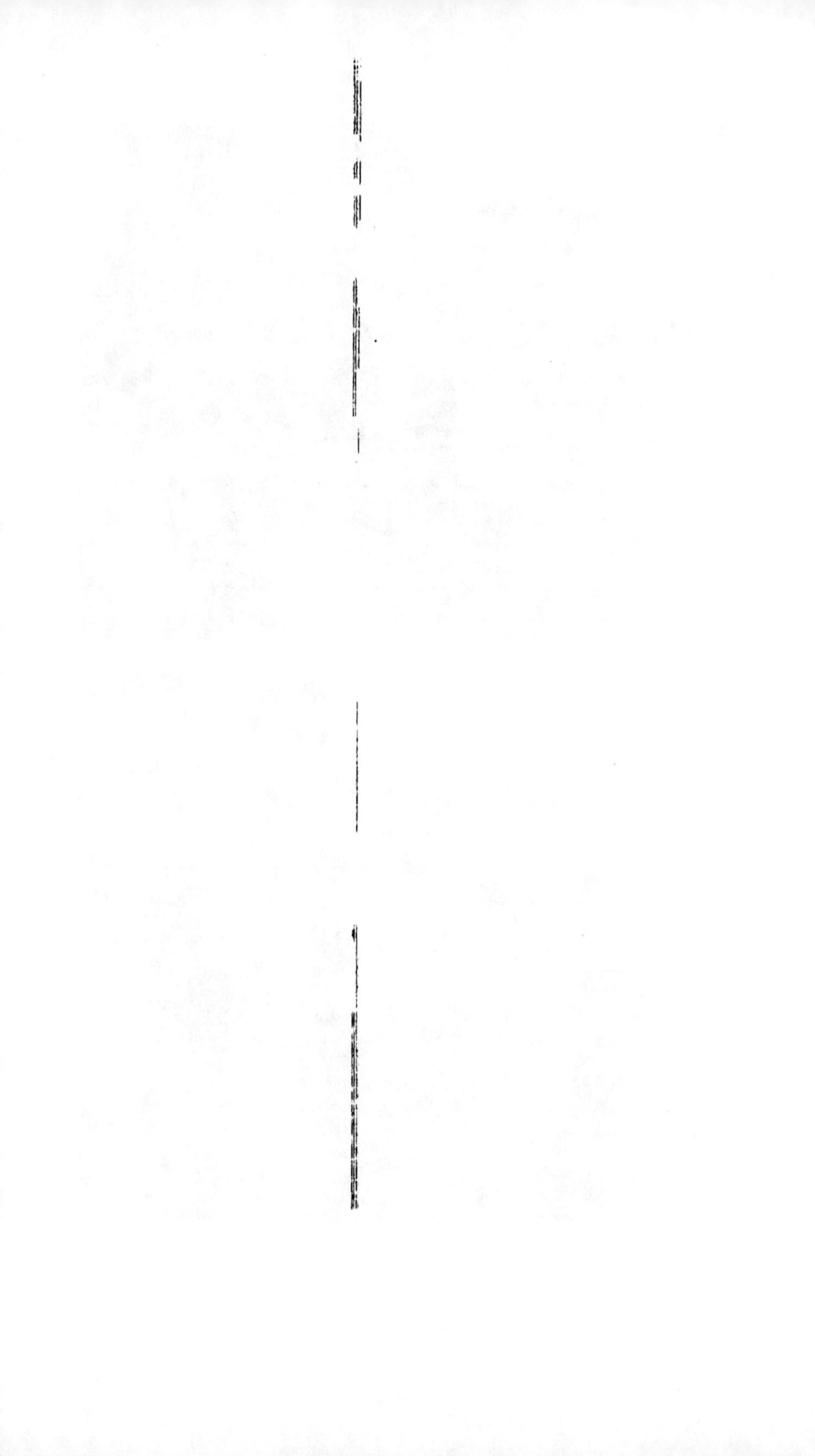

Typical lava flow in sagebrush desert on east side of Madeline Plains, Lassen County, California. Territory of *Ham-mah'-we* tribe.

Ham-mah'-we man, Chief, Jack Williams. Likely Valley, South Fork Pit River, Modoc County, California.

Ham-mah'-we man, Old Pete. Likely Valley, South Fork Pit River, Modoc County, California.

Ham-mah'-we woman. Old Sally at her home, Likely Valley Rancheria, South Fork Pit River, Modoc County, California.

Coniferous forest near Burney, Shasta County, California. Territory of the *At-soo-kā'-e* tribe.

At-soo-kā'-e family. Hat Creek, Shasta County, California.

Lava dike and scattered Ponderosa pines about 7 miles north-northwest of the desert flat known as Grasshopper Valley. Lassen County, California. Territory of the *Ap-woo'-ro-kä'e* tribe.

Ap-woo'-ro-kā'e tribe (Dixie Valley tribe). Robert Rivis and wife.

Ap-woo'-ro-kā'e man, Robert Rivis. Dixie Valley tribe. Note the round head—extreme brachycephalic type.

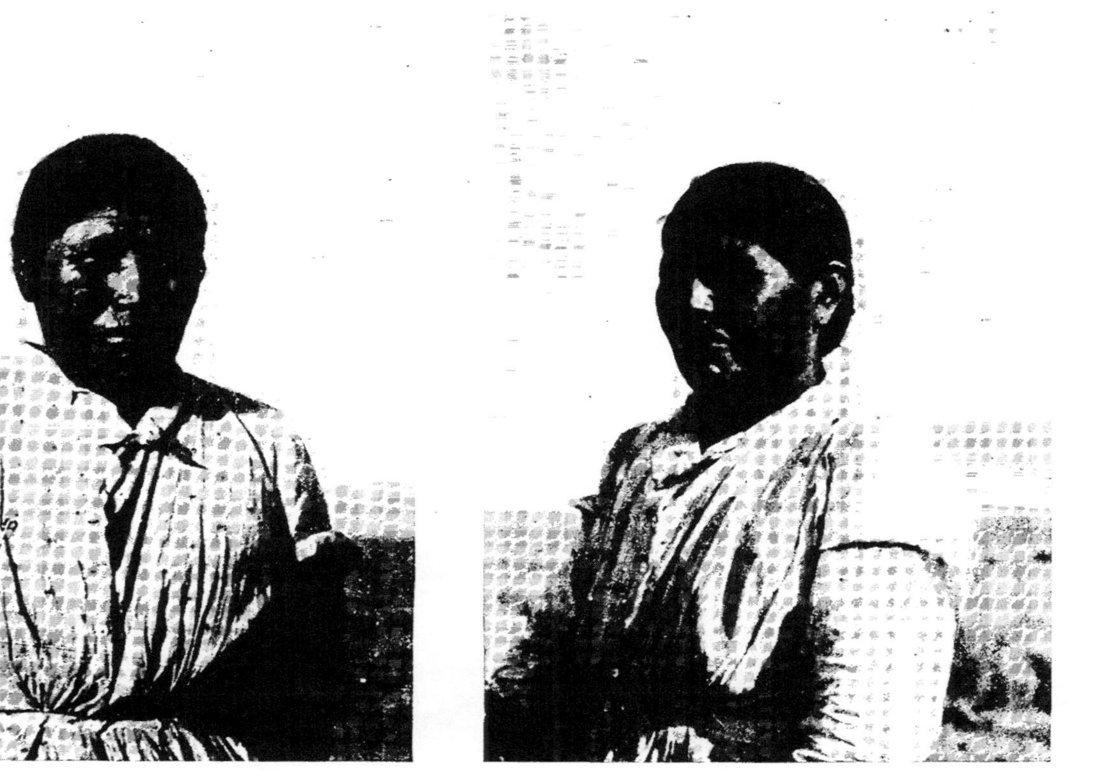

Ap-woo'-ro-kā'e woman, Mrs. Robert Rivis. Dixie Valley tribe.

SMITHSONIAN MISCELLANEOUS COLLECTIONS
VOLUME 78, NUMBER 4

SOLAR ACTIVITY AND LONG-PERIOD WEATHER CHANGES

BY
HENRY HELM CLAYTON

(Publication 2875)

CITY OF WASHINGTON
PUBLISHED BY THE SMITHSONIAN INSTITUTION
SEPTEMBER 30, 1926

The Lord Baltimore Press
BALTIMORE, MD., U. S. A.

SOLAR ACTIVITY AND LONG-PERIOD WEATHER CHANGES

By HENRY HELM CLAYTON

CONTENTS

PREFACE

The results presented in this paper are a continuation of those presented in a previous paper, No. 6, Vol. 77, of the Smithsonian Miscellaneous Collections. This investigation of the relation of weather to changes in solar radiation was made possible by a grant for that purpose to the Smithsonian Institution by Mr. John A. Roebling.

In the preparation of the data I have been assisted by Mr. Eliot C. French, Miss Hazel V. Miller and Miss M. Isabel Robinson.

1. HIGH AND LOW SOLAR RADIATION AND ASSOCIATED TEMPERATURES. MONTHLY VALUES

In the preceding papers of this series, the discussion of the relation of solar radiation to weather has been confined largely to short period solar changes, shown by the day to day values. It was only in these short period changes that there was a sufficient mass of data for statistical handling. In the case of a few very large and very small individual values of the monthly means of solar radiation, it was shown[1] that there was a distinct relation to world-wide meteorological conditions, but it was considered desirable to ascertain to what extent the average result of many smaller monthly departures from the mean showed a systematic response to variations in solar output.

Monthly mean values of solar radiation between 1.910 and 1.930 gram calories per square centimeter were taken as low values, and monthly mean values above 1.950 (all but two of which lay between 1.950 and 1.960) were taken as high values. The mean monthly departures of temperature from the normal were then obtained for a number of widely separated stations in North America, for the interval from two months before the occurrence of the solar values to twelve months following. This was done separately for high solar values and for low solar values, and for the winter half-year and the summer half-year. A correction for the influence of changes of longer period was then made by getting the average of the 15 monthly mean temperature departures in each case, and deducting this average from the individual means. The final results are given in table 1.

The departures given in table 1 are not large, and do not show a sharply marked effect of the solar radiation differences on the temperature for any single month. That there is an effect, however, is indicated by a high negative correlation between the averages of temperature, for the interval 0 to 4 months accompanying and following high values of solar radiation, and for the interval 0 to 4 months accompanying and following low solar radiation. These averages are entirely independent of each other, and there is no obvious reason why they should be correlated with each other, except through their relation to solar values.

The correlation for the five months (0 to 4 months) for the opposing solar conditions are as follows: Nome, -0.72 ± 0.16; Juneau, -0.80 ± 0.12; Edmonton, -0.81 ± 0.12; St. Johns, N. F., -0.52 ± 0.24; Hatteras, -0.89 ± 0.07; Key West, -0.64 ± 0.20.

Furthermore, it will be noted that the oscillations at northern stations are opposite in phase to those at southern stations, as is shown by the plots in figure 1.

[1] Smithsonian Misc. Coll., Vol. 77, No. 6, 1925, pp. 31-37.

TABLE 1.—*Mean departures of monthly temperatures with high and low monthly means of solar radiation, years 1918-1924.**

Station	Solar radiation	Cases	Months before —2	Months before —1	0	Months after 1	2	3	4	5	6	7	8	9	10	11	12
Nome, Alaska	1.911-1.930	26	—0.7	—1.3	—1.0	—1.5	—0.6	0.0	1.3	—0.3	1.0	0.6	1.0	0.3	1.4	—1.0	0.3
	1.951-1.960	17	—0.4	—1.1	0.5	1.1	2.0	0.2	—0.7	—0.2	—1.6	—0.8	—0.5	0.0	1.4	1.2	—1.1
	Diff.		—0.3	—0.2	—1.5	—2.6	—2.6	—0.2	2.0	—0.1	2.6	1.4	1.5	0.3	0.0	—2.2	1.4
Juneau, Alaska ...	1.911-1.930	27	0.5	—0.1	0.2	—0.2	—0.4	0.0	—0.1	—0.6	0.3	—0.1	—0.1	0.5	0.5	—0.1	0.5
	1.951-1.960	17	0.9	—0.6	—0.4	0.0	0.6	—0.6	0.3	0.1	—0.2	—0.4	—0.1	0.0	—0.5	—0.1	0.6
	Diff.		—0.4	0.5	0.6	—0.2	—1.0	0.6	—0.4	—0.7	0.5	0.3	0.0	0.5	1.0	0.0	—0.1
Edmonton, Canada	1.911-1.930	27	—0.5	—0.1	0.4	—0.5	—0.7	0.6	—0.2	—0.7	—0.1	—0.1	0.0	—0.2	0.4	0.1	0.9
	1.951-1.960	17	0.9	—2.0	—1.4	0.4	0.6	—0.4	0.6	—0.1	—0.4	—0.7	0.4	1.4	0.9	—[illegible]	0.3
	Diff.		—1.4	1.9	1.8	—0.9	—1.3	1.0	—0.8	—0.6	0.3	0.6	—0.4	—1.6	—0.5	0.3	0.6
St. Johns, N. F..	1.911-1.930	16	—0.7	1.6	0.2	—1.3	—0.6	—0.6	—1.2	—0.4	0.0	0.7	0.3	0.2	0.5	0.7	1.2
	1.951-1.960	17	0.7	—0.9	—1.6	—0.8	—0.2	—0.2	0.7	0.4	—0.3	0.6	1.4	0.0	0.0	0.0	—0.4
	Diff.		—1.4	2.5	1.8	—0.5	—0.4	—0.4	—1.9	—0.8	0.3	0.1	—1.1	0.2	0.5	0.7	1.6
San Diego, Calif..	1.911-1.930	27	0.3	0.1	0.2	—0.1	0.5	0.1	—0.1	0.1	0.0	0.1	—0.5	—0.8	—0.2	0.2	0.7
	1.951-1.960	17	0.1	—0.2	0.0	0.5	0.3	0.1	—0.1	—0.8	0.2	—0.4	—0.2	0.1	0.6	—0.2	—0.7
	Diff.		0.2	0.3	0.2	—0.6	0.2	0.0	0.0	0.9	—0.2	0.5	—0.3	—0.9	—0.8	0.4	1.4
Fort Smith, Ark..	1.911-1.930	27	0.5	0.2	—0.4	0.6	0.6	—0.3	—0.7	0.6	0.6	0.1	0.2	0.5	—0.5	—0.8	—0.6
	1.951-1.960	17	0.1	—0.4	0.4	—0.3	—0.4	1.0	—0.6	0.3	—0.6	—0.5	0.0	0.1	1.2	0.7	—0.6
	Diff.		0.4	0.6	—0.8	0.9	1.0	—1.3	—0.1	0.3	1.2	0.6	0.2	0.4	—1.7	—1.5	0.0
Hatteras, N. C...	1.911-1.930	27	0.4	0.2	—0.3	0.4	—0.1	—0.3	—0.5	0.3	0.4	0.4	0.1	—0.5	—0.1	—0.2	—0.6
	1.951-1.960	17	0.3	0.3	0.3	—0.5	—0.3	0.2	0.3	—0.3	—0.4	—0.8	—0.6	—0.2	0.8	0.9	0.2
	Diff.		0.1	—0.1	—0.6	0.9	0.2	—0.5	—0.8	0.6	0.8	1.2	0.7	—0.3	—0.9	—1.1	—0.8
Key West, Fla....	1.911-1.930	27	0.2	0.2	—0.1	0.0	—0.1	—0.2	0.2	0.3	0.3	0.2	—0.1	—0.4	0.0	0.0	—0.6
	1.951-1.960	17	—0.6	0.3	0.4	0.4	0.2	0.2	0.0	0.0	—0.2	—0.3	—0.3	0.1	0.1	0.2	0.2
	Diff.		0.8	—0.1	—0.5	—0.4	—0.3	—0.4	0.2	0.3	0.5	0.5	0.2	—0.5	—0.1	—0.2	—0.8
Merida, Mexico ..	1.911-1.930	26	0.2	0.0	—0.1	0.0	—0.1	0.0	0.0	0.1	0.3	0.0	—0.2	—0.8	0.0	—0.2	—0.6
	1.951-1.960	17	0.2	0.3	0.7	0.0	0.3	0.9	0.2	—0.2	—0.2	—0.4	—0.4	—0.2	—0.4	0.0	—0.2
	Diff.		0.0	—0.3	—0.8	0.0	—0.4	—0.9	—0.2	0.3	0.5	0.4	0.2	—0.6	0.4	—0.2	—0.4

* The temperature departures in each instance, as here given, are cleared of long-period changes by deducting from them the average departure of the 15 months represented.

Figure 1 indicates that there are two pulses accompanying and following each high and low solar value, (1) a rise or a depression of temperature accompanying the high or low solar value, and (2) a similar departure about three months later. The cause for this second delayed departure from the mean is not evident, and is a

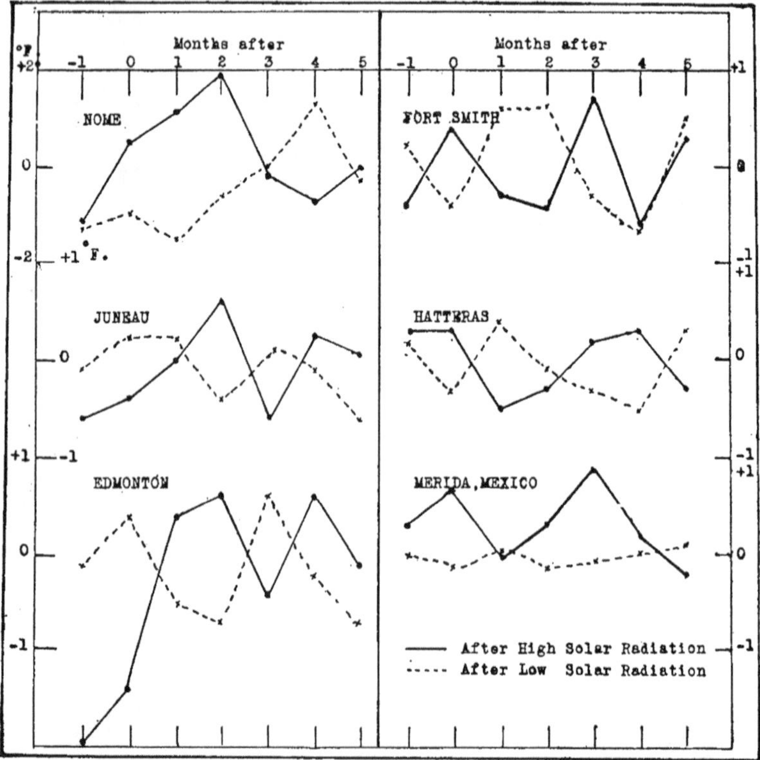

FIG. 1.—Mean departures of monthly temperatures from the average with high and with low monthly means of solar radiation.

matter for future research. The fact of its existence indicates that there can be no simple correlation between the mean monthly temperatures and mean monthly solar radiation variations.

2. THE GEOGRAPHICAL DISTRIBUTION OF WEATHER EFFECTS OF SOLAR VARIATION

(*a*) *As derived from solar radiation data.*—In order to study the geographical distribution of the differences in the weather conditions accompanying high solar radiation from those accompany-

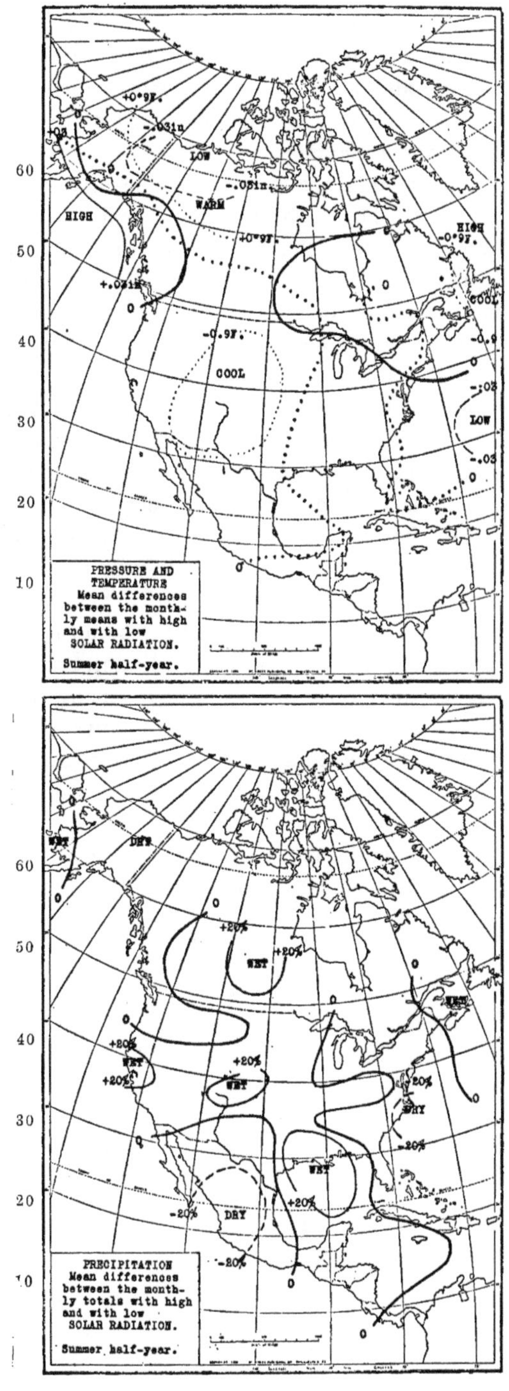
PRESSURE AND TEMPERATURE
Mean differences between the monthly means with high and with low SOLAR RADIATION.
Summer half-year.
HIGH
LOW
WARM
COOL
+0°9F.
-.03in
+.03in
-0.9F.
-0°9F.
LOW
PRECIPITATION
Mean differences between the monthly totals with high and with low SOLAR RADIATION.
Summer half-year.
WET
DRY
+20%
-20%
60
50
40
30
20
10

FIG. 2.

ing low solar radiation, the solar values were divided into low, medium, and high values. All values between 1.911 and 1.930 were called low values, and all above 1.950 were called high values. There were only two above 1.960, so that most of the high values were between 1.950 and 1.960.

The results are given in table 2, and plotted in the charts in figure 2. These charts show the distribution in weather changes accompanying a change in solar radiation equivalent to 1.3 per cent increase of the mean value. The pressure departures given in table 2 are mean departures from the normal in thousandths of an inch, the mean temperature departures from normal are in degrees and tenths Fahrenheit, and the mean precipitation is given in percentages of the normal for each station. The pressure lines in figure 2 are drawn for each .03 inch, which is the equivalent of one millibar; and temperature lines are drawn for each 1.8° F., or half that value, which make them equivalent to degrees or half degrees Centigrade.

The charts in figure 2 show that, with increased solar radiation, the pressure during the winter half-year rises in high latitudes over the continental mass of North America, and falls along the southern coast of Alaska, and probably over the ocean to the south, as well as over the central and western United States, and southward at Colon. The temperature falls over Alaska, Canada, and the northern United States, and rises south of about latitude 38° down to at least 10° south. The percentage of rainfall is greater with high solar radiation over nearly the whole of North America, down to about latitude 35° N. South of that latitude the rainfall is less, the most marked deficiency being in southern Texas and northern Mexico, while the greatest excess is in central Canada.

TABLE 2.—*Means of the monthly departures from normal of pressure, temperature and precipitation with low, medium and high monthly values of solar radiation*

Solar radiation. Calories	Winter half-year				Summer half-year			
	No. of months	Press., inches	Temp., ° F.	Precip., %	No. of months	Press., inches	Temp., ° F.	Precip., %
ALASKA, YEARS 1905 TO 1925								
Dutch Harbor, 53° 54′ N., 166° 32′ W.								
1.911-30	16	—.008	0.2	107	26	—.056	1.0	55
1.931-50	19	—.060	0.2	86	32	—.032	—0.2	91
1.951-72	14	.053	—0.4	22	20	.057	0.1	112
High-low		.061	—0.6	—85		.113	—0.9	57
Eagle, 64° 46′ N., 141° 12′ W.								
1.911-30	15	—.042	1.6	116	29	.025	—0.7	108
1.931-50	21	.004	—1.0	78	30	—.010	1.5	93
1.951-72	14	.041	—0.3	51	21	—.020	0.6	91
High-low		.083	—1.9	—65		—.095	1.3	—17
Juneau, 58° 18′ N., 134° 24′ W.								
1.911-30	16	—.009	1.2	107	29	—.006	0.1	111
1.931-50	22	—.022	—0.6	108	32	—.009	0.5	90
1.951-72	14	.062	—0.5	115	21	.006	—0.4	98
High-low		.071	—1.7	8		.012	—0.5	—13
Nome, 64° 30′ N., 165° 24′ W.								
1.911-30	14	—.012	1.1	120	28	.017	—0.8	108
1.931-50	20	.000	—0.3	105	32	—.014	0.5	86
1.951-72	14	.031	—0.5	109	18	.010	—0.6	95
High-low		.043	—1.6	—11		—.007	0.2	—13
Tanana, 65° 12′ N., 152° 0′ W.								
1.911-30	7	—.017	2.5	56	19	—.002	—1.1	97
1.931-50	16	—.016	—0.1	92	29	.001	0.2	98
1.951-72	14	.024	1.0	95	20	.001	—0.2	102
High-low		.041	—1.5	39		.003	0.9	5
Valdez, 61° 6′ N., 146° 13′ W.								
1.911-30	3	.044	2.9	99	10	.023	1.4	102
1.931-50	14	—.040	—0.7	86	26	—.006	0.2	104
1.951-72	12	.038	—0.2	104	17	—.003	—0.7	91
High-low		—.006	—3.1	5		—.026	—2.1	—11
CANADA, 1905 TO 1925								
Barkerville, 53° 2′ N., 121° 35′ W.								
1.911-30	17	.026	0.3	116	31	.003	—0.7	100
1.931-50	23	—.026	—2.0	119	39	—.001	—0.7	112
1.951-72	14	.024	0.3	105	22	.017	—0.9	100
High-low		—.002	0.0	—11		.014	—0.2	0
Charlottetown, 46° 14′ N., 63° 10′ W.								
1.911-30	17	—.008	—1.0	68	31	—.004	0.5	81
1.931-50	23	.050	2.0	73	39	.007	0.5	92
1.951-72	14	.046	—0.8	94	22	.024	—0.1	98
High-low		.054	0.2	26		.028	—0.6	17
Dawson, 64° 4′ N., 139° 20′ W.								
1.911-30	17	.023	2.3	75	30	.027	—1.1	103
1.931-50	23	—.008	—0.7	99	37	—.024	0.1	82
1.951-72	14	.037	—0.3	86	22	—.016	—0.2	94
High-low		.014	—2.6	11		—.043	0.9	—9

TABLE 2.—*Means of the monthly departures from normal of pressure, temperature and precipitation with low, medium and high monthly values of solar radiation* (continued)

Solar radiation, Calories	Winter half-year				Summer half-year			
	No. of months	Press., inches	Temp., ° F.	Precip., %	No. of months	Press., inches	Temp., ° F.	Precip., %
				CANADA (continued)				
				Edmonton, 53° 33′ N., 113° 30′ W.				
1.911-30	17	.023	1.4	115	31	.022	0.5	95
1.931-50	23	—.006	—1.0	105	38	.009	0.3	106
1.951-72	14	.033	—0.5	156	22	.010	0.3	107
High-low		.010	—1.9	41		—.012	—0.2	12
				Father Point, 48° 31′ N., 68° 19′ W.				
1.911-30	17	—.009	—1.7	98	29	—.004	—0.6	61
1.931-50	23	.009	1.7	97	37	.010	0.3	97
1.951-72	14	.040	—0.1	93	22	.013	0.2	93
High-low		.049	1.6	—5		.017	0.8	32
				Montreal, 45° 30′ N., 73° 35′ W.				
1.911-30	17	—.018	—0.1	95	31	—.012	—0.1	114
1.931-50	23	.016	2.2	98	39	.003	0.5	95
1.951-60	14	.030	0.3	92	22	.007	0.1	97
High-low		.048	0.3	—3		.019	0.2	—17
				Moose Factory, 51° 16′ N., 80° 56′ W.				
1.911-30	5	.004	0.4	..	14	.018	—0.2	..
1.931-50	11	.009	3.7	..	27	.020	—0.2	..
1.951-72	11	.019	—0.3	..	22	.025	—0.3	..
High-low		.015	—0.7	..		.007	—0.1	..
				Prince Albert, 53° 10′ N., 106° 0′ W.				
1.911-30	17	.017	—1.0	62	31	.008	0.0	67
1.931-50	23	—.015	1.9	109	39	—.009	0.1	120
1.951-72	14	.006	—1.9	137	22	.006	—0.2	90
High-low		—.011	—0.9	75		—.002	—0.2	23
				St. Johns, N. F., 47° 34′ N., 52° 42′ W.				
1.911-30	13	—.039	—1.7	107	23	—.019	1.2	95
1.931-50	20	—.006	—0.1	93	35	—.013	0.3	101
1.951-72	14	.026	—1.0	95	22	.006	—0.9	101
High-low		.065	0.7	—12		.025	—2.1	6
				Winnipeg, 49° 53′ N., 97° 7′ W.				
1.911-30	17	.016	0.4	97	30	—.005	—0.2	92
1.931-50	23	—.012	0.5	100	39	.002	0.4	98
1.951-72	14	.016	—2.4	122	22	.004	—0.6	95
High-low		.000	—2.8	25		.009	—0.4	3
				UNITED STATES, 1905 TO 1925				
				Abilene, 32° 23′ N., 99° 40′ W.				
1.911-30	17	.021	—0.6	139	29	—.001	0.5	111
1.931-50	23	.004	0.9	105	37	.001	0.8	78
1.951-72	14	—.001	0.2	105	22	.000	—0.6	105
High-low		—.022	0.8	—34		—.001	—1.1	—6
				Bismarck, 46° 47′ N., 100° 38′ W.				
1.911-30	17	.007	2.4	67	29	.004	0.0	100
1.931-50	23	—.007	2.2	87	37	—.003	1.0	96
1.951-72	14	.001	0.2	98	22	.004	—0.9	109
High-low		—.006	—2.2	31	7	.000	—0.9	9

TABLE 2.—*Means of the monthly departures from normal of pressure, temperature and precipitation with low, medium and high monthly values of solar radiation* (continued)

Solar radiation. Calories	Winter half-year No. of months	Press., inches	Temp., °F.	Precip., %	Summer half-year No. of months	Press., inches	Temp., °F.	Precip., %
UNITED STATES (continued)								
Boston, 42° 21′ N., 71° 4′ W.								
1.911-30	17	—.022	0.5	79	29	—.005	1.0	90
1.931-50	23	.025	3.3	70	37	—.002	0.7	115
1.951-72	14	.016	1.1	96	22	—.004	0.0	96
High-low		.038	0.6	17		.001	—1.0	6
Charleston, 32° 47′ N., 79° 56′ W.								
1.911-30	17	—.017	0.5	99	29	.000	—0.2	103
1.931-50	23	.021	0.9	84	37	—.004	—0.1	83
1.951-72	14	—.006	1.0	68	22	—.003	0.4	89
High-low		.011	0.5	—31		—.003	0.6	—14
Cheyenne, 41° 8′ N., 104° 48′ W.								
1.911-30	17	.025	—0.6	149	29	.014	—0.5	113
1.931-50	23	—.006	0.4	120	37	.004	—0.1	105
1.951-72	14	.006	—0.3	156	22	—.001	—1.6	120
High-low		—.019	0.3	7		—.015	—1.1	7
Chicago, 41° 53′ N., 87° 37′ W.								
1.911-30	17	.008	0.2	83	29	.009	—0.8	100
1.931-50	23	.002	2.3	101	37	.010	0.5	98
1.951-72	14	.008	0.3	98	22	.000	—0.3	85
High-low		.000	0.1	15		—.009	0.5	—15
Cincinnati, 39° 6′ N., 84° 30′ W.								
1.911-30	17	—.004	1.5	88	29	.012	—0.5	95
1.931-50	23	.024	2.5	84	37	.000	0.0	103
1.951-72	14	—.009	1.0	132	22	—.011	0.3	106
High-low		—.005	—0.5	34		—.023	0.8	11
Corpus Christi, 27° 49′ N., 97° 25′ W.								
1.911-30	17	.006	0.2	102	29	—.002	0.6	101
1.931-50	23	.012	1.1	99	37	.005	0.7	83
1.951-72	14	.002	0.9	81	22	—.003	0.5	94
High-low		—.004	0.7	—21		—.001	—0.1	—7
Eastport, 44° 54′ N., 66° 59′ W.								
1.911-30	17	—.015	—0.9	69	29	—.001	0.1	88
1.931-50	23	.029	2.1	66	37	.008	0.1	86
1.951-72	14	.045	—0.3	93	22	.011	0.5	81
High-low		.060	0.6	24		.012	0.4	—7
Galveston, 29° 18′ N., 94° 50′ W.								
1.911-30	17	.007	0.4	135	29	.006	—0.1	82
1.931-50	23	.020	0.7	92	37	.013	0.1	89
1.951-72	14	.015	1.3	79	22	.004	0.0	110
High-low		.008	0.9	—56		—.002	0.1	28
Hatteras, 35° 15′ N., 75° 40′ W.								
1.911-30	17	—.036	0.7	85	20	.000	0.1	112
1.931-50	23	.020	1.7	108	37	.000	0.1	90
1.951-72	14	—.003	1.4	77	22	—.001	0.0	79
High-low		.033	0.7	—8		—.001	—0.1	—33

TABLE 2.—*Means of the monthly departures from normal of pressure, temperature and precipitation with low, medium and high monthly values of solar radiation* (continued)

Solar radiation. Calories	Winter half-year				Summer half-year			
	No. of months	Press., inches	Temp., ° F.	Precip., %	No. of months	Press., inches	Temp., ° F.	Precip., %
				UNITED STATES (continued)				
				Helena, 46° 34′ N., 112° 4′ W.				
1.911-30	17	.018	1.0	..	29	.006	0.0	..
1.931-50	23	—.010	0.2	..	37	—.002	0.0	..
1.951-72	14	.016	0.2	..	22	.001	—0.3	..
High-low		—.002	—0.8	..		—.005	—0.3	..
				Key West, 24° 33′ N., 81° 48′ W.				
1.911-30	17	—.006	0.7	86	29	.003	0.0	98
1.931-50	23	.001	0.7	65	37	—.001	—0.1	102
1.951-72	14	.004	0.7	94	22	.000	—0.1	94
High-low		.010	0.0	8		—.003	—0.1	—4
				Little Rock, 34° 45′ N., 92° 6′ W.				
1.911-30	17	—.005	0.2	78	29	—.007	—0.4	105
1.931-50	23	.019	1.9	78	37	.004	0.4	86
1.951-72	14	—.004	1.1	136	22	—.006	0.0	105
High-low		.001	0.9	58		.001	0.4	00
				Mobile, 30° 41′ N., 88° 2′ W.				
1.911-30	17	—.009	0.8	117	29	.000	0.1	102
1.931-50	23	.011	1.5	58	37	.002	0.4	100
1.951-72	14	.001	1.8	93	22	—.001	0.5	124
High-low		.010	1.0	—24		—.001	0.4	22
				Nashville, 36° 10′ N., 86° 47′ W.				
1.911-30	17	.006	0.2	89	29	.010	—0.8	109
1.931-50	23	.034	1.7	86	37	.011	—0.2	93
1.951-72	14	.002	0.3	143	22	.004	—0.2	108
High-low		—.004	0.1	54		—.006	0.6	—1
				New York, 40° 43′ N., 74° 0′ W.				
1.911-30	17	—.020	0.2	111	29	—.005	—0.1	99
1.931-50	23	.011	2.3	72	37	.002	0.0	94
1.951-72	14	.011	0.5	101	22	.001	—0.4	96
High-low		.031	0.3	—10		.006	—0.3	—3
				North Platte, 41° 8′ N., 100° 45′ W.				
1.911-30	17	.015	0.4	109	29	.004	0.1	94
1.931-50	23	—.007	1.6	106	37	—.007	0.7	94
1.951-72	14	—.005	0.0	173	22	—.011	—0.8	112
High-low		—.020	—0.4	66		—.015	—0.9	18
				Phoenix, 33° 28′ N., 112° 0′ W.				
1.911-30	17	—.002	0.5	88	29	—.003	0.4	62
1.931-50	23	.002	0.1	81	37	.003	—0.2	105
1.951-72	14	—.007	0.3	91	22	—.004	—0.7	67
High-low		—.005	—0.2	3		—.001	—1.1	5
				Portland, Ore., 45° 32′ N., 122° 41′ W.				
1.911-30	17	.048	0.4	89	29	.005	0.6	96
1.931-50	23	—.005	—0.1	99	37	—.003	0.4	98
1.951-72	14	.022	0.3	85	22	—.006	0.6	83
High-low		—.026	—0.1	—4		—.011	0.0	—13

TABLE 2.—*Means of the monthly departures from normal of pressure, temperature and precipitation with low, medium and high monthly values of solar radiation* (continued)

Solar radiation. Calories	Winter half-year				Summer half-year			
	No. of months	Press., inches	Temp., °F.	Precip., %	No. of months	Press., inches	Temp., °F.	Precip., %
UNITED STATES (continued)								
St. Paul, 44° 58′ N., 93° 3′ W.								
1.911-30	17	.012	1.0	91	29	.004	—0.7	100
1.931-50	23	—.002	2.2	115	37	.004	—0.6	120
1.951-72	14	.005	0.3	110	22	.000	—0.8	105
High-low		—.007	—0.7	19		—.004	—0.1	5
Salt Lake City, 40° 46′ N., 111° 54′ W.								
1.911-30	17	.027	—0.3	88	29	.007	0.6	97
1.931-50	23	—.002	0.1	117	37	.000	0.5	126
1.951-72	14	.007	0.1	111	22	.000	0.0	100
High-low		—.020	0.4	23		—.007	—0.6	3
San Diego, 32° 43′ N., 117° 10′ W.								
1.911-30	17	—.003	0.9	70	29	.000	0.3	58
1.931-50	23	.012	—0.4	125	37	.006	—0.4	123
1.951-72	14	.004	—0.3	92	22	.002	—0.3	70
High-low		.007	—1.2	22		.002	—0.6	12
San Francisco, 37° 48′ N., 122° 26′ W.								
1.911-30	17	.017	0.7	90	29	.005	0.1	58
1.931-50	23	.002	—0.2	114	37	.001	0.3	116
1.951-72	14	—.003	0.3	54	22	.004	0.2	124
High-low		—.020	—0.4	—36		—.001	0.1	66
San Luis Obispo, 35° 18′ N., 120° 39′ W.								
1.911-30	17	.006	0.3	63	29	.006	—0.3	26
1.931-50	23	.013	—0.5	101	37	.010	—0.2	140
1.951-72	14	—.002	—0.4	69	22	.003	—0.4	24
High-low		—.008	—0.7	6		—.003	—0.1	—2
Santa Fe, 35° 41′ N., 105° 57′ W.								
1.911-30	17	.021	—0.9	67	29	.003	—0.1	78
1931-50	23	.007	—0.2	101	37	.005	—0.2	93
1.951-72	14	.003	0.5	102	22	—.004	—0.9	101
High-low		—.018	1.4	35		—.007	—0.8	23
Washington, 38° 54′ N., 77° 3′ W.								
1.911-30	17	—.020	0.9	111	29	.002	0.1	104
1.931-50	23	.027	2.6	76	37	—.002	—0.1	107
1.951-72	14	.003	0.8	117	22	—.004	—0.4	114
High-low		.023	—0.1	6		—.006	—0.5	10
MEXICO, 1905-1925								
Merida, 20° 58′ N., 89° 37′ W.								
1.911-30	15	—.012	0.1	106	27	—.001	0.1	90
1.931-50	21	.004	0.2	110	34	.007	—0.1	97
1.951-72	13	.008	0.2	67	18	—.005	0.0	109
High-low		.020	0.1	—39		—.004	—0.1	19
Monterey, 25° 40′ N., 100° 18′ W.								
1.911-30	16	—.001	—0.9	133	29	—.005	—0.4	121
1.931-50	21	—.005	0.0	107	35	.003	—0.1	101
1.951-72	14	.007	0.4	55	22	.003	—0.3	97
High-low		.008	—1.3	—78		.008	0.1	—32

TABLE 2.—*Means of the monthly departures from normal of pressure, temperature and precipitation with low, medium and high monthly values of solar radiation* (continued)

Solar radiation. Calories	Winter half-year				Summer half-year			
	No. of months	Press., inches	Temp., ° F.	Precip., %	No. of months	Press., inches	Temp., ° F.	Precip., %
			CENTRAL AMERICA, 1905-1925					
			San Salvador, 13° 44′ N., 89° 9′ W.					
1.911-30	11	..	—0.6	92	19	..	0.5	91
1.931-50	20	..	0.1	122	25	..	0.5	100
1.951-72	12	..	0.9	78	17	..	1.2	97
High-low		..	1.5	—14		..	0.7	6
			Colon, 9° 23′ N., 79° 23′ W.					
1.911-30	13	.001	0.3	80	29	.001	0.3	101
1.931-50	19	.003	0.5	97	35	—.001	0.5	103
1.951-72	13	—.004	0.5	70	18	—.008	0.5	92
High-low		—.005	0.2	—10		—.009	0.2	—9
			BERMUDA AND JAMAICA, 1905-1925					
			Hamilton, 32° 17′ N., 64° 46′ W.					
1.911-30	12	—.020	—1.0	82	26	.020	—0.3	84
1.931-50	21	.018	0.2	99	37	.001	0.0	102
1.951-72	14	—.001	1.0	116	22	—.025	—0.6	95
High-low		.019	2.0	—34		—.045	—0.3	11
			Port au Prince, 18° 34′ N., 72° 22′ W.					
1.911-30	17	.006	0.0	87	31	.006	0.2	96
1.931-50	23	—.009	—0.2	133	39	—.002	0.0	98
1.951-72	14	.004	0.7	67	22	—.004	0.4	96
High-low		—.002	0.7	—20		—.010	0.2	0

The charts in figure 2 for the summer half-year show a distribution of the differences of pressure and temperature almost the opposite of that of the winter half-year, north of 50° latitude. The pressure is lower with increased solar radiation over Alaska and northern Canada, and higher over the oceans. The temperature is higher over northern Alaska and Canada, but it is lower over a large part of the United States, just as it is for the winter half-year. The precipitation map shows a deficiency of rainfall in Alaska and northwestern Canada, but there is an excess over central Canada and a large part of the United States. There is a deficiency in Mexico as was found in the winter half-year.

(*b*) *As derived from sun-spot data.*—The solar radiation numbers cover only a few years, but Dr. Abbot has shown that there is a relation between the monthly sun-spot numbers and the monthly means of solar radiation, so that when the radiation values are arranged in the order of increasing magnitude, the average of the sun-spot numbers also shows a progressive increase.[1] Hence, by means of the Wolf and Wolfer sun-spot numbers, the investigation of the influence of solar radiation can be carried back to the earliest meteorological observations.

At a few stations in the United States the meteorological observations extend back to more than a century, but at most of the stations in the net which I have used for the North American Continent they cover periods varying from about 40 to 60 years. I decided to limit the investigation to the period beginning with 1856, and to use all the available data.

[1] Smithsonian Misc. Coll., Vol. 77, No. 5, p. 21. Also Monthly Weath. Rev., May, 1926.

TABLE 3.—*Means of the monthly values of pressure, temperature and precipitation with low, medium and high monthly values of the Wolf and Wolfer sun-spot numbers*

Sun spot number	Winter half-year: No. of months	Press., inches	Temp., °F.	Precip., %	Summer half-year: No. of months	Press., inches	Temp., °F.	Precip., %
				ALASKA, 1880-1923				
				Dutch Harbor, 53° 54′ N., 166° 32′ W.				
0-20	51	..	0.3	102	46	..	0.5	106
21-50	47	..	—0.3	96	41	..	0.0	112
Over 50	53	..	0.2	101	53	..	—0.5	85
High-low		..	—0.1	—1		..	—1.0	21
				Eagle, 64° 46′ N., 141° 12′ W.				
0-20	65	—.002	1.6	91	68	.003	—1.4	105
21-50	39	—.039	0.7	104	35	—.004	1.0	100
Over 50	41	.043	—1.4	106	36	—.003	2.0	96
High-low		.045	—3.0	15		—.006	3.4	—9
				Juneau, 58° 18′ N., 134° 24′ W.				
0-20	80	.026	0.4	102	78	—.012	—0.2	96
21-50	36	—.038	0.0	83	44	—.022	0.6	100
Over 50......	44	.002	—0.4	104	38	.024	0.0	96
High-low		—.024	—0.8	2		.036	0.2	0
				Kodiak, 57° 47′ N., 152° 22′ W.				
0-20	8	.033	1.2	80	9	.009	—0.6	91
21-50	17	—.023	0.5	101	13	—.002	0.3	102
Over 50	24	—.002	—1.0	105	25	—.003	0.0	99
High-low		—.035	—2.2	25		—.012	0.6	8
				Nome, 64° 30′ N., 165° 24′ W.				
0-20	32	—.025	3.2	100	38	.012	0.0	99
21-50	26	—.042	0.5	113	28	—.025	0.8	94
Over 50	35	.042	—1.8	92	31	—.002	—0.7	103
High-low		.067	—5.0	—8		—.014	—0.7	3
				Tanana, 65° 12′ N., 152° 0′ W.				
0-20	39	—.001	2.3	86	40	.000	—0.9	100
21-50	42	—.071	—0.8	100	38	.002	0.3	89
Over 50	52	.059	—1.2	110	39	.007	0.7	114
High-low		.060	—3.5	24		.007	1.6	14
				St. Paul Island, 57° 15′ N., 170° 10′ W.				
0-20	10	—.071	3.4	96	15	.001	0.6	115
21-50	18	.036	0.4	115	16	.001	—0.1	106
Over 50	43	.028	—0.8	98	38	—.002	—0.5	89
High-low		—.099	—4.2	2		—.003	—1.1	—26
				CANADA, 1873-1923				
				Barkerville, 53° 2′ N., 121° 35′ W.				
0-20	94	—.010	0.7	96	94	.016	—0.2	104
21-50	61	—.027	—0.5	95	60	.002	0.6	90
Over 50	67	.011	—0.5	108	61	.029	—0.2	101
High-low		.021	—1.2	12		.013	0.0	—3
				Bella Coola				
0-20	50	..	1.2	107	57	..	—0.2	98
21-50	38	..	—0.1	97	37	..	0.4	97
Over 50	40	..	—0.9	109	36	..	0.0	101
High-low		..	—2.1	2		..	0.2	3

TABLE 3.—*Means of the monthly values of pressure, temperature and precipitation with low, medium and high monthly values of the Wolf and Wolfer sun-spot numbers* (continued)

Sun spot number	Winter half-year				Summer half-year			
	No. of months	Press., inches	Temp., ° F.	Precip., %	No. of months	Press., inches	Temp., ° F.	Precip., %
CANADA (continued)								
Charlottetown, 46° 14′ N., 63° 10′ W.								
0-20	104	.002	1.6	95	99	—.003	0.0	114
21-50	72	—.014	—0.1	110	73	.008	0.2	88
Over 50	76	.005	—1.1	103	68	—.013	—0.3	94
High-low		.003	—2.7	8		—.010	—0.3	—20
Dawson, 64° 4′ N., 139° 20′ W.								
0-20	34	.017	1.3	98	37	.029	—0.9	91
21-50	35	.022	0.3	95.	34	.003	0.5	105
Over 50	44	.048	—1.6	101	40	—.016	0.4	99
High-low		.031	—2.9	3		—.045	1.3	8
Edmonton, 53° 33′ N., 113° 30′ W.								
0-20	104	.002	1.6	95	99	—.003	0.0	114
21-50	72	—.014	—0.1	110	73	.008	0.2	88
Over 50	76	.005	—1.1	103	68	—.013	—0.3	94
High-low		.003	—2.7	8		—.010	—0.3	—20
Father Point, 48° 31′ N., 68° 19′ W.								
0-20	121	—.012	0.0	108	117	.005	0.0	108
21-50	83	.015	0.3	91	83	.004	—0.1	103
Over 50	87	.007	—0.7	90	90	—.002	0.1	98
High-low		.019	—0.7	—18		—.007	0.1	—10
Kamloops, 50° 41′ N., 120° 29′ W.								
0-20	72	.006	0.2	98	69	—.002	—0.4	109
21-50	58	—.016	0.4	111	60	—.008	0.6	92
Over 50	57	.025	—0.6	96	56	.010	0.1	100
High-low		.019	—0.8	—2		.012	0.5	—9
Massett, 53° 58′ N., 132° 9′ W.								
0-20	51	..	0.6	90	57	..	—0.3	102
21-50	38	..	—0.2	102	36	..	0.8	96
Over 50	44	..	—0.5	101	41	..	0.1	98
High-low		..	—1.1	11		..	0.4	—4
Moose Factory, 51° 16′ N., 80° 56′ W.								
0-20	67	—.015	0.8	..	51	.016	—0.2	..
21-50	53	.025	0.0	..	51	—.012	0.0	..
Over 50	69	—.001	—0.7	..	67	—.001	—0.2	..
High-low		.014	—1.5	..		—.017	0.0	..
Montreal, 45° 30′ N., 73° 35′ W.								
0-20	127	—.010	0.3	103	127	.003	0.2	105
21-50	88	.018	0.8	90	87	—.003	0.2	96
Over 50	91	.002	—0.7	94	92	.005	—0.1	99
High-low		.012	—1.0	—9		.002	—0.3	—6
Prince Albert, 53° 10′ N., 106° 0′ W.								
0-20	103	.005	0.5	93	99	.003	0.1	113
21-50	68	—.007	0.4	119	72	.009	—0.1	97
Over 50	72	.017	—0.6	88	71	.004	0.1	95
High-low		.012	—1.1	—5		.001	0.0	—18

TABLE 3.—*Means of the monthly values of pressure, temperature and precipitation with low, medium and high monthly values of the Wolf and Wolfer sun-spot numbers* (continued)

Sun spot number	Winter half-year				Summer half-year			
	No. of months	Press., inches	Temp., ° F.	Precip., %	No. of months	Press., inches	Temp., ° F.	Precip., %
				CANADA (continued)				
				Qu'appelle, 50° 30′ N., 103° 47′ W.				
0-20	104	—.004	1.1	107	103	—.002	0.2	101
21-50	73	—.006	1.3	106	74	.010	0.0	112
Over 50	78	.003	—0.7	97	81	.003	0.0	90
High-low		.007	—1.8	—10		—.005	0.2	—11
				Sable Island, 43° 57′ N., 60° 6′ W.				
0-20	54	—.013	0.3	96	59	—.008	0.4	98
21-50	40	.029	0.5	101	38	—.009	—0.1	100
Over 50	44	—.008	—0.8	103	41	.008	—0.5	102
High-low		.005	—1.1	7		.016	—0.9	4
				SW Point, Anticosti, 49° 24′ N., 63° 33′ W.				
0-20	86	—.024	0.5	104	84	.001	0.2	98
21-50	68	.021	—0.3	99	66	—.004	—0.1	94
Over 50	82	.002	—0.3	98	84	.004	—0.2	107
High-low		.026	—0.8	—6		.033	—0.4	9
				St. Johns, N. F., 47° 34′ N., 52° 42′ W.				
0-20	126	—.024	—0.2	92	123	—.022	0.2	100
21-50	90	.025	—0.1	99	86	—.003	—0.1	97
Over 50	101	—.004	—0.1	107	101	.021	—0.1	103
High-low		.020	0.1	15		.043	—0.3	—3
				Toronto, 43° 40′ N., 79° 24′ W.				
0-20	131	—.012	0.6	103	130	.003	0.4	99
21-50	89	.017	0.6	93	82	—.006	0.2	96
Over 50	93	—.001	—0.7	96	95	—.010	—0.3	110
High-low		.011	—1.3	—7		—.013	—0.7	11
				Winnipeg, 49° 53′ N., 97° 7′ W.				
0-20	128	.007	1.0	80	128	.003	0.7	102
21-50	91	—.008	0.1	123	90	.005	—0.1	102
Over 50	93	.003	—0.7	111	94	.001	—0.4	93
High-low		—.004	—1.7	31		—.002	—1.1	—9
				UNITED STATES, 1856-1923				
				Abilene, 32° 23′ N., 99° 40′ W.				
0-20	98	.002	—0.3	126	97	—.001	0.1	107
21-50	63	—.003	0.6	87	70	.000	0.0	97
Over 50	71	.004	—1.3	93	67	—.002	—0.1	92
High-low		.002	—1.0	33		—.001	—0.2	—15
				Bismarck, 46° 47′ N., 100° 38′ W.				
0-20	131	.006	0.1	88	127	—.003	0.4	104
21-50	85	—.005	—0.4	112	170	.001	—0.2	104
Over 50	87	.004	—1.5	97	94	.000	—0.4	94
High-low		—.002	—1.6	9		.003	—0.8	—10
				Boston, 42° 21′ N., 71° 4′ W.				
0-20	159	—.010	—0.4	102	156	.002	0.1	99
21-50	126	.019	0.2	101	134	.001	0.1	107
Over 50	161	—.001	—1.3	95	157	—.004	—0.2	110
High-low		.009	—0.9	—7		—.006	—0.3	11

TABLE 3.—*Means of the monthly values of pressure, temperature and precipitation with low, medium and high monthly values of the Wolf and Wolfer sun-spot numbers* (continued)

Sun spot number	Winter half-year No. of months	Press., inches	Temp., ° F.	Precip., %	Summer half-year No. of months	Press., inches	Temp., ° F.	Precip., %
UNITED STATES (continued)								
Charleston, 32° 47′ N., 79° 56′ W.								
0-20	160	—.007	—0.1	104	156	.002	0.3	102
21-50	126	.009	0.3	100	134	.000	0.3	91
Over 50	161	.001	0.1	101	156	—.005	0.4	96
High-low		—.006	0.2	—3		—.007	0.1	—6
Cheyenne, 41° 8′ N., 104° 48′ W.								
0-20	131	.007	0.0	114	128	—.002	0.3	97
21-50	94	—.008	—0.1	98	91	.002	—0.2	111
Over 50	93	—.002	0.0	92	95	.002	—0.4	101
High-low		—.009	0.0	—22		.004	—0.7	4
Cincinnati, 39° 6′ N., 84° 30′ W.								
0-20	131	—.009	0.2	96	127	—.002	0.2	101
21-50	107	.121	0.3	86	100	.006	—0.1	91
Over 50	93	—.001	—0.7	99	95	—.003	—0.8	99
High-low		.008	—0.9	3		—.001	—1.0	—2
Denver, 35° 45′ N., 105° 0′ W.								
0-20	131	.008	—0.1	104	127	.000	0.5	96
21-50	92	—.004	0.0	90	92	.002	—0.3	109
Over 50	93	—.003	0.1	107	88	.000	—0.3	102
High-low		—.011	0.2	3		.000	—0.8	6
Detroit, 42° 20′ N., 83° 3′ W.								
0-20	159	—.004	—0.4	106	152	.002	—0.1	106
21-50	123	.012	0.6	97	125	.004	0.5	100
Over 50	152	—.002	—0.1	98	151	—.002	0.1	97
High-low		.002	0.3	—8		—.004	0.2	—9
Eastport, 44° 54′ N., 66° 59′ W.								
0-20	130	—.014	0.0	104	126	.001	0.2	102
21-50	90	.019	0.7	100	89	.000	0.0	90
Over 50	89	.003	0.6	89	92	.003	0.3	104
High-low		.017	0.6	—15		.002	0.1	2
El Paso, 31° 47′ N., 106° 30′ W.								
0-20	111	.002	0.3	93	106	.000	0.2	100
21-50	82	—.001	0.0	101	82	—.001	0.0	88
Over 50	87	.004	—0.1	107	90	.000	—0.1	105
High-low		.002	—0.4	14		.000	—0.3	5
Galveston, 29° 18′ N., 94° 50′ W.								
0-20	130	—.006	0.0	112	128	—.006	0.0	105
21-50	100	.004	—0.1	87	92	.005	—0.1	95
Over 50	93	.007	0.2	87	94	.002	0.0	96
High-low		.013	0.2	—25		.008	0.0	—9
Hatteras, 35° 15′ N., 75° 40′ W.								
0-20	131	—.007	0.2	103	127	.005	0.1	102
21-50	86	.007	0.1	104	85	—.001	0.0	100
Over 50	87	—.001	—0.5	91	90	—.007	—0.1	99
High-low		.006	—0.7	—12		—.012	—0.2	—3

TABLE 3.—*Means of the monthly values of pressure, temperature and precipitation with low, medium and high monthly values of the Wolf and Wolfer sun-spot numbers* (continued)

Sun spot number	Winter half-year: No. of months	Press., inches	Temp., ° F.	Precip., %	Summer half-year: No. of months	Press., inches	Temp., ° F.	Precip., %
				UNITED STATES (continued)				
				Helena, 46° 34′ N., 112° 4′ W.				
0-20	103	.007	0.1	92	100	—.003	—0.5	97
21-50	82	—.006	0.4	98	82	.005	0.3	95
Over 50	87	—.002	—0.4	106	90	.001	—0.1	103
High-low		—.009	—0.5	14		.004	0.4	6
				Key West, 24° 33′ N., 81° 48′ W.				
0-20	152	—.005	0.1	108	144	.002	—0.2	96
21-50	121	—.005	0.3	110	124	—.003	0.2	111
Over 50	152	.001	0.3	94	151	—.001	0.6	102
High-low		.006	0.2	—14		—.003	0.8	6
				Little Rock, 34° 45′ N., 92° 6′ W.				
0-20	105	.001	—0.3	88	100	.005	0.0	104
21-50	82	—.001	0.3	101	82	—.001	0.2	92
Over 50	87	.003	—0.1	108	90	—.007	—0.3	103
High-low		.002	0.2	20		—.012	—0.3	—1
				Marquette, 46° 34′ N., 87° 24′ W.				
0-20	125	.002	0.4	94	125	.002	0.4	106
21-50	88	.003	0.4	101	86	.003	0.0	89
Over 50	93	—.004	—0.9	99	95	—.004	—0.6	98
High-low		—.006	—1.3	5		—.006	—1.0	—8
				Mobile, 30° 41′ N., 88° 2′ W.				
0-20	131	—.003	—0.1	96	127	.002	0.1	97
21-50	92	.004	—0.1	103	92	.004	0.0	91
Over 50	93	.001	0.1	114	95	—.006	0.0	101
High-low		.004	0.2	18		—.008	—0.1	4
				Modena, 37° 48′ N., 113° 54′ W.				
0-20	60	.007	..	92	60	.001	..	83
21-50	42	.003	..	99	45	—.003	..	108
Over 50	45	.000	..	109	41	.004	..	104
High-low		—.007	..	17		.003	..	21
				Nashville, 36° 10′ N., 86° 47′ W.				
0-20	131	—.001	0.0	98	127	.002	0.1	105
21-50	92	.007	0.2	106	92	.003	0.1	95
Over 50	102	—.001	—0.2	97	95	—.008	—0.3	98
High-low		.000	—0.2	—1		—.010	—0.4	—7
				New York, 40° 43′ N., 74° 0′ W.				
0-20	160	—.012	—0.1	105	155	.000	0.4	102
21-50	126	.017	0.3	102	134	—.005	0.0	100
Over 50	160	—.002	—0.6	106	154	—.002	0.0	99
High-low		.010	—0.5	1		—.002	—0.4	—3
				North Platte, 41° 8′ N., 100° 45′ W.				
0-20	135	.007	0.0	93	128	—.002	0.5	95
21-50	86	—.004	0.3	101	85	.007	—0.2	102
Over 50	87	.003	—0.6	104	90	.001	—0.4	94
High-low		—.004	—0.6	11		.003	—0.9	—1

TABLE 3.—*Means of the monthly values of pressure, temperature and precipitation with low, medium and high monthly values of the Wolf and Wolfer sun-spot numbers* (continued)

Sun spot number	Winter half-year: No. of months	Press., inches	Temp., ° F.	Precip., %	Summer half-year: No. of months	Press., inches	Temp., ° F.	Precip., %
				UNITED STATES (continued)				
				Omaha, 41° 16′ N., 95° 56′ W.				
0-20	131	.004	0.5	96	127	—.002	0.4	92
21-50	92	—.002	0.1	101	92	.005	—0.4	107
Over 50	93	.000	—0.4	104	95	.000	—0.3	99
High-low		—.004	—0.9	8		.002	—0.7	7
				Phoenix, 33° 26′ N., 112° 0′ W.				
0-20	122	.002	0.5	91	117	.002	0.4	87
21-50	83	.002	0.0	88	83	—.001	0.2	93
Over 50	87	.001	—0.4	123	90	—.002	—0.5	95
High-low		—.001	—0.9	32		—.004	—0.9	8
				Portland, Ore., 45° 32′ N., 122° 41′ W.				
0-20	131	.007	0.3	97	127	.000	0.1	111
21-50	92	—.014	—0.1	102	92	.000	0.1	88
Over 50	93	.003	—0.4	98	95	—.001	—0.2	95
High-low		—.004	—0.7	1		—.001	—0.3	16
				Red Bluff, 40° 10′ N., 122° 15′ W.				
0-20	117	.005	0.2	101	113	—.003	0.3	98
21-50	82	—.007	0.0	101	82	—.006	0.1	83
Over 50	87	.003	—0.4	92	90	.004	—0.3	101
High-low		—.002	—0.6	—9		.007	—0.6	3
				St. Louis, 38° 38′ N., 90° 12′ W.				
0-20	161	.001	—0.3	94	156	.002	0.2	94
21-50	126	.003	0.2	109	134	.002	—0.4	107
Over 50	161	—.003	—0.6	100	156	—.005	—0.3	93
High-low		—.004	—0.3	6		—.007	—0.5	—1
				St. Paul, 44° 58′ N., 93° 3′ W.				
0-20	143	.004	0.1	94	139	—.002	—0.1	94
21-50	111	—.004	0.0	108	113	.004	—0.9	94
Over 50	146	—.002	—0.7	102	146	—.002	—0.6	108
High-low		—.006	—0.8	8		.000	—0.5	14
				Salt Lake City, 40° 46′ N., 111° 54′ W.				
0-20	131	.006	0.5	102	127	—.001	0.5	94
21-50	86	—.004	—0.2	99	85	—.003	—0.2	104
Over 50	87	.002	—0.6	97	90	.002	0.0	104
High-low		—.004	—1.1	—5		.003	—0.5	10
				San Diego, 32° 43′ N., 117° 10′ W.				
0-20	161	—.002	0.4	101	156	.002	0.7	94
21-50	126	.002	0.0	90	134	—.002	0.8	91
Over 50	160	.002	—0.1	104	151	—.001	0.8	98
High-low		.004	—0.5	3		—.003	0.1	4
				San Francisco, 37° 48′ N., 122° 26′ W.				
0-20	150	—.001	0.3	103	139	—.004	0.3	70
21-50	126	—.002	—0.1	102	132	—.002	0.1	113
Over 50	139	.004	—0.1	94	136	.007	0.2	108
High-low		.005	—0.4	—9		—.011	—0.1	18

TABLE 3.—*Means of the monthly values of pressure, temperature and precipitation with low, medium and high monthly values of the Wolf and Wolfer sun-spot numbers* (continued)

Sun spot number	Winter half-year No. of months	Press., inches	Temp., °F.	Precip., %	Summer half-year No. of months	Press., inches	Temp., °F.	Precip., %
UNITED STATES (continued)								
San Luis Obispo, 35° 18′ N., 120° 39′ W.								
0-20	75	.001	0.3	105	76	—.001	0.0	68
21-50	56	.004	—0.2	109	59	—.002	—0.1	72
Over 50	52	.000	—0.1	101	47	.003	0.3	65
High-low		—.001	—0.4	—4		.004	0.3	—3
Santa Fe, 35° 41′ N., 105° 57′ W.								
0-20	155	.004	0.1	99	146	.001	0.7	101
21-50	119	—.004	0.0	108	129	—.005	0.6	102
Over 50	147	.007	0.3	91	137	.003	1.3	94
High-low		.003	0.2	—8		.002	0.6	—7
Spokane, 47° 40′ N., 117° 25′ W.								
0-20	104	.013	0.2	96	99	.003	0.0	103
21-50	76	—.015	0.3	104	77	—.006	0.4	104
Over 50	87	.001	—0.3	97	89	.000	0.0	96
High-low		—.012	—0.5	1		—.003	0.0	7
Washington, 38° 54′ N., 77° 3′ W.								
0-20	159	—.010	—0.2	105	140	.005	0.0	124
21-50	129	.018	0.1	96	123	.002	—0.3	105
Over 50	136	.000	—0.6	104	137	—.004	—0.2	98
High-low		.010	—0.4	—1		—.009	—0.2	—26
MEXICO, 1878-1924								
Leon, 21° 7′ N., 101° 41′ W.								
0-20	80	.007	0.0	99	80	.006	—0.3	105
21-50	42	.007	—0.2	131	41	.000	—0.2	100
Over 50	47	—.010	—0.2	76	47	—.013	0.2	94
High-low		—.017	—0.2	—23		—.019	0.5	—11
Mazatlan, 23° 12′ N., 106° 25′ W.								
0-20	68	—.006	0.4	107	64	—.004	0.6	70
21-50	61	.000	0.3	102	62	.007	0.3	142
Over 50	60	—.007	0.4	107	65	—.008	0.7	70
High-low		—.001	0.0	0		—.004	0.1	0
Merida, 20° 58′ N., 89° 37′ W.								
0-20	61	—.007	0.3	97	63	.006	0.3	117
21-50	51	.000	0.0	98	50	—.007	0.0	88
Over 50	38	.010	0.2	106	36	.003	0.2	93
High-low		.017	—0.1	9		—.003	—0.1	—24
Mexico City, 19° 26′ N., 99° 8′ W.								
0-20	114	.000	0.3	98	111	.000	0.0	106
21-50	81	.004	0.0	130	78	.000	0.0	100
Over 50	87	.000	0.0	79	89	.000	0.2	89
High-low		.000	—0.3	—19		.000	0.2	—17

TABLE 3.—*Means of the monthly values of pressure, temperature and precipitation with low, medium and high monthly values of the Wolf and Wolfer sun-spot numbers* (continued)

Sun-spot number	Winter half-year				Summer half-year			
	No. of months	Press., inches	Temp., ° F.	Precip., %	No. of months	Press., inches	Temp., ° F.	Precip., %
MEXICO (continued)								
Monterey, 25° 40′ N., 100° 18′ W.								
0-20	67	.021	—0.5	97	67	.008	—0.5	90
21-50	50	.008	0.3	108	56	.000	0.2	115
Over 50	45	.012	0.5	85	40	.007	0.0	100
High-low		.009	—1.0	—12		—.001	0.5	10
Oaxaca, 16° 4′ N., 96° 43′ W.								
0-20	37	.000	—0.3	114	34	.003	—0.3	86
21-50	52	.004	0.2	90	51	.000	0.0	108
Over 50	48	.000	0.0	94	47	.007	0.2	107
High-low		.000	0.3	—20		.004	0.5	21
Puebla, 19° 2′ N., 98° 12′ W.								
0-20	72	—.004	0.3	75	71	.000	0.0	97
21-50	55	.003	—0.2	168	52	.003	0.0	111
Over 50	76	—.003	—0.3	77	81	—.007	—0.2	96
High-low		.001	—0.6	2		—.007	—0.2	—1
CENTRAL AMERICA, 1863-1920								
Colon, 9° 23′ N., 79° 23′ W.								
0-20	38	.001	0.4	94	46	.000	0.5	90
21-50	30	—.004	0.3	98	26	.006	0.0	105
Over 50	27	.006	—0.1	95	27	.000	—0.1	99
High-low		.005	—0.5	1		.000	—0.6	9
San Salvador, 13° 44′ N., 89° 9′ W.								
0-20	34	.000	—0.1	110	38	.000	—0.1	85
21-50	19	—.001	0.1	141	18	.000	0.0	102
Over 50	19	.001	0.2	158	25	—.001	0.1	113
High-low		.001	0.3	48		—.001	0.2	28
BERMUDA AND WEST INDIES, 1882-1920								
Christiansted, Virgin Islands								
0-20	105	..	0.1	97	101	..	0.0	101
21-50	68	..	0.0	107	63	..	0.0	111
Over 50	63	..	—0.2	100	101	..	0.0	71
High-low		..	—0.3	3		..	0.0	—30
Port au Prince, 18° 34′ N., 72° 22′ W.								
0-20	76	—.001	0.3	100	79	.001	0.2	97
21-50	53	—.002	0.3	103	52	.000	0.2	99
Over 50	68	.002	—0.5	99	67	—.001	—0.3	100
High-low		.003	—0.8	—1		—.002	—0.5	3
Hamilton, Bermuda, 32° 17′ N., 64° 46′ W.								
0-20	72	.000	0.1	98	73	.009	—0.1	100
21-50	60	.018	0.4	101	59	.006	0.3	105
Over 50	70	—.005	—0.4	101	67	.006	0.0	96
High-low		—.005	—0.5	3		—.003	0.1	—4
San Juan, Porto Rico, 18° 29′ N., 66° 7′ W.								
0-20	54	.000	0.3	101	56	.003	0.0	104
20-50	37	—.003	0.1	106	34	—.001	0.1	100
Over 50	44	.002	—0.4	94	43	—.004	0.0	92
High-low		.002	—0.7	—7		—.007	0.0	—12

The monthly sun-spot numbers were divided into low, medium, and high numbers, and the dates of occurrence were tabulated. The values from 0 to 20 were called low, those from 21 to 50 medium, and those over 50 high. The mean departures from normal pressure and temperature, and the mean percentages of the normal rainfall, were then obtained for each of the three divisions of sun spots. This was done separately for the winter half-year, and for the summer half-year, and the results are given in table 3.

The number of months included in each average varied according to the length of time data were available from the station used, but in some cases it was 160, or more, for temperature and precipitation, as shown by the tabulated data for Boston, Charleston, and New York. The mean departures from normal pressure are given in thousandths of an inch, the mean departures of temperature in degrees and tenths Fahrenheit, and the mean precipitation in percentages of the normal. The differences between the mean values found with high sun-spot numbers and with low sun-spot numbers are also given. These differences are shown plotted in figure 3 in the same manner as were the differences between the means in the case of high and low solar-radiation values. The lines for pressure are drawn for .03 inch, equal to one millibar, and the lines for temperature are drawn for 1.8° F., or for half that amount, equal to one degree, or to a half degree Centigrade.

(*c*) *The findings of the two investigations compared.*—The two sets of charts in figures 2 and 3 depend on entirely different measures of solar variation, and are largely for different periods of time. Yet they show a striking similarity. In winter, accompanying increased solar activity, both studies reveal higher pressure over Alaska and Canada, and lower pressure along southern Alaska, and in the western United States. The temperature is lower in both cases over Alaska, Canada, and the northern United States, and warmer in the southern states, and southward to Colon. The precipitation is also in excess over the region where the temperature is lower, and in defect in Texas and Mexico. Both investigations show in the summer a reversal of pressure, as compared to the winter, north of about latitude 50°. A lower pressure prevails in summer over northwestern Canada and higher temperature in the same region. Higher pressure prevails along the north Pacific and the north Atlantic coast of North America, between latitude 50° and 60°. Lower temperature is found over the interior of the United States, both during the winter and summer half-year. The similarity of the rainfall distribution during the

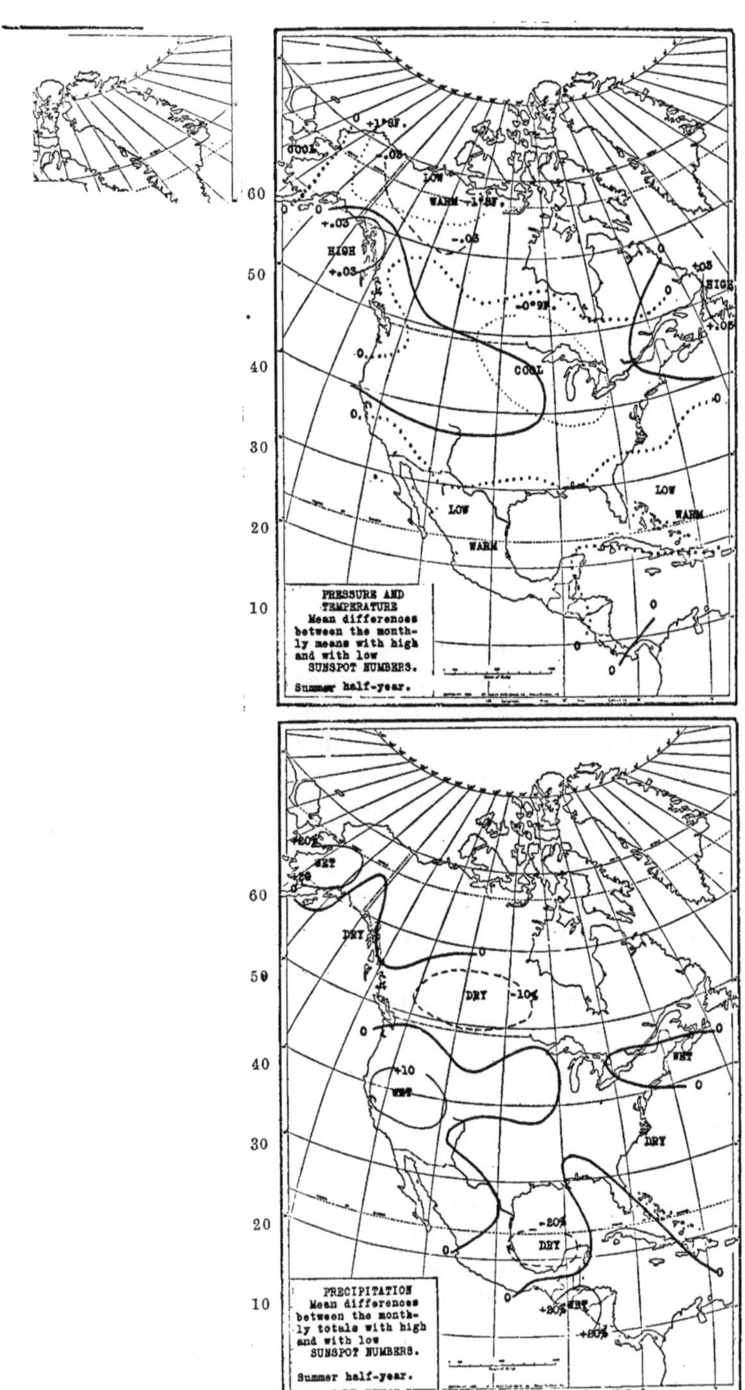
PRESSURE AND TEMPERATURE
Mean differences between the monthly means with high and with low SUNSPOT NUMBERS.
Summer half-year.
PRECIPITATION
Mean differences between the monthly totals with high and with low SUNSPOT NUMBERS.
Summer half-year.
COOL
WARM
HIGH
LOW
WET
DRY
-10%
+10
-20%
+20%

FIG. 3.

summer months is less marked, but there is evidently a relation between the two sets of data.

TABLE 4.—*Differences between the means of pressure and temperature for high and low solar radiation and for high and low sun-spot numbers*

	Winter half-year				Summer half-year			
	Pressure		Temperature		Pressure		Temperature	
Station	Solar rad.	Sun spots	Solar rad.	Sun spots	Solar rad.	Sun spots	Solar rad.	Sun spots
Dutch Harbor ..	.061	..	—0.6	—0.1	.113	..	—0.9	—1.0
Eagle	.083	.045	—1.9	—3.0	—.095	—.006	1.3	3.4
Juneau	.071	—.024	—1.7	—0.8	.012	.036	—0.5	0.2
Nome	.043	.067	—1.6	—5.0	—.007	—.014	0.2	—0.7
Tanana	.041	.060	—1.5	—3.5	.003	.007	0.9	1.6
Valdez-Kodiak .	—.006	—.035	—3.1	—2.2	—.026	—.012	—2.1	0.6
Barkerville	—.002	.021	0.0	—1.2	.014	.013	—0.2	0.0
Charlottetown .	.054	.003	0.2	—2.7	.028	—.010	—0.6	—0.3
Dawson	.014	.031	—2.6	—2.9	—.043	—.045	0.9	1.3
Edmonton	.010	.003	—1.9	—2.7	—.012	—.010	—0.2	—0.3
Father Point ..	.049	.019	1.6	—0.7	.017	—.007	0.8	0.1
Etc. for 47 stations	...	...	..	..	...	...	..	..
See Tables 8 and 9.								

In order to determine numerically the correlation between the differences in the means of pressure and temperature, found with high and with low values of solar radiation, and those found with high and with low sun-spot numbers, the differences given in tables 2 and 3 were tabulated in the manner illustrated in table 4. The correlations between the two classes of differences, one for solar radiation measurements and the other for sun spots, were then computed without further corrections. There are 45 stations given in tables 2 and 3 for which the means of pressure were computed for both the values of solar radiation and for sun spots, and 47 for which the means of temperature were computed. Valdez and Kodiak were treated as one station.

The correlation coefficients found for the four sets of differences are as follows:

Winter half-year, for pressure, 0.56 ± 0.07; for temperature, 0.62 ± 0.06
Summer half-year, for pressure, 0.45 ± 0.08; for temperature, 0.50 ± 0.07

It is possible to doubt the accuracy of the work, but it seems impossible for anyone to suppose that these two independent sets of correlations could be the result of chance. Fairly interpreted, they mean that higher solar-radiation values prevail at times of numerous sun spots, and that definite geographically located weather changes attend changes in the solar activity, whichever measure of it we employ.

3. THE GEOGRAPHICAL MARCH OF WEATHER EFFECTS DEPENDING ON THE INTENSITY OF SOLAR ACTIVITY

The observed values of solar radiation and sun spots are not numerous enough for the accompanying temperature departures to give smooth curves, when they are subdivided into numerous grades, and mean temperatures obtained for each grade, but the results for a few widely separated stations are given in table 5.

The means in table 5 do not show a steady progress from high to low values, but they do show that on the North American Continent, with very low solar radiation and low sun-spot numbers, the temperature departures above normal are greatest in high latitudes; that they are greatest in middle latitudes with medium solar radiation and sun-spot numbers; and greatest in the subtropical regions of southern Mexico and Central America with high solar radiation and high sun-spot numbers. On the other hand, temperatures below normal occur in high latitudes, and also in the equatorial region, with high solar radiation and high sun-spot numbers.

On account of the paucity of solar radiation measurements, it seemed worth while also to study the distribution of the departures of the monthly means of the weather elements, with very low and very high sun-spot numbers. Accordingly, the means of the monthly departures from normal of pressure and temperature, and the percentages of normal precipitation, were worked out for sun-spot numbers 0 to 5, and for those over 70, for the same stations as given in table 2. Means were obtained for the winter half-year and for the summer half-year, separately. The results are given in table 6 and are shown graphically in figures 4 and 5. In the separate charts in these figures the pressure lines are drawn for each .003 inch, equal to one millibar, and the temperature lines are drawn for 1.8° F., equal to one degree C., or were drawn for half that amount.

TABLE 5.—*Mean departures from normal temperature for six grades of solar radiation during the years 1918 to 1925 and for seven grades of sun-spot numbers during the years 1870 to 1923 Winter half-year*

Solar radiation	No. of cases	Nome	Dawson	Edmonton	St. Paul	Little Rock	San Salvador	Colon
		Stations						
1.911-1.920	7	**3.3**° F.	4.1° F.	**2.6**° F.	—0.8° F.*	—2.1° F.	—0.4° F.	0.6° F.
1.921-1.930	10	—0.5	1.1	0.5	**2.2**	1.8	—0.8*	0.1
1.931-1.940	10	2.5	—2.7	—2.4	1.5	**2.0**	—0.4	0.1
1.941-1.950	13	—2.2	0.9	0.1	**2.7**	1.8	0.5	**0.8**
1.951-1.960	12	0.4	0.2	—0.7	0.4	1.1	**0.8**	0.6
1.961-1.972	2	—6.2*	—3.1*	0.6	—0.5	0.8	..	—0.1
Sun-spot numbers		**Nome**	**Dawson**	**Edmonton**	**St. Paul**	**Little Rock**	**Oaxaca**	**Colon**
0- 5	..	**3.6**	**2.3**	1.8	—0.1	—0.8*	—0.1	0.5
6-10	..	—1.2	0.6	**2.6**	**1.7**	0.3	—0.4*	—0.5
11-20	..	..	**2.6**	0.1	**2.0**	—0.1	0.4	..
21-30	..	..	1.2	—1.2	0.4	**1.2**	**0.4**	**0.6**
31-50	..	0.0	**1.0**	0.7	0.0	—0.4	0.2	0.5
51-70	..	—1.4	—1.5	—1.2	—0.5	—0.2	—0.3	—0.2
Over 70	..	—2.6*	—2.7*	—2.3*	—1.0*	—0.3	**0.4**	—0.6*

Heavy faced type show maximum and asterisks show minimum values. The number of cases are not given for the sun spots because they are different for each station according to the length of record. Oaxaca, Mexico, is substituted for San Salvador in the sun spots, because the record at San Salvador covers a short period. On the other hand no records were available from Oaxaca since 1918.

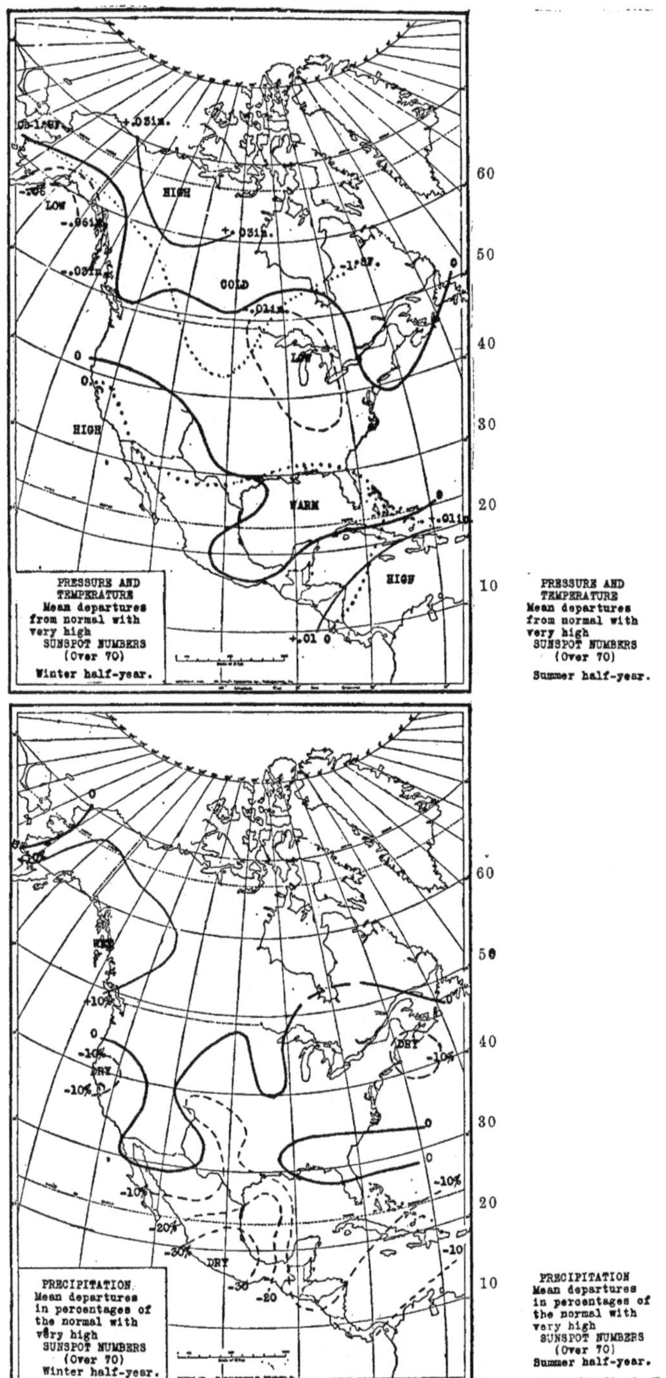
PRESSURE AND TEMPERATURE
Mean departures from normal with very high SUNSPOT NUMBERS (Over 70)
Winter half-year.
PRESSURE AND TEMPERATURE
Mean departures from normal with very high SUNSPOT NUMBERS (Over 70)
Summer half-year.
HIGH
LOW
COLD
WARM
+.03in.
+.01in.
60
50
40
30
20
10
PRECIPITATION
Mean departures in percentages of the normal with very high SUNSPOT NUMBERS (Over 70)
Winter half-year.
PRECIPITATION
Mean departures in percentages of the normal with very high SUNSPOT NUMBERS (Over 70)
Summer half-year.
DRY
+10%
-10%
-20%
-30%

FIG. 4.

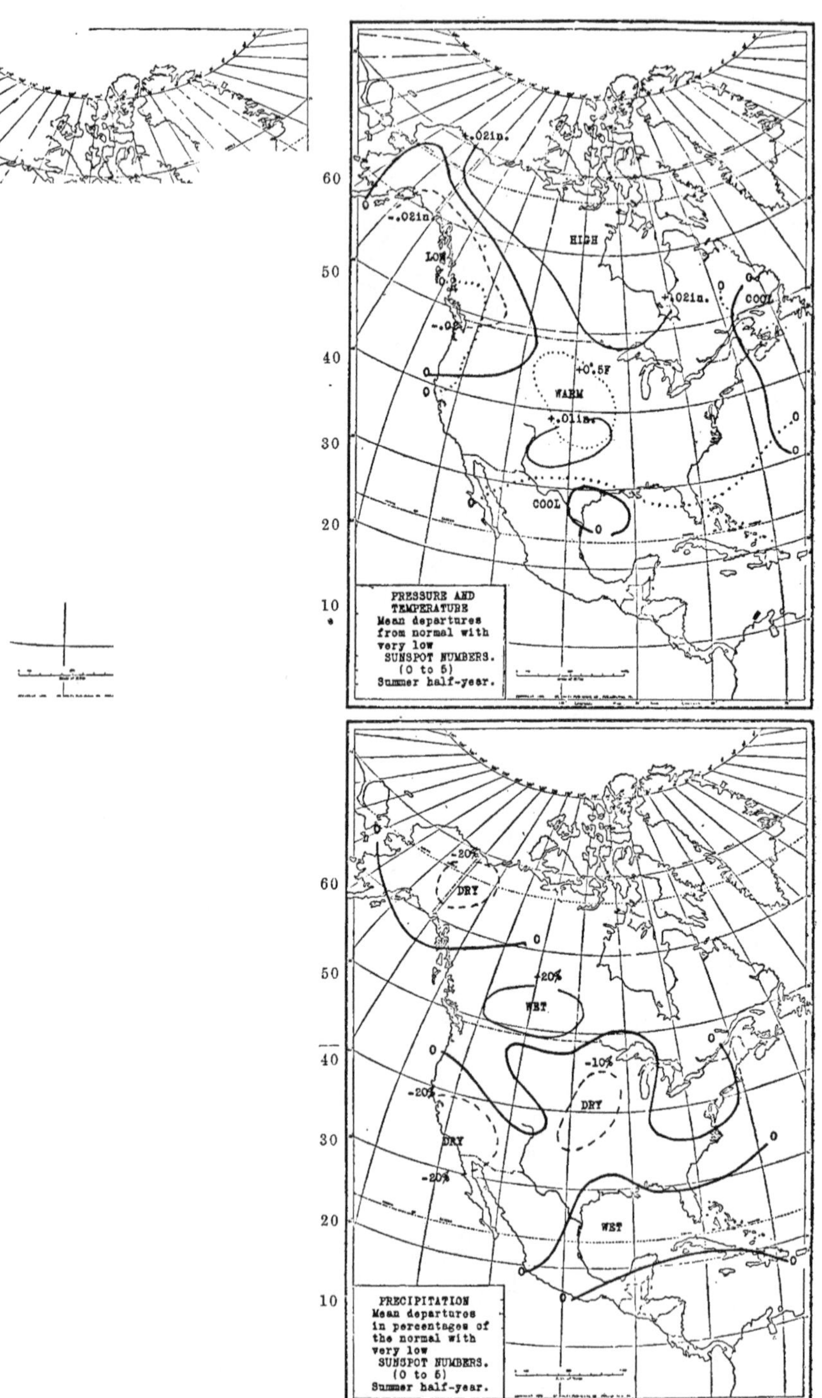

60
50
40
30
20
10
+.02in.
-.02in.
HIGH
LOW
+.02in.
COOL
-.02
+0.5F
WARM
+.01in.
COOL
PRESSURE AND TEMPERATURE
Mean departures from normal with very low SUNSPOT NUMBERS. (0 to 5) Summer half-year.
-20%
DRY
+20%
WET
-10%
DRY
-20%
DRY
-20%
WET
PRECIPITATION
Mean departures in percentages of the normal with very low SUNSPOT NUMBERS. (0 to 5) Summer half-year.

Fig. 5.

TABLE 6.—*Means of the monthly departures from normal pressure, temperature and precipitation for very low and very high monthly sun-spot numbers*

Sun-spot number	Winter half-year Cases	Press., inches	Temp., ° F.	Precip., %	Summer half-year Cases	Press., inches	Temp., ° F.	Precip., %
			ALASKA					
			Dutch Harbor					
0-5	33	..	0.2	109	22	..	0.6	109
Over 70	24	..	0.1	96	28	..	0.0	87
			Eagle					
0-5	36	—.002	1.9	91	15	.006	—1.2	96
Over 70	17	.004	—1.5	133	20	—.004	1.1	91
			Juneau					
0-5	41	—.041	0.5	109	36	—.031	0.3	94
Over 70	17	—.030	—1.2	119	20	.026	—0.3	100
			Kodiak					
0-5	..	..	..	..	..	..	..	..
Over 70	11	—.062	—1.4	116	16	—.001	—0.4	93
			Nome					
0-5	19	..	3.6	79	19	..	2.7	79
Over 70	14	—.035	—2.6	70	19	..	—0.7	120
			Tanana					
0-5	20	—.003	2.7	93	19	—.001	0.6	91
Over 70	24	.016	—1.0	130	19	.007	—0.3	128
			St. Paul Islands					
0-5	..	..	..	..	6	.073	1.0	102
Over 70	20	.033	—2.3	92	27	—.007	0.5	90
			CANADA					
			Barkerville					
0-5	45	.002	0.7	94	41	—.021	—0.5	115
Over 70	26	.009	—0.9	118	32	.048	—0.1	98
			Bella Coola					
0-5	27	..	0.8	103	27	..	0.0	103
Over 70	15	..	—1.1	111	20	..	—0.2	94
			Dawson					
0-5	20	.026	2.3	101	20	.075	—0.5	76
Over 70	17	.029	—2.7	103	20	—.010	0.2	94
			Calgary					
0-5	48	—.002	1.3	105	41	—.022	—0.1	133
Over 70	43	.009	0.2	119	40	.003	0.5	70
			Charlottetown					
0-5	53	—.033	1.0	90	48	—.002	0.3	89
Over 70	39	.008	—1.0	99	46	.007	0.1	98
			Edmonton					
0-5	49	.019	1.8	93	41	.003	0.3	130
Over 70	35	—.007	—2.3	106	43	—.004	—0.1	91

TABLE 6.—*Means of the monthly departures from normal pressure, temperature and precipitation for very low and very high monthly sun-spot numbers* (continued)

Sun-spot number	Winter half-year				Summer half-year			
	Cases	Press., inches	Temp., °F.	Precip., %	Cases	Press., inches	Temp., °F.	Precip., %
CANADA (continued)								
Father Point								
0-5	57	—.031	0.5	106	47	.003	—0.1	107
Over 70	36	.014	—1.4	85	45	.011	0.1	95
Kamloops								
0-5	36	.018	0.3	94	31	—.013	—0.5	122
Over 70	21	.036	—1.4	97	29	.021	0.0	104
Massett								
0-5	27	..	0.9	87	28	..	—0.2	86
Over 70	17	..	—1.2	117	21	..	0.1	81
Montreal								
0-5	54	—.020	0.4	107	51	.010	0.1	104
Over 70	39	.007	—1.1	94	45	.012	0.0	106
Moose Factory								
0-5	33	—.003	—1.9	..	31	.037	—0.5	..
Over 70	28	—.004	—1.1	..	37	.020	0.0	..
Prince Albert								
0-5	49	.013	0.3	90	41	.011	0.4	127
Over 70	29	.011	—2.0	84	39	.035	0.1	84
Qu'appelle								
0-5	49	.005	0.9	110	41	.008	0.1	122
Over 70	35	—.012	—2.2	100	43	.009	0.0	81
Sable Island								
0-5	30	—.028	0.3	95	28	—.008	0.6	89
Over 70	17	—.005	—1.0	96	21	.014	—0.4	107
St. John's, N. F.								
0-5	53	—.028	0.4	93	49	—.024	0.4	100
Over 70	17	—.005	—1.0	96	21	.014	—0.4	107
S. W. Point, Anticosti								
0-5	41	—.041	0.6	108	36	—.007	—0.4	93
Over 70	36	.006	—0.5	101	44	.018	0.2	105
Toronto								
0-5	57	—.016	0.4	97	52	.012	0.5	105
Over 70	39	—.003	—1.2	95	46	—.007	—0.1	122
Winnipeg								
0-5	57	.022	1.3	74	52	.024	0.9	104
Over 70	39	—.015	—1.9	112	46	.007	—0.6	86

TABLE 6.—*Means of the monthly departures from normal pressure, temperature and precipitation for very low and very high monthly sun-spot numbers* (continued)

Sun-spot number	Winter half-year				Summer half-year			
	Cases	Press., inches	Temp., ° F.	Precip., %	Cases	Press., inches	Temp., ° F.	Precip., %
UNITED STATES								
Abilene								
0-5	49	.007	—0.5	109	41	.010	0.9	86
Over 70	26	—.008	—0.1	90	30	—.003	0.2	115
Bismarck								
0-5	57	.017	0.2	75	51	.005	0.7	111
Over 70	36	—.005	—2.3	102	45	.010	—0.7	90
Boston								
0-5	66	—.034	—0.1	104	65	.001	0.1	100
Over 70	78	.001	—1.0	87	89	—.004	—0.1	114
Charleston								
0-5	..	—.012	—0.6	102	..	.006	0.3	94
Over 70	..	—.006	0.1	106	..	—.005	0.5	88
Cheyenne								
0-5	57	.017	—0.4	117	52	.006	0.8	86
Over 70	39	—.015	—0.3	82	46	.002	—1.3	99
Cincinnati								
0-5	57	—.007	—0.7	99	51	.005	0.7	102
Over 70	39	—.011	—1.1	98	46	—.010	—0.2	98
Corpus Christi								
0-5	48	.007	—0.4	105	42	.009	—0.2	105
Over 70	28	.001	0.0	86	37	—.003	—0.2	79
Denver								
0-5	57	.019	—0.6	112	51	.006	0.8	100
Over 70	39	—.019	0.3	95	46	.001	—1.1	93
Detroit								
0-5	67	—.010	—1.0	99	64	.009	0.1	118
Over 70	76	—.009	—0.4	99	84	—.001	—0.2	99
Eastport								
0-5	57	—.032	0.2	68	52	.000	0.1	96
Over 70	36	.004	—1.1	89	45	.008	—0.4	99
El Paso								
0-5	52	.006	—0.2	101	43	.005	0.4	95
Over 70	36	.005	—0.3	107	45	.001	—0.4	105
Galveston								
0-5	57	.004	—0.5	113	52	—.001	—0.1	116
Over 70	39	.000	0.2	78	46	.003	—0.2	98
Hatteras								
0-5	57	—.015	—0.1	98	52	.004	0.3	97
Over 70	36	—.005	—0.7	85	45	—.006	—0.1	106

TABLE 6.—*Means of the monthly departures from normal pressure, temperature and precipitation for very low and very high monthly sun-spot numbers* (continued)

Sun-spot number	Winter half-year Cases	Press., inches	Temp., °F.	Precip., %	Summer half-year Cases	Press., inches	Temp., °F.	Precip., %
UNITED STATES (continued)								
Helena								
0-5	49	.010	0.7	93	41	—.003	0.2	98
Over 70	36	—.011	—1.3	117	45	.003	—0.5	104
Key West								
0-5	64	—.004	—0.1	104	60	.003	—0.2	101
Over 70	76	—.001	0.4	93	84	.000	0.7	97
Little Rock								
0-5	49	.002	—0.8	97	41	.011	0.6	124
Over 70	36	—.004	—0.3	99	45	—.006	—0.2	100
Marquette								
0-5	53	.003	—0.1	101	51	.014	0.5	110
Over 70	39	—.017	—1.7	86	46	—.005	—0.8	99
Mobile								
0-5	57	.002	—0.7	100	52	.009	0.2	94
Over 70	39	—.004	0.0	122	46	—.004	—0.2	96
Nashville								
0-5	57	.002	—0.7	102	52	.010	0.5	98
Over 70	48	—.010	—0.7	96	46	—.012	—0.7	96
New York								
0-5	67	—.025	—0.2	109	65	.000	0.4	103
Over 70	77	.001	—0.6	97	86	—.005	0.1	99
North Platte								
0-5	57	.021	—0.5	109	52	.005	1.0	93
Over 70	36	—.009	—1.8	96	45	.005	—1.0	95
Omaha								
0-5	57	.008	—0.1	102	52	—.001	1.3	83
Over 70	39	—.009	—1.1	112	46	.001	—0.7	95
Phoenix								
0-5	57	.004	0.3	106	47	.001	0.1	80
Over 70	36	—.001	—0.7	137	45	.001	—0.5	95
Portland, Ore.								
0-5	57	.018	0.4	94	52	—.001	.0.1	133
Over 70	39	.011	—0.2	102	46	.004	—0.4	86
Red Bluff								
0-5	57	.010	0.7	92	48	.001	—0.4	83
Over 70	36	.008	—0.3	84	45	.009	—0.3	84
Salt Lake City								
0-5	57	.015	0.4	89	52	.001	0.5	108
Over 70	36	—.001	—1.1	103	45	.006	—0.5	99

TABLE 6.—*Means of the monthly departures from normal pressure, temperature and precipitation for very low and very high monthly sun-spot numbers* (continued)

Sun-spot number	Winter half-year Cases	Press., inches	Temp., ° F.	Precip., %	Summer half-year Cases	Press., inches	Temp., ° F.	Precip., %
UNITED STATES (continued)								
St. Louis								
0-5	67	.001	—1.0	96	65	.009	0.8	91
Over 70	78	—.013	—0.7	97	88	—.012	—0.4	91
St. Paul								
0-5	60	.008	—0.1	87	56	.005	0.1	97
Over 70	74	—.013	—1.0	101	87	.001	—0.9	111
San Diego								
0-5	67	.000	0.4	93	65	.006	0.7	86
Over 70	78	.005	0.0	100	88	.004	0.8	93
San Francisco								
0-5	62	.005	0.8	97	51	.002	0.2	70
Over 70	62	.011	0.1	90	71	.011	0.0	71
San Luis Obispo								
0-5	38	.001	0.4	100	33	.002	—0.3	79
Over 70	18	.005	0.4	98	23	.010	0.3	50
Santa Fe								
0-5	64	.009	—0.1	107	60	.009	1.1	109
Over 70	71	.001	0.9	81	82	.002	1.6	87
Washington								
0-5	66	—.025	—0.4	106	65	.003	0.2	105
Over 70	65	.003	—0.8	97	77	—.008	—0.1	100
MEXICO								
Leon								
0-5	42	.008	0.0	98	35	.008	—0.2	101
Over 70	17	—.024	0.4	67	24	—.024	0.2	69
Mazatlan								
0-5	27	—.004	0.7	72	22	—.004	1.4	62
Over 70	24	.012	0.5	77	29	.004	0.4	58
Merida								
0-5	31	.004	0.2	100	28	.016	—0.5	112
Over 70	9	.000	0.5	94	16	.012	—0.2	108
Mexico City								
0-5	55	.000	0.4	132	47	.004	—0.2	119
Over 70	36	.000	0.2	57	44	.004	0.2	80
Monterey								
0-5	31	.012	—1.6	100	26	.012	—1.1	93
Over 70	16	.012	0.2	90	21	.016	0.0	71
Oaxaca								
0-5	12	.004	—0.2	111	8	.024	0.4	79
Over 70	20	.000	0.7	57	23	.004	1.2	122

TABLE 6.—*Means of the monthly departures from normal pressure, temperature and precipitation for very low and very high monthly sun-spot numbers* (continued)

Sun-spot number	Winter half-year				Summer half-year			
	Cases	Press., inches	Temp., ° F.	Precip., %	Cases	Precip., inches	Temp., ° F.	Precip., %
				WEST INDIES				
				Christiansted, Virgin Islands				
0-5	49	..	0.0	107	43	..	0.0	105
Over 70	24	..	—0.2	105	29	..	0.2	89
				Hamilton, Bermuda				
0-5	34	—.001	—0.1	111	32	—.001	—0.3	101
Over 70	28	—.004	—0.8	101	37	.010	0.0	95
				Port-au-Prince, Haiti				
0-5	37	.001	0.4	100	33	—.006	0.0	94
Over 70	27	.004	—0.7	90	37	—.001	—0.5	101
				San Juan, Porto Rico				
0-5	30	.002	0.4	106	27	.003	—0.1	112
Over 70	17	.011	—0.8	89	23	—.002	0.0	90
				Colon, Panama				
0-5	27	—.001	0.5	85	22	.003	0.6	88
Over 70	12	.014	—0.6	89	17	—.001	—0.3	104

It is seen from the charts in figure 4 that, with very high sun-spot numbers in winter, the pressure averages above normal over the larger part of Alaska and Canada. It averages below normal over southern Alaska and the Aleutian Islands, over the Great Lakes, and along the Atlantic coast of North America. It averages colder than normal over practically the whole of Canada and the United States, and warmer than normal in Mexico, and along the Gulf Coast of the United States. The low pressure over the Great Lakes probably endures only so long as the lakes remain unfrozen.

In summer, with very high sun-spot numbers, the pressure averages low in northern Canada, and the temperature averages above normal. High pressure is found along the North Pacific coast, and to the north of Newfoundland. A second area of low pressure is found in the Gulf States of the United States. With very high solar radiation, as in July, 1917, these two areas of low pressure in Canada and the United States unite to form one. The temperature averages lower than normal over the larger part of the United States.

With high sun-spot numbers in winter, the average precipitation is in excess over nearly all of Canada and Alaska. The greatest excess is found on the North Pacific coast. It is in defect over the eastern and southern part of the United States, and over Mexico. In summer, the average rainfall is in defect over Canada and the western part of the United States, and in excess over Alaska, the North Atlantic and Gulf States of the United States, and in Central America.

The distribution of the average departures from normal with very low sun-spot numbers is shown in figure 5. It is seen that the pressúre in winter averages low over northern and eastern Canada, and the temperature averages above normal. The pressure averages high on the Pacific coast, and over the United States west of the 85th meridian; while the temperature averages below normal in the central and eastern United States. The precipitation averages below normal over Canada and the northern United States, and in excess over a large part of the southern United States, in eastern Mexico, and in Central America. In summer, it averages dry in the central United States, and wet along the Gulf coast, and in southern Canada.

4. THE ANNUAL MARCH OF WEATHER EFFECTS DEPENDING ON SOLAR VARIATIONS

In order to determine more accurately the character of the annual period in the relation of sun spots to weather, the means of the departures from normal of pressure, temperature, and precipitation

TABLE 7.—*The annual period in the influence of sun spots on pressure, temperature, and precipitaition*

	Sun spots over 50				Sun spots 0-5			
Month	Cases	Mean press., inches	Mean temp., ° F.	Mean precip., %	Cases	Mean press., inches	Mean temp., ° F.	Mean precip., %
				Dawson				
Jan.	6	.172	—3.1	96	9	—.017	—1.5	93
Feb.	7	.044	—1.1	110	8	.031	4.9	122
Mar.	10	.043	—1.1	103	7	.091	1.0	90
Apr.	7	—.009	1.7	66	9	.010	—3.1	115
May	5	—.012	—0.3	125	7	.011	—0.3	76
June	5	—.010	—0.5	120	8	.035	0.0	79
July	8	—.019	0.3	87	8	.038	—1.1	102
Aug.	8	—.010	0.5	107	10	.039	—0.9	92
Sept.	7	—.037	0.5	101	9	.037	—0.3	82
Oct.	9	.013	—0.6	103	9	—.016	4.1	73
Nov.	6	—.040	0.4	88	9	.032	—0.1	80
Dec.	7	.073	—4.6	100	8	—.003	—0.1	121
Year	85	.018	—0.6	100	101	.024	0.1	94
				Prince Albert				
Jan.	10	.098	—4.5	100	18	—.016	0.4	95
Feb.	13	.028	0.9	62	15	.017	—0.2	89
Mar.	15	—.030	1.4	100	17	.035	0.1	108
Apr.	12	—.008	—0.6	110	19	.005	0.6	92
May	10	—.002	—0.4	96	16	—.012	0.5	122
June	11	—.003	0.0	86	17	.008	0.5	102
July	13	.019	0.3	84	15	.005	—0.4	121
Aug.	12	—.001	0.8	111	16	.006	—0.4	127
Sept.	13	.016	0.2	83	16	.003	—0.5	116
Oct.	14	.011	—1.1	102	19	.000	0.4	80
Nov.	9	—.008	2.6	59	18	.010	—0.4	107
Dec.	11	.033	—3.5	95	16	—.015	2.6	78
Year	143	.011	—0.2	91	202	.004	0.2	104
				St. Paul				
Jan.	14	.024	—2.6	87	22	.003	0.2	103
Feb.	18	—.005	0.0	115	19	.015	0.1	72
Mar.	19	—.014	0.0	111	21	.004	—0.9	51
Apr.	17	.001	1.0	110	24	—.002	0.4	91
May	12	—.015	—1.0	126	20	.000	—0.4	85
June	14	.000	—0.4	91	21	.001	0.0	98
July	18	.002	—0.8	104	20	—.006	0.4	83
Aug.	16	.004	—0.8	111	21	—.012	—0.4	96
Sept.	18	—.007	0.0	106	21	.009	—1.2	107
Oct.	17	—.005	—0.2	103	24	.000	0.2	100
Nov.	13	—.004	—0.5	101	23	.003	0.8	88
Dec.	12	—.001	—1.5	92	22	.007	0.0	108
Year	188	—.002	—0.7	105	258	.022	0.0	94

TABLE 7.—*The annual period in the influence of sun-spots on pressure, temperature, and precipitation* (continued)

Month	Sun spots over 50				Sun spots 0-5			
	Cases	Mean press., inches	Mean temp., ° F.	Mean precip., %	Cases	Mean press., inches	Mean temp., ° F.	Mean precip., %
				El Paso				
Jan.	12	.004	—0.3	118	18	.014	0.5	73
Feb.	16	.009	0.6	100	15	—.007	0.1	143
Mar.	18	.000	—0.5	84	18	—.003	—0.1	83
Apr.	16	—.005	0.1	86	20	.006	—0.1	110
May	12	—.003	—1.2	200	16	.006	0.9	66
June	14	.005	0.6	51	17	—.001	—0.1	134
July	16	—.002	0.4	95	15	—.001	—0.3	95
Aug.	14	.007	—0.6	125	16	.008	0.8	89
Sept.	18	—.001	—0.2	94	16	—.003	0.4	104
Oct.	17	.004	0.1	109	19	—.005	0.4	82
Nov.	12	.015	—0.4	130	18	.002	0.3	100
Dec.	12	.003	—0.3	118	17	.012	0.3	65
Year	177	.002	—0.1	106	205	.001	0.3	97
				Charleston				
Jan.	14	—.005	—0.8	106	22	.008	—0.4	105
Feb.	18	—.013	1.1	98	19	—.014	0.0	94
Mar.	19	—.001	0.2	107	21	—.014	—0.8	102
Apr.	17	—.015	0.4	85	24	.000	0.2	120
May	12	—.003	0.5	86	20	.008	0.4	118
June	14	—.002	0.3	110	21	.001	0.2	106
July	18	—.002	0.6	100	20	.007	0.6	92
Aug.	16	—.003	0.4	96	21	—.003	0.4	80
Sept.	18	—.003	0.1	97	21	.000	0.2	95
Oct.	17	.012	0.8	84	24	—.012	—0.3	107
Nov.	13	.015	—0.5	106	23	—.013	0.4	104
Dec.	12	.001	—0.6	105	22	.001	0.7	109
Year	188	—.001	0.2	98	258	—.003	0.1	103
				Boston				
Jan.	14	.014	—1.2	110	22	—.014	—1.1	102
Feb.	18	—.013	—0.9	103	19	—.023	—0.5	90
Mar.	19	—.020	—0.8	79	20	.010	—0.2	109
Apr.	17	—.018	—0.8	98	24	.003	0.1	112
May	12	—.026	—0.5	107	20	.012	0.3	85
June	14	.028	—0.2	119	21	—.013	0.2	92
July	18	.002	—0.1	95	20	.007	0.4	108
Aug.	16	—.002	0.1	109	21	—.007	0.1	106
Sept.	18	—.007	0.3	128	21	.008	—0.3	89
Oct.	17	.011	—0.9	97	24	—.012	—0.4	98
Nov.	13	.014	—1.6	79	23	—.021	0.0	106
Dec.	12	—.004	—2.2	105	22	—.002	—0.3	105
Year	188	—.002	—0.7	102	257	—.004	—0.1	100

were computed for each of the twelve months of the year, at a few widely scattered stations in North America. The results are given for high and low sun-spot numbers in table 7.

The means in table 7 show clearly that there is an annual period, and a semi-annual period, in the relation of sun spots to weather. Throughout the continental part of Canada and the United States the greatest plus departures of pressure, and the greatest minus departures of temperature, occur in December or January with high sun-spot numbers, and there is a tendency toward the opposite departures in summer. There is, however, evidently a semi-annual period combined with the annual in which the highest pressures and lowest temperature tend to occur in December-January and June-July, and the opposite about March and September. With low sun-spot numbers the trend is in the opposite direction, but is not so marked. These periods could be brought out more clearly by harmonic analysis or by numerical smoothing.

5. SUMMARY OF PRECEDING RESULTS

The results of these studies indicate that there is a real relation between weather conditions and the monthly means of solar radiation and monthly sun-spot numbers, but in the average the amounts of the changes in pressure, temperature, and precipitation are not large. Either there are large disturbing causes, or, as seems more probable, the phase of the effect is not constant at any one place, being sometimes positive and sometimes negative according to some law not yet fully disclosed.

6. THE ELEVEN YEAR SUN-SPOT PERIOD AND OTHER PERIODS IN WEATHER PHENOMENA

To anyone who examines the meteorological records, it is evident that there is no sharply defined eleven-year period in the weather elements in any part of the world. Sir Gilbert Walker computed correlation coefficients between the annual pressures and temperatures in various parts of the world and the annual sun-spot numbers. He found a systematic distribution of the positive and negative coefficients over certain areas, but no high correlations.

Plots show that the weather elements are much more variable than the sun spots. There is a two to four-year oscillation in the weather, which is not evident in the sun-spot curve. In order to compare the two, it is necessary to eliminate the short period oscillations, just as it is necessary to eliminate the oscillations due to ordinary waves on the surface of the ocean, in order to study the tidal oscillation due to the moon. This elimination is effected in a certain type of tide-recording machine by means of a small opening which does not permit the water to enter and leave fast enough to record the rapid fluctuations, but responds to the slow rising and falling of the tides. An analogous result may be obtained by the numerical process of smoothing recorded observations, a method which is frequently used in meteorological research.

It is difficult to determine in advance the amount of smoothing of the annual meteorological means needed in order to compare them with annual sun-spot changes; but, from a study of the plotted curves, I decided that three-year means would eliminate the most striking of the short-period oscillations. Accordingly, I computed overlapping three-year means of pressure for a large number of stations scattered over the world, and computed correlation coefficients between the mean pressures and the annual sun-spot numbers for the length of time covered by the records in each case. Coefficients were computed for the same year, and for one, two, and three years following the sun-spot observations, in order to ascertain whether any lag occurred in the relation with the meteorological changes.

These computations were begun a number of years ago when I was in Buenos Aires and I was materially assisted by Mr. Nils Hessling in their preparation. The results are given in table 8. The first two columns give the position of the stations, and the third the number of years of observation. It is seen from this table that at many of the stations the highest correlation coefficients were found for the same year as the sun-spot observations, while at others there was a lag of one year; but at no station was there an indication of a lag greater than one year. The results for the year of the sun-spot observations are plotted in figure 6, in which lines are drawn indicating areas of 0.50 or more correlation, and of zero correlation. The continuous lines inclose areas of positive correlation, and broken lines inclose areas of negative correlation.

It is seen that in the equatorial belt there is a large area where the negative correlations exceed −0.50, extending from western Brazil across Africa and the Indian Ocean and out into the Pacific. Within this area there are found negative correlations of −0.59 at Cuyaba, −0.59 at Quixeramobim, of −0.61 at Recife, Brazil, of −0.54 at Zanzibar, Africa, of −0.54 at Bombay, and −0.88 at Singapore. The high positive correlations are found within the areas of normal high pressure in temperate latitudes; with coefficients of 0.60 at Sydney,

FIG. 6.—Correlation between sun spots and 3-year means of pressure.

TABLE 8.—*Coefficients of correlation between sun spots and atmospheric pressure (means of 3 consecutive years)*

Places	Lat.	Long.	No. of years	Years following obs. of sun spots 0	1	2
North America:						
Victoria, Canada	48° 24′ N.	123° 19′ W.	18	—0.11	+0.11	.....
Winnipeg, Canada	49° 53′ N.	97° 7′ W.	41	+0.22	+0.18	—0.04
Toronto, Canada	43° 29′ N.	79° 23′ W.	69	+0.03	0.00	.....
Montreal, Canada	45° 30′ N.	73° 35′ W.	42	+0.31	+0.20	+0.11
Sydney, Canada	46° 10′ N.	60° 10′ W.	32	+0.60	+0.53	.....
Helena, U. S. A.	46° 34′ N.	112° 4′ W.	36	—0.16	—0.13	—0.03
Duluth, U. S. A.	46° 47′ N.	92° 6′ W.	45	—0.02	—0.30	—0.30
Denver, U. S. A.	39° 45′ N.	105° 0′ W.	38	—0.17	—0.06	+0.06
Washington, U. S. A.	38° 54′ N.	77° 3′ W.	38	—0.38	+0.20	.....
Nashville, U. S. A.	36° 10′ N.	86° 47′ W.	38	+0.10	—0.02	—0.10
San Diego, U. S. A.	32° 43′ N.	117° 10′ W.	38	+0.10	+0.21	+0.33
Galveston, U. S. A.	29° 18′ N.	94° 50′ W.	38	+0.41	+0.51	+0.49
Mobile, U. S. A.	30° 41′ N.	88° 2′ W.	45	+0.02	+0.15	+0.30
Key West, U. S. A.	24° 34′ N.	81° 49′ W.	38	+0.35	+0.43	+0.44
Mexico, Mex.	19° 26′ N.	99° 8′ W.	32	+0.25	+0.13	.....
North Atlantic:						
Jacobshaven	69° 13′ N.	51° 2′ W.	45	+0.24	+0.18	.....
Godthaab	64° 11′ N.	51° 46′ W.	33	+0.35	+0.21	—0.03
Beruford	64° 40′ N.	14° 19′ W.	43	—0.15	—0.10	—0.12
Stykisholm	65° 5′ N.	22° 46′ W.	61	—0.12	—0.19	—0.24
Bermuda	32° 18′ N.	64° 47′ W.	23	—0.09	—0.06	.....
Ponta Delgada	37° 45′ N.	25° 41′ W.	39	+0.19	+0.25	+0.13
Jamaica	18° 6′ N.	76° 42′ W.	17	—0.28	—0.19	.....
Barbadoes	13° 8′ N.	59° 40′ W.	21	+0.33	+0.40	+0.28
Europe:						
Vardö	70° 22′ N.	31° 8′ E.	45	+0.04	+0.30	+0.43
Tromso	69° 39′ N.	18° 58′ E.	45	+0.14	+0.09	+0.17
Archangel	64° 33′ N.	40° 32′ E.	32	—0.09	+0.02	+0.19
Petrograd	59° 56′ N.	30° 16′ E.	56	+0.23	+0.37	+0.31
Moscow	55° 46′ N.	37° 40′ E.	36	+0.02	+0.07	+0.27
Valencia	51° 56′ N.	10° 15′ W.	39	—0.22	—0.30	—0.22
Greenwich	51° 29′ N.	0	56	—0.09	—0.27	—0.25
Lisbon	38° 43′ N.	9° 9′ W.	54	—0.23	—0.22	...
Perpignan	42° 42′ N.	2° 53′ E.	51	—0.21	—0.23	—0.19
Basel	47° 33′ N.	7° 35′ E.	60	+0.16	—0.10	.....
Vienna	48° 15′ N.	16° 21′ E.	58	+0.13	—0.06	.....
Athens	37° 58′ N.	23° 43′ E.	46	+0.10	+0.24	+0.12
Asia:						
Ekaterinburg	56° 50′ N.	60° 38′ E.	53	+0.34	+0.28	.....
Barnaul	53° 20′ N.	83° 47′ E.	53	—0.06	—0.11	—0.20
Yeniseisk	58° 27′ N.	92° 6′ E.	35	+0.51	+0.49	+0.19
Nertchinsk	51° 19′ N.	119° 37′ E.	50	+0.05	+0.17	.....
Scutari	40° 0′ N.	29° 3′ E.	40	+0.31	+0.28	+0.31
Taschenkut	45° 21′ N.	32° 29′ E.	35	+0.26	+0.15	—0.01
Tokio	35° 41′ N.	139° 45′ E.	37	—0.02	+0.01	.....
Nagasaki	32° 44′ N.	129° 51′ E.	28	+0.32	+0.14	—0.11
Bushire	28° 59′ N.	50° 49′ E.	30	—0.25	—0.23	.....
Zi-Ka-Wei	31° 12′ N.	119° 6′ E.	35	+0.04	—0.03	.....
Hong Kong	22° 18′ N.	114° 11′ E.	26	0.00	—0.06	.....
Agra	27° 10′ N.	78° 5′ E.	37	0.00	—0.03	—0.11
Aden	12° 45′ N.	45° 3′ E.	27	—0.18	—0.11	.....
Bombay	18° 54′ N.	72° 49′ E.	65	—0.54	—0.61	—0.52

TABLE 8.—*Coefficients of correlation between sun spots and atmospheric pressure (means of 3 consecutive years)* (continued)

Places	Lat.	Long.	No. of years	Years following obs. of sun spots 0	1	2
Asia (continued):						
Calcutta	22° 32′ N.	88° 20′ E.	57	—0.41	—0.48	—0.43
Madras	13° 4′ N.	80° 14′ E.	70	—0.45	—0.47	—0.34
Rangoon	16° 46′ N.	96° 12′ E.	36	—0.27	—0.16	.....
Colombo	6° 56′ N.	79° 52′ E.	42	—0.51	—0.63	—0.58
Batavia	6° 11′ S.	106° 50′ E.	43	—0.49	—0.56	—0.46
North Pacific:						
Honolulu	21° 18′ N.	157° 50′ W.	33	+0.11	+0.35	0.23
South America:						
Quixeramobim	5° 16′ S.	39° 15′ W.	23	—0.30	—0.59	—0.59
Recife	8° 4′ S.	34° 52′ W.	29	—0.61	—0.56	—0.36
Bahia	12° 54′ S.	38° 24′ W.	17	—0.14	—0.40	.....
Cuyabá	15° 36′ S.	56° 6′ W.	17	—0.59	—0.52	—0.35
Rio de Janeiro	22° 54′ S.	43° 10′ W.	57	—0.29	—0.16	.....
Santiago	33° 27′ S.	70° 41′ W.	51	+0.52	+0.60	+0.52
Córdoba	31° 25′ S.	64° 12′ W.	42	+0.26	+0.40	+0.39
Buenos Aires	34° 36′ S.	58° 22′ W.	42	+0.26	+0.47	+0.48
Punta Arenas	53° 10′ S.	70° 54′ W.	31	+0.12	+0.03	+0.03
South Atlantic:						
St. Helena	15° 55′ S.	5° 43′ W.	23	—0.18	—0.25	—0.17
South Georgia	54° 14′ S.	36° 33′ W.	15	+0.83	+0.58	+0.17
Orcadas	60° 42′ S.	44° 42′ W.	18	+0.43	+0.39	+0.33
Africa:						
Abassia	30° 5′ N.	31° 17′ E.	40	—0.21	—0.15	—0.10
Zanzibar	6° 10′ S.	39° 11′ E.	30	—0.54	—0.37	—0.05
Durban	29° 51′ S.	30° 30′ E.	32	—0.17	—0.10	.....
Cape Town	33° 56′ S.	18° 29′ E.	61	—0.60	—0.67	—0.52
Indian Ocean:						
Singapore	1° 15′ N.	103° 51′ E.	34	—0.88	—0.77	—0.43
Mauritius	20° 6′ S.	57° 53′ E.	37	—0.27	—0.43	—0.41
Australia:						
Port Darwin	12° 28′ S.	130° 51′ E.	30	—0.39	—0.45	.....
Carnarvon	24° 54′ S.	113° 39′ E.	26	—0.17	—0.39	—0.52
Perth	31° 57′ S.	115° 52′ E.	27	—0.30	—0.19	.....
Albany	35° 2′ S.	117° 52′ E.	33	—0.13	—0.19	—0.17
Adelaide	34° 57′ S.	138° 35′ E.	55	—0.53	—0.58	—0.45
Sydney (N. S. W.)	33° 52′ S.	151° 12′ E.	53	—0.10	—0.17	—0.21
Hobart	42° 53′ S.	147° 20′ E.	29	—0.17	—0.15	—0.14
South Pacific:						
Apia	13° 48′ S.	171° 46′ W.	36	+0.16	+0.22	+0.21

Canada, of 0.51 at Yeniseisk, Russia, of 0.52 at Santiago, Chile, and of 0.83 at the South Georgias; but the observations at the latter station cover only a short period, and the coefficient will probably be lower for a longer interval. These correlations are large enough to be significant, and indicate that the eleven-year period is of sufficient importance to be considered in the long-period changes in certain regions. There are areas of negative correlation in the North Pacific, in the region of the Great Lakes, and in the North Atlantic near Iceland; but the correlation coefficients are not high.

If the annual means of pressure are examined at stations near the same latitude north and south of the equator, similarly situated in relation to the belts of positive and negative correlations outlined in

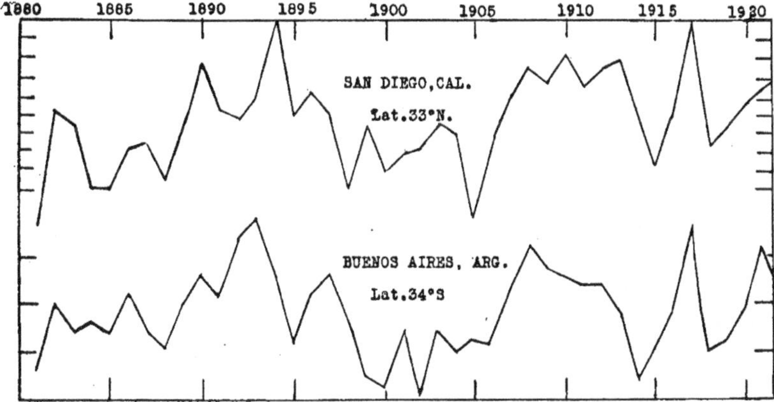

FIG. 7.—Mean annual atmospheric pressure.

figure 6, it is found that the annual pressure changes show a striking similarity. This similarity is illustrated in figure 7 by a comparison of the annual means of pressure at Buenos Aires and San Diego. The similarity of the pressure changes at these widely separated stations, in opposite hemispheres, is evidence that the pressure changes are controlled by world-wide conditions, and not by local causes.

In the United States and Canada, the correlation of the sun spots with the three-year means of pressure is not high, and in order to study in what way the long-period changes in this part of the earth were related to solar changes, the three-year means of pressures, for a large number of stations, were plotted and compared with the sun-spot curve.

Figure 8 shows a plot of the three-year means of pressure at Chicago and St. Louis, and also a plot of the three-year means of summer rainfall at Cordoba, Argentina, which shows that after 1887, at least, oscillations of the same nature were taking place in both hemispheres. Preceding 1887, the Cordoba oscillation was inverted to the northern one.

At Chicago there were maxima of pressure in 1889, in 1900, in 1912, and in 1921, which approximated to an eleven-year period inverted to the sun-spot curve. But there are also other maxima showing a combination of the eleven-year period with oscillations of another order. These secondary maxima come out more strikingly at St. Louis. By referring to figure 6 it is seen that St. Louis is near a line of zero

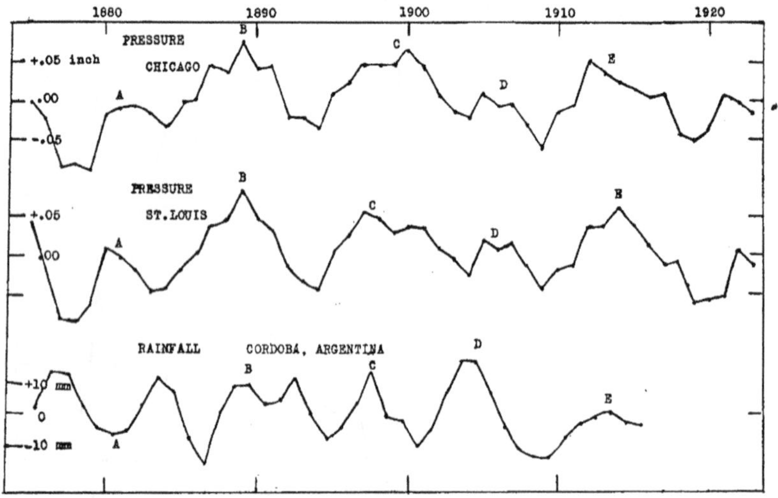

FIG. 8.—Three-year means of pressure and rainfall.

correlation with sun spots. The maxima of pressure at St. Louis are nearly equally spaced, and the time interval between the maxima appears to be about one-third of a double sun-spot period, or of an interval of 22.5 years, which Hale now inclines to believe is the real sun-spot period. Maxima of pressure at St. Louis coincided with minima of sun spots in 1889 and in 1913; but, instead of there being one intervening maximum, there were two maxima. These facts lead to a consideration of oscillations of pressure in the atmosphere which are harmonics of the sun-spot period.

When annual means of pressure are plotted, they show that the eight-year period tends to divide into two periods of about four years, of unequal strength, so that alternate maxima are higher. Sir Norman Lockyer was one of the first to call attention to this period of about four years, suggesting that it was a fraction of the sun-spot period, and was connected with similar changes in the amount of the prominences on the sun. F. H. Bigelow arrived at a similar conclusion in investigating the weather changes of the United States, and called attention to other periods which were fractions of the sun-spot period. Sir Napier Shaw in his book on "The Air and its Ways," p. 176. shows that the yield of wheat for England may be represented from 1885 to 1905 with remarkable fidelity by a combination of six harmonic terms of an eleven-year period. Dinsmore Alter has recently made an extensive study of periodicity in various parts of the world, and arrives at the conclusion that most of the periodic terms are harmonics of the sun-spot period, which he puts at 22.5 years.

Evidence that there are harmonic oscillations of weather in short periods was given by me in the American Journal of Science, March, 1894, and the Meteorologische Zeitschrift, 1895, p. 22. Recently Otto Myrbach has accumulated data bearing on the same point. (Ann. d. Hydr. u. s. w., 1926, Vol. IV.) The researches of Clough lead to somewhat similar conclusions.

This is a subject demanding further research in order to explain how these periodic oscillations arise, why they vary in intensity from time to time, and to determine whether they are related to solar changes of the same kind. It is not yet certain that the eight-year period, for example, is simply one-third of a 22.5-year period, or an harmonic of a much longer period, for there appear to be periodic oscillations of about 2 years, 4 years, 8 years, 16 to 18 years, and 33 to 35 years, which may be parts of one series.

The period of about two years was very marked in the United States during the years 1874 to 1881, when I made an investigation of it. The oscillations are shown in figure 9, reproduced from the American Meteorological Journal of August, 1884. The continuous curves in this diagram were plotted from the progressive averages of successive twelve monthly means of pressure, at several stations in the United States. The curves show an oscillation slightly longer than two years, with a long period swing indicated by the dotted curves.

The departures of the means of 12 months from the means of two years at these four stations, together with those from eight other stations treated in the same manner, furnished the data for the charts in figure 10. The lines in the charts show the departures at the time of the minimum of the period in the central United States (see plot for St. Paul, in figure 9). The broken lines show values below normal for each .010 inch, and the continuous lines show values above normal of the same amount. The charts show that the center of the greatest minus departures which was near Chicago in December, 1875, had moved westward to North Dakota in March, 1880, and the

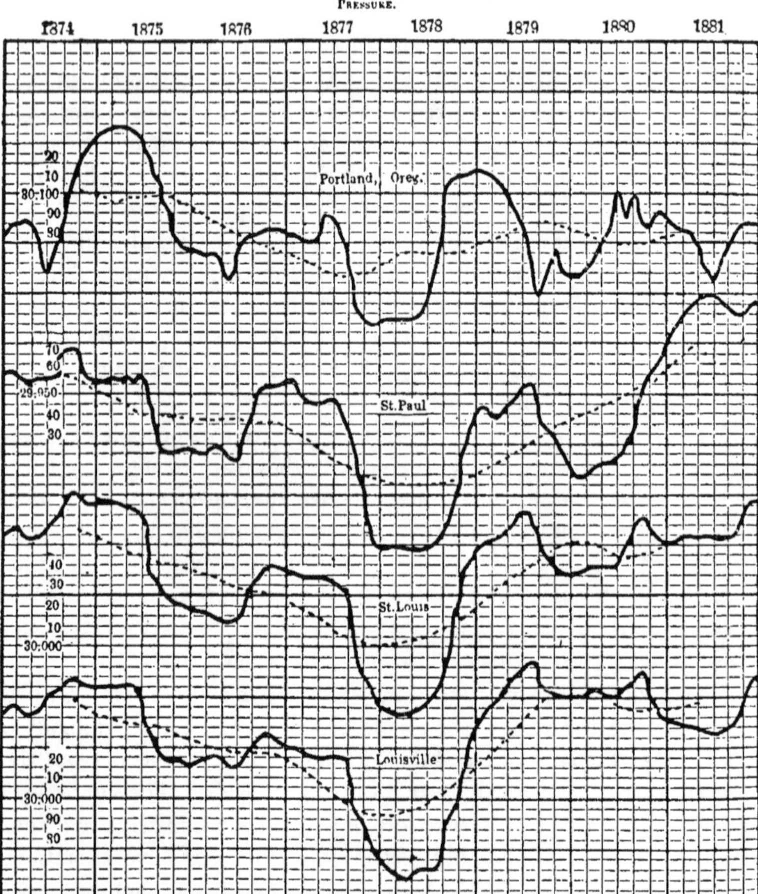

Fig. 9.—Twelve months means of pressure showing oscillations of pressure of slightly over two years duration. (See plot for St. Paul.)

phase of the period had inverted at eastern stations. This is an important fact; for it becomes evident that in investigating these periodic oscillations, one must consider the progressive motion of these centers, and study not merely single stations where the phase of the period is likely to invert, but must deal with a network of stations covering a large area, the whole world if possible.

7. FORECASTS OF NEW YORK TEMPERATURE FOR FIVE DAYS IN ADVANCE

The forecasts of temperature for New York for three, four, and five days in advance were continued during 1925 up to December 1, which thus completed two full years.[1] These forecasts were based on observed solar conditions, in combination with the temperatures observed at two or three stations in the United States.

The forecasts for five days in advance were selected for verification, because, in my opinion, it is impossible to forecast successfully daily temperatures so far in advance, without the aid of solar conditions. The correlation of the daily departures from normal temperature at New York, with similar departures at western stations in the United States, five days earlier, give correlation coefficients of practically zero, as determined from observations covering several months.

The verifications were made as in the preceding year by means of averages. As agreed on in advance with Dr. Abbot, predictions of five degrees above normal were to be considered forecasts of high temperature, those between +4 and −4 were to be considered normal, and those below minus five degrees were to be considered forecasts of low temperature. The forecasted temperatures for five days in advance, during the year ending December 1, 1925, were divided into these three classes, and the average departures of the maximum temperatures from normal on the days for which the forecasts were made are as follows:

TEMPERATURE

Forecasted	No. of cases	Mean observed
High	59	+2°1
Normal	188	+0.5
Low	83	—0.1
Difference		2.2

The difference between the mean temperature following forecasts of high temperature and that following forecasts of low temperature is 2°2 F. in the right direction, and with the mean observed value for normal predictions standing intermediate. The magnitude of this difference is, I think, a measure of success. If the forecasts had been without any basis, this difference would have been near zero; if perfectly successful, it would have been nearly four times as large.

[1] See Smithsonian Misc. Coll., Vol. 77, No. 6, 1925, pp. 54-59.

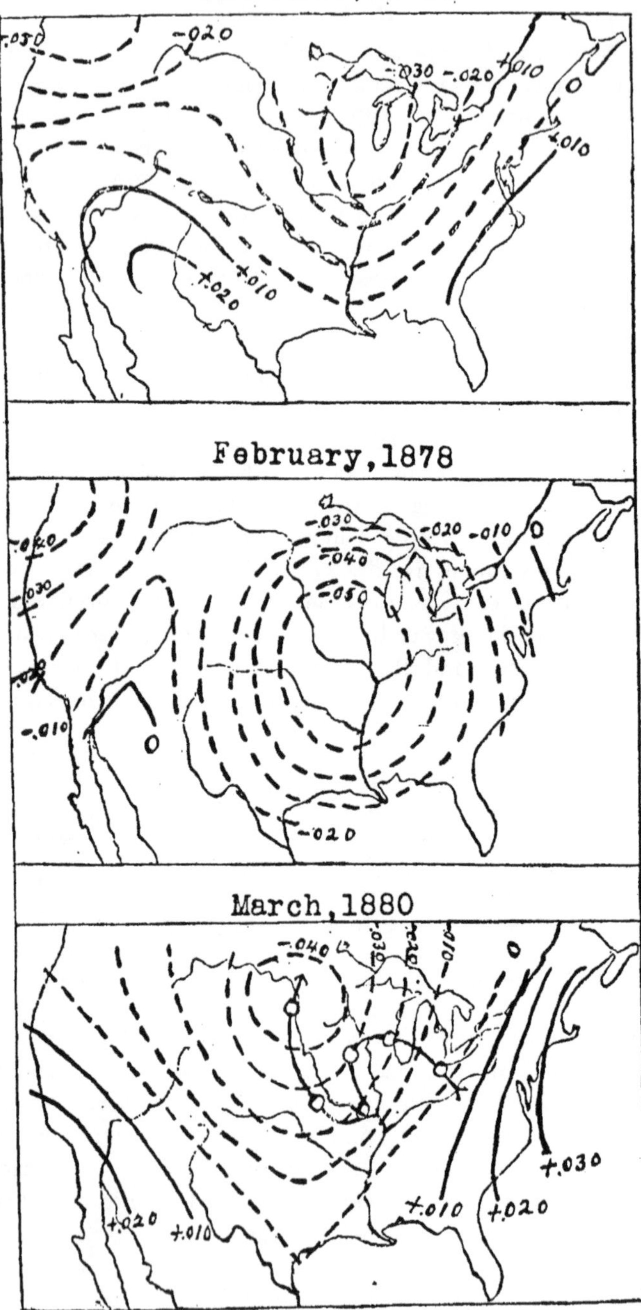

FIG. 10.—Centers of greatest minus departure in period of slightly over two years, showing movement of the centers of oscillation.

The forecasted temperatures for five days in advance, and the observed temperatures from July 10 to September 3, are shown by means of plots in figure 11. These curves fairly indicate, I think, the character of the successes and failures. In some cases the observed maximum or minimum of temperature occurred a day later, or a day earlier, than predicted, and in one or two cases the expected rise or fall of temperature did not occur; but in most cases there was a peak or depression of temperature at or very near the times forecasted. The breaks in the dotted curve representing the forecasts were due to Sundays when no forecasts were made.

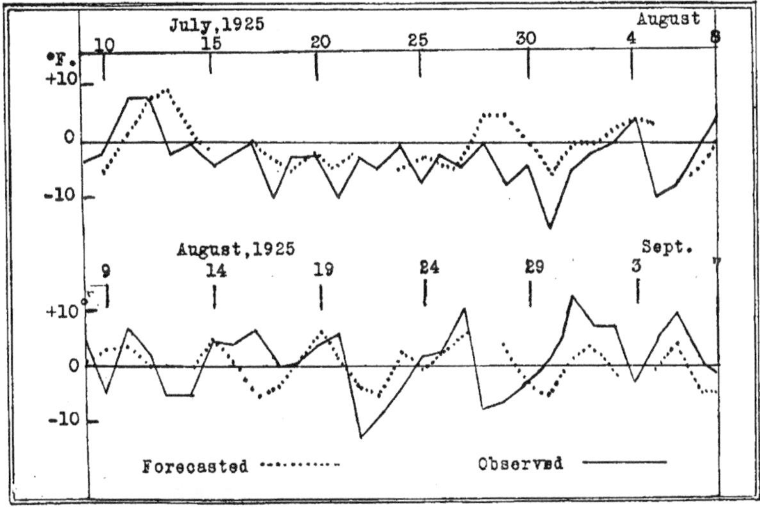

FIG. 11.—The temperature at New York as forecasted five days in advance and the observed values.

8. REPLY TO CRITICISM

In preceding papers of this series a large amount of evidence was presented to show that there are systematic and opposing variations in the weather conditions in different parts of the world, correlated with variations in solar radiation, as measured by the Astrophysical Observatories of the Smithsonian Institution. Recently, however, Prof. C. F. Marvin, Dr. H. H. Kimball, and Mr. H. W. Clough have raised the question whether the apparent solar variation may not be due largely, if not entirely, to errors of observation, (Monthly Weather Review, 1925, Vol. 53, pp. 285, 303 and 343.) Fortunately, the maintenance of two observing stations by the Smithsonian Institution permits a determination of the relative values of the variability, as compared with the probable errors of observation, to be made with great accuracy, provided that the two stations are independent of each other.

(*a*) *Tests of the reality of solar variation based on numerical analysis.*—In comparing solar and meteorological data, my first work dealt with observed values of solar radiation from Mt. Wilson, California, later with those from Calama, Chile, and finally with the combined observations from Montezuma, Chile, and Harqua Hala, Arizona. In order to compare the observations at Calama and Mount Wilson, I arranged the observations at Calama in a series of steps separated by 0.010 calorie, as shown in table 9, and for each class at Calama counted the frequency with which simultaneous values occurred in different classes at Mount Wilson.

TABLE 9.—*Comparison of Solar Radiation Measurements at Calama, Chile, and Mount Wilson, Calif., Years 1918-1920*

Values at Calama	1.920-9	1.930-9	1.940-9	1.950-9	1.960-9
Simultaneous values at Mount Wilson:					
1.900-9	1	1	0	0	0
1.910-9	**4**	1	1	2	0
1.920-9	**3**	**3**	1	2	1
1.930-9	0	**4**	**8**	3	0
1.940-9	1	1	**5**	**8**	1
1.950-9	1	4	2	**7**	**3**
1.960-9	1	0	1	6	**3**
1.970-9	1	0	1	0	1

If there were no relation between the measurements at the two stations the observations would be scattered through the different classes at random. The tabulation shows that this is not the case, but that for each group of observations at Calama there is a maximum occurrence near the same value at Mount Wilson, and a progressive displacement of the maximum frequency, as the solar radiation values increase from 1.920-9 to 1.960-9. The most natural conclusion is that the observers were measuring the same phenomenon, and that this phenomenon showed a range from grade 1.920-9 to 1.960-9, or more than two per cent of the mean solar radiation value. There was no marked secular change during this interval, so that the whole of this variation is attributable to short-period changes. The fact that the maxima tended to come at a slightly lower level at Mt. Wilson shows that there was some constant difference in level between the two, which may well have been due to a difference in the calibration of the instruments, or other similar cause.

The scatter of the observations on each side of the maximum frequency is a measure of the errors of observation. In order to determine the probable error of the observations, I obtained all the differences between the pairs of simultaneous observations, 110 in all, and found that they were distributed as shown in table 10.

TABLE 10.—*Distribution of the Differences in Solar Radiation Values Observed Simultaneously at Calama and Mt. Wilson*

Mean difference, thousandths of a calorie	−70	−60	−50	−40	−30	−20	−10	0	10	20	30	40	50	60	70
Frequencies ...	1	1	0	3	4	13	23	28	14	6	4	4	6	0	3

In counting the number of observations for −10, for example, all the observations between −6 and −14 were used; for zero, all the observations between −5 and +4 were included; and for +10 all

TABLE 11.—*Comparison of Solar Radiation Measurements at Montezuma and Harqua Hala, Years 1920-1924*

Values at Montezuma	1.890-1	1.900-9	1.910-9	1.920-9	1.930-9	1.940-9	1.950-9	1.960-9
Values at Harqua Hala:								
1.870-9	1	..	..	1	1	..	..	..
1.880-9	..	..	1	1	1	..	..	..
1.890-9	**3**	2	1	6	..	..	..	..
1.900-9	..	3	11	10	..	..	..	..
1.910-9	1	5	**20**	21	10	4	2	..
1.920-9	1	**10**	18	**35**	18	4	2	1
1.930-9	1	4	7	23	**25**	3	8	..
1.940-9	..	..	4	5	11	**10**	**8**	3
1.950-9	1	..	..	3	6	8	6	**3**
1.960-9	..	..	..	..	3	3	1	2
1.970-9	..	..	..	..	..	..	3	1

observations between +5 and +14 were taken. As the distribution of these numbers evidently follows the normal law of distribution of errors of observation, they were reduced to percentages, and a curve of best fit was drawn through them. From this curve the probable error of the differences is found to be ±0.0121 calorie. Since this value is made up by the combined errors of observation at Mt. Wilson and Calama, the probable error at one station is $\frac{0.0121}{\sqrt{2}}$, which gives a value of ±0.0086 for the observations at one station, assuming the errors at the two stations to be equal. Or if we assume, as is probable, that they were somewhat larger at Mt. Wilson, we may take the probable error there as ±0.010, and at Calama, 0.007. The probable error obtained in the usual way from the mean square of the differences also gives ±0.009 as the probable error at one station.

Turning to the more recent measurements at Montezuma, in northern Chile, and Harqua Hala, in Arizona, for the interval from October, 1920, to November, 1924, table 11 gives for each class of observations at Montezuma the frequency of occurrence of different values at Harqua Hala.

In forming this table all "unsatisfactory" values were discarded except where they were marked *U*+.

When observations were made in one grade at Montezuma there was a maximum frequency in exactly the same grade at Harqua Hala from 1.890 to 1.970, excepting in grades 1.900-9, 1.950-9, and 1.960-9, where there were only slight displacements. There seems but one explanation of this fact, namely, that the two observers were measuring changes in solar radiation, which progressed from 1.890 to 1.970. This difference is equivalent to a change of four per cent. The scatter of the observations indicates errors of measurement. The number of observations in each grade between 1.910 and 1.950 is sufficiently great, so that they can be converted into percentages, and normal curves of error drawn through them.

These curves are shown in figure 12. From these plots the probable error of the measurements in each grade was determined.

The results agree very closely in each grade in giving a probable error of approximately 0.0085 calorie. This is the combined errors of the measurements at both stations, and needs to be divided by $\sqrt{2}$ to give the probable error at each individual station, which is thus found to be 0.006. This value agrees very closely with Dr. Abbot's value of 0.0065 found from the whole mass of observations.

In my paper in the Smithsonian Miscellaneous Collections, Vol. 77, No. 6, p. 2, it is shown that the observed probable solar variability from July, 1918, to September, 1922, was ± 0.011. That is, there were as many deviations from the median value exceeding that amount as there were below it. But the observed probable solar variability is determined by the combined effect of the true probable solar variability and the probable errors of observation. Having obtained the probable error of the observations, as shown above, I think that we are in a position to compute the true probable solar variability. Let *tv* represent the true probable solar variability, then since ± 0.011 is the observed probable solar variability, and ± 0.006 the probable error of the observations, we have:

$$(tv)^2 + (0.006)^2 = (0.011)^2$$

$$tv = 0.0092$$

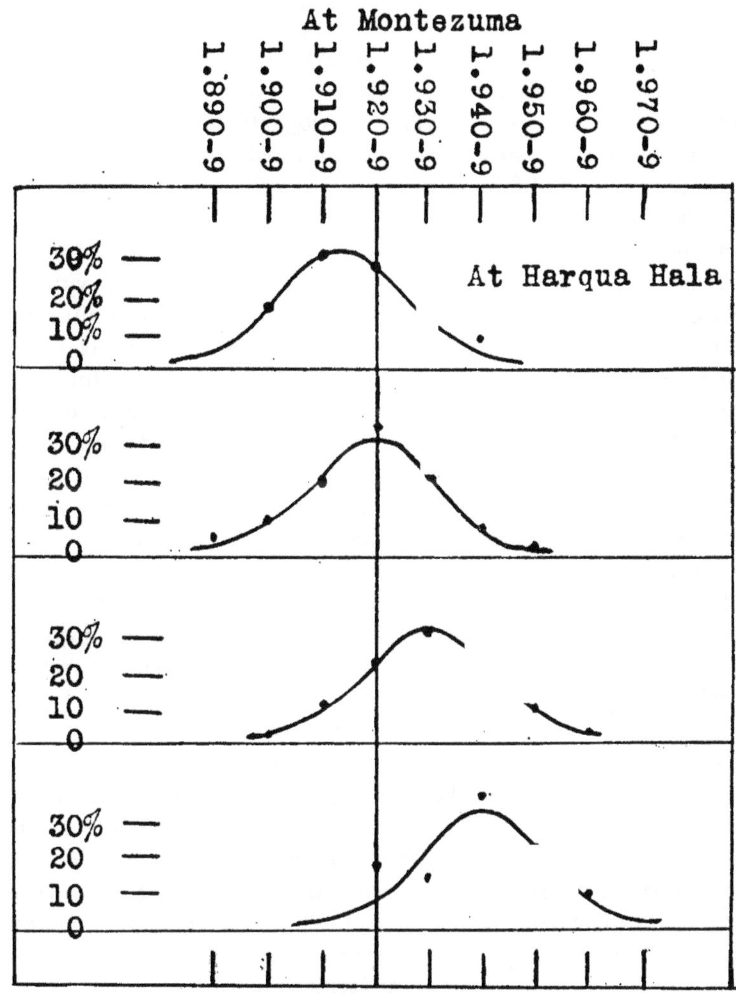

Fig. 12.—Frequency of occurrence of different values of solar radiation at Harqua Hala corresponding to simultaneous observations at Montezuma.

The arithmetical mean variability of true solar observations, unaffected by accidental error, for the interval July, 1918, to September, 1922, would therefore be:

$$\frac{0.0092}{0.845} = 0.0109 \text{ calorie}$$

(*b*) *Other confirmations of the probable reality of the solar radiation variations reported by the Smithsonian Institution.*—In addition to the comparison of observations at two stations, there are the following evidences of solar variation furnished by various classes of researches and by different types of workers. From measurements with the bolometer, Dr. Abbot has found that when the solar radiation increases, the ratio between the intensity of the short-wave radiation and the long-wave radiation increases. This is in accord with the well-known fact that when a body increases in temperature the proportion of short-wave radiation increases, so that the body becomes first red, then yellow, and finally blue, as the temperature continues to rise. Recently Dr. Pettit of Mt. Wilson, by spectroscopic means has measured the relative intensity of solar radiation in the green and ultra-violet. This ratio shows a wide variability of something like 80 per cent, which he has correlated with changes in the mean monthly values of solar radiation, finding a high correlation between the two.[1]

Dr. L. A. Bauer has found a close relation between the mean annual interdiurnal variability of solar radiation and certain magnetic effects, which for the years 1919 to 1924 give a correlation reaching 0.97.[2]

I found in an average of 200 cases that there is a sharp maximum of solar radiation coinciding with the times of maximum of faculæ on the sun, as shown by the published observations of the Greenwich Observatory. For the months of April to September of the years 1918-1921 there were 121 cases, and the mean maximum of solar radiation varied from the mean value of preceding and following days to the extent of nine times the probable error of the mean.

I found also that there was a marked depression of solar radiation when sun spots and their attendant faculæ crossed the central area of the sun. In this case the depression of the mean solar radiation, below the mean of the values obtained when the spots were near the limb of the sun, was seven times the probable error of the means. These results agree with preliminary ones found by Dr. Abbot.

[1] Pub. Ast. Soc. Pacific, February, 1926, Vol. XXXVIII, No. 221, p. 21.

[2] Terrestrial Magnetism, December, 1925, Vol. XXX, No. 4, p. 205.

From March to May, 1920, Mr. F. E. Fowle found a high correlation between the flocculi crossing the central disc of the sun and simultaneous solar radiation values.

From results on days of nearly equal atmospheric conditions, Dr. Abbot has found that pyrheliometric observations alone confirm closely the variations in solar-constant values, and show close correlation with sun-spot numbers.[1]

Other evidence might be cited, but those given seem sufficient to prove the reality of solar variability.

(*c*) *Solar variability and weather: The reality of their correlation.*—Granted solar variability, the question arises, are these variations correlated with terrestrial weather conditions more closely than could be explained by chance coincidence?

I used the observations at Mt. Wilson for a study of the correlation between solar radiation and pressure and temperature in Argentina. While Mt. Wilson values are less accurate than later ones, they are, as Dr. Abbot has said, useful in the form of means of many days.[2] In one comparison, I took all of the highest values of solar radiation between the years 1909 and 1918, over 50 in number, and determined the average values of solar radiation for each of the 30 days following and for the five days preceding. Thus I formed a table of 36 columns having as many lines as high values. But owing to failures to observe on some days, all the columns contained gaps excepting the column for zero day. Thus the number of cases varied somewhat, but averaged about 35. I then obtained in a similar way averages of the temperatures for each of the corresponding days at Buenos Aires. After allowing an interval of three days for a lag in the effect, the mean temperature march showed a correlation of 0.66 with the mean march of solar radiation over the 36-day interval.

Exclusive of zero day, the mean values of solar radiation over the 36-day interval ranged from 1.930 to 1.952. As determined above, the probable error of a Mt. Wilson observation is 0.010 calorie. Hence, if there had been no real solar change, the probable variation of the mean of 35 values would have been $\frac{0.010}{\sqrt{35}}=0.0017$. The observed range is hence more than 12 times the probable error of any of the 36 individual means.

For the year 1916 I correlated 10-day means of solar radiation with 10-day means of temperature at various stations in Argentina and

[1] Monthly Weath. Rev., May, 1926.

[2] Smithsonian Misc. Coll., 1925, Vol. 77, No. 5, p. 3.

obtained correlations exceeding -0.80 (in one case, -0.82 ± 12) between the 10-day mean temperature and the 10-day mean solar-radiation values. The range of the mean solar-radiation values in this case is 0.032 gram calorie. Assuming an average of seven values for each 10-day mean, the probable error of such a mean is $\pm \frac{0.010}{\sqrt{7}} = 0.0038$. Here the observed range in mean values is about nine times their probable error.

These computations may incline my critic in Nature of November 20, 1925, to view more favorably the reality of the relations of the solar changes to meteorological changes, which were among the results of my former papers.

(*d*) *Revision of a former evidential result.*—In computing the values given in table 8 of my paper " Solar Radiation and Weather " (Smithsonian Miscellaneous Collections, Vol. 77, No. 6, p. 27), I used observed maximum temperatures, but Mr. R. H. Weightman called my attention to the fact that the data were not distributed equally among the months, and for that reason the influence of the annual period was not eliminated, and the resulting differences were too large. To correct for this difference in level, I have in each case obtained the average of all of the mean values for the 15 days from two days before to 12 days after the day of solar observation, and deducted this average from each of the mean values. The residuals are given in the lines marked *a*, in table 12.

In order to eliminate the influence of the annual period in another way I recomputed the means. For this purpose I used the maximum temperatures given in the daily weather maps of the United States Weather Bureau for the 12 hours between 8 a. m. and 8 p. m. each day, and from these obtained the departures from the normals of the days on which the observations were made. These daily normals were derived from the monthly normals by interpolation, taking the monthly normal as the mean temperature of the 16th day of the month, except in February when the 14th day was used. Using the daily departures from normal thus obtained means were obtained for the interval from two days before to 12 days after high and low solar values.

The results show that, even after eliminating the annual period in this manner, the mean temperatures during the entire period covered by the observations were lower with high solar radiation than with low. This difference in level I attribute to long-period changes, and it was corrected for in the same way as described previously, namely, by getting the average of each set of mean values for the 15 days covered by the observations, and deducting this average from each of the mean values. The results are given in the lines marked *b* in table 12.

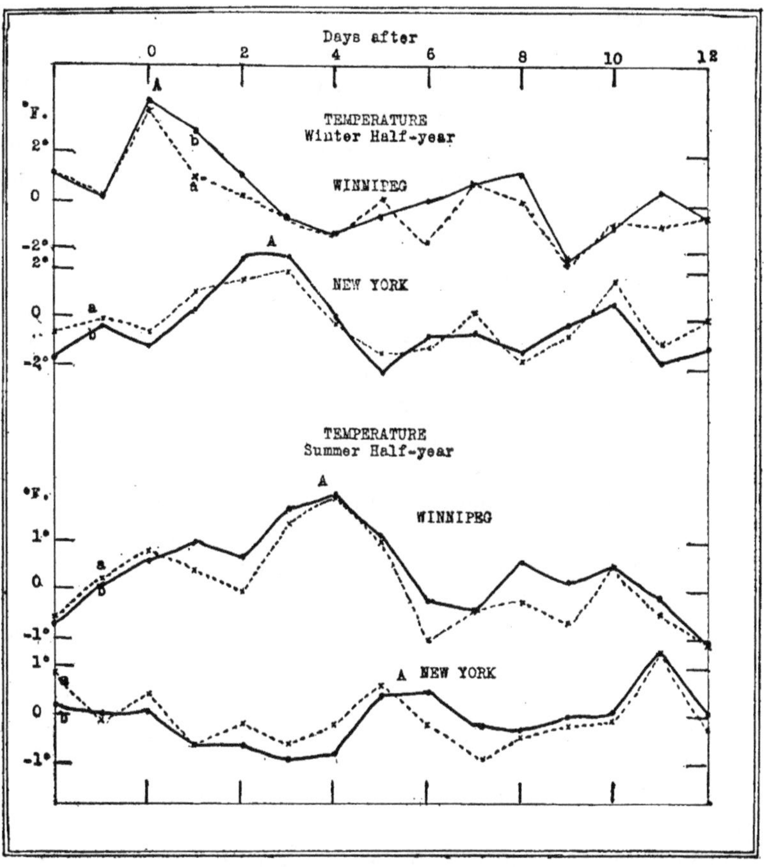

FIG. 13.—Mean differences between temperatures with high and with low solar radiation values, years 1918-1922.

The differences between the values accompanying high and low solar radiation are plotted in figure 13. The corrected differences by the first method marked *a* in table 12 are plotted with dotted lines. The means derived from the departures from normal temperature, marked *b* in table 12, are plotted with a continuous line. It is seen from the diagram that these two sets of mean values follow the same course. The minor differences arise largely, if not entirely, from a few changes in dates in the revised data. These curves also are of much the same form as those shown in figure 19 of Smithsonian Miscellaneous Collections, Vol. 77, No. 6. The difference is largely a difference in level, brought about by a complete elimination of long-period effects. The evidence of real weather changes depending on solar variation, and their lag as between different stations, remains unimpaired.

The maxima at New York occur later than at Winnipeg, and the maxima at both stations occur about three days later in summer than in winter. This lag between winter and summer probably results from a displacement of the centers of action. Allowing for the lag, and using the values *b* in table 12, the winter departures at Winnipeg show a correlation with the summer departures of 0.87 ± 0.07.

(*e*) *Do the solar variation and weather correlations have permanency?*—Another criticism of the results previously published is that investigations for successive periods of time were not made and compared. Such comparisons have, however, been made, but not hitherto published. Some of them are now given in tables 13 and 14.

Table 13 was computed several years ago, and gives a comparison of the mean temperatures at Buenos Aires following high values of solar radiation for two intervals, (1) for the years 1909 to 1918, and (2) for the years 1919 to 1920. They are for the winter half-year. Up to 1918 no solar radiation measurements were available for the summer half-year of the southern hemisphere. The results in table 13 show that the means of the departures of temperature for the two periods follow almost identical courses. The correlation between the two for the 13 days covered by the observations is 0.73 ± 0.09.

The work of Sr. Hoxmark and the researches of Sr. Julio Bustos Navarrete indicate that these influences of the solar radiation changes on the pressure and temperature of Chile and Argentina continue to the present time.

Table 14 was computed more recently, and shows for the winter half-year a comparison of the mean temperature at Winnipeg following high solar values for two intervals, (1) for the time July, 1918, to December, 1919, and (2) for the time January, 1920, to March, 1922. The values given in the table are departures from the average of the 11 days. The correlation between the two sets of values is high for the entire period of 11 days covered by the observations, but is highest for the interval 0 to 5 days, coinciding with and immediately following the maximum of solar radiation.

TABLE 12.—*Mean departures of the daily maximum temperatures for the interval from two days before to twelve days after high and low values of solar radiation*

		Days before			Days after											
Days		—2	1	0	1	2	3	4	5	6	7	8	9	10	11	12
				Solar radiation below 1.931—Winter half-year												
Winnipeg	*a*	2°5 F.	1°1	2°8	0°6	0°7	0°1	—0°1	0°3	—1°2	0°6	—0°8	—2°2	—1°8	—1°6	—1°5
	b	1.1	—0.1	2.2	1.5	0.5	—0.4	—0.3	0.3	—0.2	0.3	0.1	—1.9	—1.2	—0.2	—1.3
New York	*a*	0.4	0.9	1.2	1.5	0.9	1.0	—0.2	—1.4	—1.6	0.0	—1.4	—0.2	0.6	—1.1	—0.9
	b	—1.2	0.1	—0.2	0.9	1.8	1.9	0.4	—1.4	—0.6	0.4	—0.4	0.0	0.7	—1.3	—1.0
				Solar radiation above 1.960—Winter half-year												
Winnipeg	*a*	1.3	0.7	—0.9	—0.4	0.4	0.9	1.5	0.0	0.6	—0.6	—0.8	0.4	—1.0	—0.6	—0.9
	b	0.2	0.0	—1.7	—1.1	0.3	0.6	1.3	1.2	0.5	—0.3	—0.8	0.7	0.1	—0.2	—0.4
New York	*a*	1.0	1.0	1.8	0.4	—0.7	—1.0	0.0	0.0	—0.4	—0.3	0.4	—0.5	—0.9	—0.1	—0.9
	b	0.2	0.1	0.7	0.3	—1.0	—1.0	—0.1	0.5	—0.2	0.6	0.7	0.0	—0.3	0.1	—0.1
				Solar radiation below 1.931—Summer half-year												
Winnipeg	*a*	—0.2	—0.6	—0.5	—0.5	—0.2	0.2	0.6	0.2	—0.4	—0.6	0.2	0.3	0.5	0.9	0.7
	b	0.2	0.0	—0.4	—0.3	—0.1	0.2	0.6	—0.1	—0.4	—0.8	0.1	0.2	0.1	0.3	0.2
New York	*a*	—0.3	—0.2	0.0	0.0	—0.3	0.1	0.1	0.2	0.0	—0.5	—0.5	—0.3	0.3	0.7	0.2
	b	0.0	0.4	0.4	0.0	—0.3	0.0	0.0	0.0	0.0	—0.4	—0.6	—0.4	0.0	0.9	0.3
				Solar radiation above 1.960—Summer half-year												
Winnipeg	*a*	0.4	—0.8	—1.3	—0.9	—0.2	—1.2	—1.4	—0.8	0.7	—0.1	0.4	1.0	0.0	1.5	2.3
	b	1.3	0.3	—0.6	—0.9	—0.4	—1.1	—1.0	—0.8	0.2	0.0	—0.1	0.4	0.0	0.8	1.7
New York	*a*	—1.2	—0.1	—0.4	0.6	—0.1	0.7	0.3	—0.4	0.2	0.4	—0.1	—0.1	0.6	—0.5	0.3
	b	—0.3	0.3	0.2	0.5	0.2	0.8	0.7	—0.3	—0.4	—0.3	—0.4	—0.5	—0.2	—0.8	0.1
				Differences between values below 1.931 and above 1.960—Winter half-year												
Winnipeg	*a*	1.2	0.4	3.7	1.0	0.3	—0.8	—1.6	0.3	—1.8	1.2	0.0	—2.6	—0.8	—1.0	—0.6
	b	0.9	—0.1	3.9	2.6	0.8	—1.0	—1.6	—0.9	—0.7	0.6	0.9	—2.6	—1.3	0.0	—0.9
New York	*a*	—0.6	—0.1	—0.6	1.1	1.6	2.0	—0.2	—1.4	—1.2	0.3	—1.8	0.3	—1.5	—1.0	0.0
	b	—1.4	0.0	—0.9	0.6	2.8	2.9	0.5	—1.9	—0.4	—0.2	—1.1	0.0	1.0	—1.4	—0.9
				Differences [illegible] [illegible] [illegible] 1.931 [illegible] [illegible] 1.960—Summer half-[illegible]												
Winnipeg	*a*	—0.6	0.2	0.8	0.4	0.0	1.4	2.0	1.0	—1.1	—0.5	—0.2	—0.7	0.5	—0.6	—1.6
	b	—1.1	—0.3	0.2	0.6	0.3	1.3	1.6	0.7	—0.6	—0.8	0.2	—0.2	0.1	—0.5	—1.5
New York	*a*	0.9	—0.1	0.4	—0.6	—0.2	—0.5	—0.2	0.6	0.2	—0.9	—0.5	—0.3	—0.3	1.2	—0.1
	b	0.3	0.1	0.2	—0.5	—0.5	—0.8	—0.7	0.3	[illegible]	—0.1	—0.2	0.1	0.2	1.7	0.2

NOTE.—The [illegible] used in [illegible] this table are the [illegible] between 8 a. m. and 8 p. m. [illegible] from the daily [illegible] of the U. S. [illegible] Bureau. The [illegible] are [illegible] the means of the 15 days covered by the [illegible] in [illegible] These are [illegible] in degrees and [illegible] Fahrenheit. The [illegible] marked *a* were [illegible] by using the [illegible] marked *b* were [illegible] subtracting the daily normals from the [illegible] values and dealing with the [illegible] normal.

TABLE 13.—*Mean departures of temperature at Buenos Aires following high solar values—Winter half-year*

Years	Solar values	Cases	0	Mean departures * Days after solar maximum: 1	2	3	4	5	6	7	8	9	10	11	12
1909-1918	1.990 and over	50	—1°.1 F.	—0.2	—0.4	—1.1	—1.8	—0.2	0.2	0.4	0.9	1.3	0.9	0.2	0.2
1919-1920	1.970 and over	33	—0.7	0.0	—0.2	—0.5	—1.1	—1.1	—0.2	2.0	1.1	1.3	1.3	0.9	0.7

NOTE.—These departures are from the means of the 13 days covered by the observations and are in degrees and tenths F.

TABLE 14.—*Mean departures of temperature at Winnipeg following high and low solar radiation values—Winter half-year*

Years	Solar values	Cases	Mean departures *: Days before: —2	—1	0	1	Days after: 2	3	4	5	6	7	8
1918-1919	1.960 and over	48	2°.1 F.	—1.8	—2.9	—1.7	0.4	1.6	2.1	1.6	0.7	—0.2	—1.2
1920-1922	do	60	0.7	1.2	—0.9	—0.8	—0.2	—0.2	0.6	0.8	0.3	—0.5	—0.6
1918-1919	Below 1.931	30	0.4	—0.4	0.7	1.0	0.3	—0.6	—1.0	—0.3	—0.2	0.0	—0.2
1920-1922	do	35	0.3	—1.1	2.3	0.7	—0.5	—1.5	—1.0	—0.5	—1.5	—0.7	—1.0

* These departures are from the means of the 11 days covered by the observations and are in degrees and tenths of degrees F.

For the six days (0 to 5 days) the correlation coefficients for the years 1918-1919 and 1920-1922 as between the two intervals, are as follows:

For temperatures following high solar values.......r =	0.88 ± 0.07
For temperatures following low solar values........r =	0.81 ± 0.10
For temperatures following high solar values compared with those following low solar values, for the years 1918-1919 ....r =	—0.87 ± 0.07
For temperatures following high solar values compared with those following low solar values, for the years 1920-1922.........................r =	—0.68 ± 0.15

Each of these sets of values are independent of each other, and the high correlations strongly support the conclusion that they are closely related with each other. It should be noted, however, that these correlations are between means, and not between individual observations.

Some able meteorologists, like Sir Napier Shaw,[1] while not denying the facts presented in the previous paper, object to some of the conclusions drawn from them. No one can feel more strongly than I do the great difficulty of correctly interpreting the complex physical processes of the atmosphere; but working hypotheses are as necessary to an investigator as is the compass to a navigator, although an occasional investigator thinks he is working without an underlying hypothesis. I regard my interpretation of observed phenomena as working hypotheses to be modified, or abandoned for better interpretations, as facts accumulate. Doubtless there are some who judge results entirely by the working hypotheses used, and accept or reject the facts entirely on this basis.

This may be illustrated by the story of an early discoverer of meteoric stones, who, having seen them fall, recovered some fragments, and took them to a philosopher. The philosopher looked at them and said, " My friend how do you suppose stones could get up into the sky? " " I don't know," replied the discoverer, " perhaps they were thrown out from a volcano." " A volcano! " said the philosopher, " There isn't a volcano within a thousand miles of here. Poof! it is impossible. Your seeing them fall is purely imaginary," and refused further to examine the evidence.

9. A PARTIAL SUMMARY OF THE EVIDENTIAL RESULTS IN THIS PAPER

As it has seemed to me that heretofore critics have been apt to overlook many of the evidences favorable to solar variation and its

[1] Meteorol. Mag., February, 1926, p. 7.

influence on weather, perhaps because these were too numerous and extensive to be mentally digested, I draw together, in the following table 15, 20 of the correlation coefficients which have been given above. Besides these, there are many other evidential results in this paper, but given in other forms.

TABLE 15.—*Some evidential correlation coefficients*

Nature of the correlation	Value	Probable error
Between monthly mean temperature 0 to 4 months succeeding high and low months of solar radiation of the years 1905 to 1925.		
Stations:		
Nome	—0.72	±0.16
Juneau	—0.80	0.12
Edmonton	—0.81	0.12
St. John's, N. F.	—0.52	0.24
Hatteras	—0.89	0.07
Key West	—0.64	0.20
Between monthly mean differences of temperature and of pressure accompanying respectively ranges of solar radiation of the years 1918-1925, and ranges of sun-spot numbers of the years 1856-1923.		
Pressure, winter half-year	0.56	0.07
Pressure, summer half-year	0.45	0.08
Temperature, winter half-year	0.62	0.06
Temperature, summer half-year	0.50	0.07
In definite geographical areas between pressures and sun-spot range.		
4 temperate zone positive correlations exceeding	0.50	
9 tropical zone negative correlations exceeding	—0.50	
Between the mean marches of temperature and solar radiation for 30 days (1-30) during which high solar radiation maxima occurred on the sixth day.		
For temperatures at Buenos Aires, 3 days after	0.66	0.07
Between 10-day means of solar radiation and of Argentine temperatures of the year 1916, June-October	—0.80	0.12
Between mean marches of departures of temperature at Winnipeg over ranges of 12 days accompanying a large range of solar radiation. As between summer and winter effects	0.87	0.07
Between the mean marches of temperature at Buenos Aires following high solar radiation. As between the results of 1909-1918 and those of 1919-1920	0.73	0.09
Between the mean marches of temperature at Winnipeg, 0 to 5 days following high and low solar values. As between results of 1918-1919 and 1920-1922.		
High values	0.88	0.07
Low values	0.81	0.10
As between high values and low values.		
For the interval 1918-1919	—0.87	0.07
For the interval 1920-1922	—0.68	0.15

SMITHSONIAN MISCELLANEOUS COLLECTIONS
VOLUME 78, NUMBER 5

THE DISTRIBUTION OF ENERGY OVER THE SUN'S DISK

(WITH ONE PLATE)

BY
C. G. ABBOT

(PUBLICATION 2876)

CITY OF WASHINGTON
PUBLISHED BY THE SMITHSONIAN INSTITUTION
OCTOBER 12, 1926

The Lord Baltimore Press
BALTIMORE, MD., U. S. A.

THE DISTRIBUTION OF ENERGY OVER THE SUN'S DISK

By C. G. ABBOT

(With One Plate)

Abstract.—Moll, Burger, and van der Bilt have attacked the accuracy and usefulness of Smithsonian solar disk drift curves. They describe them as omitting the region (from 95 to 100 per cent out on the solar radius) most important for solar theory; as affected by large instrumental error; and as subject to fatal systematic error due to tardy response of the bolometric apparatus.

Abbot states that since the Smithsonian observations were undertaken merely to test suspected variability of distribution from epoch to epoch, only differential accuracy was required. No attempt was made to carry the work over into the difficult region between 95 and 100 per cent of the solar radius, because it was unnecessary for this purpose. He fears that to reach demonstrable accuracy to 1 per cent in this region of the curves will meet insuperable difficulties.

Abbot points out that their statement regarding accidental error rests on one curve made at Washington City, sea-level, station, prior to 1908, though the work went on under highly satisfactory conditions for eight years afterwards at Mount Wilson. Photographs proving its general excellence are available.

He points out that the amount of systematic error claimed by the authors depends on the actual degree of quickness of response of the Smithsonian apparatus; that photographic evidence shows that this was 1.95 seconds; that such error tends to raise the following limb of the curve above the true values, though lowering the advancing limb beneath them, and thus tends to be eliminated in their mean; that where, as in Smithsonian observations, the measurements show negligible differences between the two limbs, the error is presumably negligible; that in different years, receiving instruments of unequal quickness of response were used without corresponding differences of result in the sense indicated by the Dutch authors' criticisms. He admits that very near the limbs the curves would not have furnished trustworthy indications. A determination of the error near the limbs is given. This indicates that at 95 and 92 per cent the

Smithsonian results would differ from the true curve on account of lag by 0.28 per cent, and 0.26 per cent, respectively.

On returning from a six months' expedition, I find the paper of Moll, Burger, and van der Bilt.[1] They take exception to the Smithsonian experiments on the distribution of energy over the sun's disk from three points of view. First, that our measurements were rarely extended beyond 95 per cent on the sun's radius. Second, that our findings were expressed with more places of figures than the experiments justified. This criticism they support by reproducing one of our early curves. Third, that owing to the tardiness of response of our bolometric apparatus, our curves differ very sensibly from true representations of the distribution sought.

That the reader may clearly understand what is in question, I recall for him that we formed a large image of the sun by a reflecting telescope. Stopping the clockwork, this image drifted its own diameter in about two minutes. The arrangements were such that this drifting solar image marched centrally and horizontally across a short vertical slit. From the ray which passed through the slit, a certain wave-length was selected by means of a spectroscope and brought to focus upon the strip of the bolometer of far more than hairlike thinness. The curve of bolometer temperatures corresponding to the intensities of the selected wave-length in the solar image was automatically recorded in the shape of an inverted U.

We observed on both solar limbs and took their mean values. We were accustomed to cut off the recording light from the galvanometer mirror at the instants when the sun's image visibly reached the slit and departed from it. At intervals we also inserted shutters which produced zeros of radiation on the records and permitted accurate examination of the behavior of the bolometer. The curves were inevitably a little wider than corresponded to the astronomical width of the sun in terms of the rate of motion of the plate. This is because of the time required for the galvanometer to descend to zero after the following limb of the sun had crossed the slit.

Our habit of measuring was to compute from astronomical and plate-speed data the widths corresponding to definite proportions of the solar radius; to adjust these places symmetrically to the central axis of the U-shaped curve: and to measure heights on both advancing and following limbs at these places. All results were finally compared to the mean form of distribution curve for the year 1913

---

[1] Bul. Astron. Inst. Netherlands, Vol. III, No. 91, Dec. 18, 1925.

as a standard. The curves, as I shall show, were symmetrical to within negligible limits up to 95 per cent of the solar radius.

We made our experiments at Washington prior to 1908, and at Mount Wilson from 1913 to 1920, every summer. Several different bolometers, several different galvanometers, and several different optical systems were used by us. Evidences of secular variations of distribution were found. Extensive discussion of the methods, sources of error and results are given in Volumes II, III, and IV of the Annals of the Astrophysical Observatory of the Smithsonian Institution.

The Dutch authors describe briefly their own experiments in which they used a thermopile whose time of lag in attaining thermal equilibrium is not stated. They, indeed, refer to a description in another paper in which several instruments of considerable quickness are described, but as one may infer from their figure 2, the sluggishness of the actual instrument used appears to have been very great. They employed a 3-centimeter solar image formed by a lens, a slit whose width in proportion to the image equalled ours, and a device adjusted to produce a uniform drift of the image across the slit in 14 minutes. They give no data as to the degree of uniformity of their galvanometer scale or the width of the curves compared with the computed width. They made observations during part of one month at the Gornergrat. Of the results, they say:

We were able to get trustworthy values of the distribution of energy along about 99 per cent of the sun's radius, against Abbot's 95 per cent.

Our measurements do not claim to give results of the highest precision obtainable. We think that our values of the energy are trustworthy to about one-hundredth part of their value at the sun's center.

Our values for the common 95 per cent exceed those of Abbot. The differences attain a maximum amount of about 2½ per cent at a distance of about 8 per cent from the sun's limb. . . . . It is easy to explain this discrepancy in a satisfactory way as a consequence of the insufficient quickness of Abbot's instruments compared with the speed of the solar image. It is not so easy to explain the fact that the discrepancies are less at a distance of 5 per cent from the sun's limb. Probably Abbot has been under the influence of a preconceived opinion, viz.: That the energy at the sun's limb must, from a finite value, abruptly fall to zero.

In looking at Abbot's curves, of which figure 1 is a specimen, a peculiarity attracts the attention: they show a certain skewness or absence of symmetry.

A first glance at his curve shows that it is not smooth, a fact which we ascribe to disturbances, probably of the galvanometer. One wonders how it was possible to derive, from curves like this, reliable results, and to give the data in four figures.

Again they say, in regard to their figure 2, given to compare 14-minute drifts with 2-minute drifts:

> Evidently the latter has been seriously affected by the slowness of our instruments. Now, since these were doubtlessly much quicker in response than Abbot's (which is evident from the absence of any perceptible skewness in our curves), his curves, which were all recorded with the sun's image at its normal speed, must have undergone a considerable deformation. No wonder that our final measurements led to data quite different from those given by Abbot.

I regard the authors' insinuations regarding our work as unfairly derogatory. Especially do I deprecate their implication that our results at 95 per cent were anything but direct computations from the measurements. That they are unbiased results from direct measurements the authors could have ascertained from Volumes III and IV of our Annals, but it will also appear plainly in certain illustrative examples below.

Again, one would hardly have expected that the authors would base a severe condemnation of our entire research on a single curve made prior to 1908 amid the murky atmosphere and rumbling vehicles of the city of Washington.[2] For we afterwards carried on the work for eight successive summers, 1913 to 1920, and made many thousand drift curves under fine conditions at Mount Wilson. Except when the sun itself presented irregularities of distribution, our curves are in general of great smoothness and symmetry. This prevails notwithstanding that our curves are on a higher scale of ordinates than those which the Dutch authors show. I am sending to the authors photographic prints which prove the prevailing smoothness and symmetry of our curves, and reproducing the same as the accompanying plate 1.

As to whether the Dutch authors have obtained or will obtain a higher degree of accuracy than we did in determining the distribution of energy over the sun's disk, we must await the more detailed experiments and descriptions which they promise before we can form a conclusion. The matter, indeed, will be exceedingly difficult to demonstrate. Certainly a degree of accuracy which they "think" extends to about 1 per cent cannot decide as between the results of two researches whose maximum discrepancy they inform us is 2½ per cent.

This is the more obvious when we reflect that our results, to whose error they would attribute the whole discrepancy, were made during more than 10 different years, at two different stations, with four different bolometric outfits, and with three different optical systems.

[2] See plate XXVIII, Annals Smithsonian Astrophys. Obs., Vol. II, 1908.

They were made on solar images of 40 and 20 centimeters diameter, as compared with three centimeters used by the authors. They indicated differences from day to day and from year to year in the solar energy distribution. The authors, indeed, do not even state with which year of our observations they have made their comparison, and seem to suppose that the distribution of radiation over the sun's disk is invariable.

The Smithsonian observers approached this research from a totally different point of view than the Dutch authors. With us it was incidental to our general study of the variation of the sun. We supposed that the sun's variations of short-interval and of long-interval might be associated with changes of the transparency or of temperature of his outer envelopes. Such alterations might, as we thought, reveal themselves by modifications of the distribution of intensity of radiation along the east and west diameter. We even hoped that the correlation of change of solar-constant with change of distribution would prove so close that we could substitute for the (at that time) tedious solar-constant observations the easy drift observations. Accordingly, we sandwiched in between solar-constant bolographs, on nearly every day of observation at Washington and at Mount Wilson, sets of drift curves at several wave-lengths. At Washington we made three drifts for each wave-length on each day. At Mount Wilson the conditions were so much better that we contented ourselves, except in the year 1920, with two.

The reasons which induced us to limit our measurements to 95 per cent out on the sun's radius were that we did not need to go farther out to show secular changes in distribution, and that we conceived that the boiling of the atmosphere, the intensity of sky light, and the extremely rapid change of intensity at the sun's limbs introduced factors of such uncertainty that the measurements farther out would be of little value for indicating such small changes from day to day and year to year as we were searching for.

In short, we did not undertake to test theories of the sun's constitution by distribution experiments, or try to obtain results suitable for that purpose, though we were, of course, glad if the measurements later proved adaptable to that inquiry. This is the problem which the Dutch authors set for themselves. For its solution they desire accurate values out to 99 per cent of the sun's radius. I am tempted to refer them to the words of Ahab:[3] "And the King of Israel answered and said, Tell him, Let not him that girdeth on his harness

[3] I Kings 20, 11.

boast himself as he that putteth it off." There are yet great difficulties before them in arriving at 1 per cent demonstrable accuracy out to 99 per cent of the sun's radius. Full details, quantitative investigation of errors, the effect of altering the experimental means and an investigation of solar variability will be demanded to support such claims.

In the remainder of my remarks, I wish to defend our results from the theoretical objection made by the authors that, owing to the quick march of the solar drift and the slow response of the bolometric outfit, our curves are sensibly deformed as far back as 95 per cent and even 92 per cent of the sun's radius. They support this objection by printing [as their fig. 2] two curves taken with their own apparatus on drifts respectively of 2-minute and 14-minute speeds. They do not show these curves on both advancing and following limbs. Yet they seem to leave their readers to infer that the *quick* drift curve is the *lower* on *both* sides of the sun. This is, of course, not so. If the receiving instrument lags behind in attaining thermal equilibrium, it will be *below* the true curve on the advancing limb, and *above* the true curve on the *following* limb. Hence, taking the mean of measurements on the two limbs tends to eliminate the error.

The elimination of error by this device cannot be perfect and it is highly desirable to use apparatus acting so quickly that the difference between the two limbs is negligible. I give in illustration a number of sets of measurements of our curves on the two limbs for different wave-lengths, different years, and different bolometers. These values are exactly as obtained and measured many years ago in our Mount Wilson work. The results are neither better nor worse than hundreds of others which I might have quoted.

It is possible to determine approximately the magnitude of the error which the Dutch authors fasten upon our results. For this purpose, consider first the effect of inserting the shutter before the slit as photographically recorded on all of our plates. I find by measurements of several such records that the trace starts to fall very steeply without preliminary slow gathering of motion, and runs to zero in 1.95 seconds according to the following schedule:

Fraction of whole time of falling	0	0.1	0.2	0.3	0.4	0.5	0.6	0.7	0.8	0.9	1.0
Fraction of whole fall	0	.05	.12	.20	.31	.44	.61	.74	.85	.92	1.0

The upward march when the shutter is removed is substantially identical.

Original Measurements on Smithsonian Drift Curves

various years, bolometers, and wave-lengths

Measurements in Centimeters

Solar Limb	Date	Bolometer	Wave Length	Per cent of radius from center										
				0	20	40	55	65	75	82.5	87.5	92	95	97
Preceding........	1914, Aug. 11	Fine strip in air	0.4265	7.03	6.98	6.70	6.11	5.71	5.17	4.61	4.24	3.60	3.12	2.71
Following........	" "	"	"	—	6.90	6.53	6.10	5.73	5.27	4.65	4.19	3.60	3.13	2.63
F minus P........	" "	"	"	—	—0.08	—0.17	—0.01	+0.02	+0.10	+0.04	—0.05	0.00	+0.01	—0.08
Per cent (F—P).	" "	"	"	—	—1.1	—2.5	—0.2	+0.3	+1.9	+0.9	—1.2	0.0	+0.3	—3.0
Preceding........	" "	"	0.5955	14.48	14.24	13.83	13.21	12.59	11.76	10.92	10.20	9.30	8.60	7.80
Following........	" "	"	"	—	14.28	13.78	13.22	12.66	11.83	11.00	10.29	9.30	8.53	7.73
F minus P........	" "	"	"	—	+0.04	—0.05	+0.01	+0.07	+0.07	+0.08	+0.09	0.00	—0.07	—0.07
Per cent (F—P).	" "	"	"	—	+0.3	—0.4	+0.1	+0.6	+0.6	+0.7	+0.9	0.0	—0.8	—0.9
Preceding........	1916, July 28	Coarse strip in vacuo	0.4265	13.78	13.52	12.90	12.07	11.32	10.19	9.02	8.00	6.90	6.00	—
Following........	" "	"	"	—	13.52	12.84	11.97	11.12	10.02	8.95	8.00	6.90	6.00	—
F minus P........	" "	"	"	—	0.00	—0.06	—0.10	—0.20	—0.17	—0.07	0.00	0.00	0.00	—
Per cent (F—P).	" "	"	"	—	0.0	—0.5	—0.9	—1.7	—1.7	—0.8	0.0	0.0	0.0	—
Preceding........	" "	"	0.5955	16.30	16.02	15.49	14.82	14.12	13.22	12.28	11.49	10.39	9.38	—
Following........	" "	"	"	—	16.04	15.50	14.80	14.12	13.21	12.23	11.42	10.32	9.32	—
F minus P........	" "	"	"	—	+0.02	+0.01	—0.02	0.00	—0.01	—0.05	—0.07	—0.07	—0.06	—
Per cent (F—P).	" "	"	"	—	+0.1	+0.1	—0.2	0.0	—0.1	—0.4	—0.6	—0.7	—0.6	—
Preceding........	1920, Aug. 30	Fine strip in vacuo	0.4265	16.07	15.83	15.08	14.00	13.00	11.73	10.49	9.40	8.20	7.15	—
Following........	" "	"	"	—	15.83	14.94	13.90	13.00	11.67	10.40	9.38	8.20	7.10	—
F minus P........	" "	"	"	—	0.00	—0.14	—0.10	0.00	—0.06	—0.09	—0.02	0.00	—0.05	—
Per cent (F—P).	" "	"	"	—	0.0	—0.9	—0.7	0.0	—0.5	—0.9	—0.2	0.0	—0.7	—
Preceding........	" "	"	0.5955	11.25	11.15	10.85	10.30	9.90	9.21	8.60	7.90	7.30	6.70	—
Following........	" "	"	"	—	11.11	10.78	10.36	9.92	9.20	8.60	7.97	7.33	6.72	—
F minus P........	" "	"	"	—	—0.04	—0.07	+0.06	+0.02	—0.01	0.00	+0.07	+0.03	+0.02	—
Per cent (F—P).	" "	"	"	—	—0.4	—0.6	+0.6	+0.2	—0.1	0.0	+0.9	+0.3	+0.3	—

As the time of complete fall or rise is about 1/65 of the time required for a complete drift on the day I investigated, namely, August 21, 1920, I gave the Dutch authors a slight advantage by calling the time of full bolographic response to a new stimulus 1/60 of the time for a complete solar drift.

Our next object is to estimate the effect of this degree of sluggishness upon the true drift curve. Not knowing positively the true curve near the limb of the sun, I have used the Dutch authors' preliminary result as our best approximation for it. As a sample, I have chosen their curve for wave-length 0.5 microns. This I plotted on our great sheets of millimeter cross section paper, on a scale of 12,000 millimeters of abscissae to the solar diameter, and 500 millimeters corresponding to the ordinate of the Dutch authors' curve at the sun's center, taken as 1,000 in what follows.

Next, recalling that Mr. Fowle, who measured all of our drift curves, was accustomed to place the curve symmetrically, and to measure to the computed abscissae corresponding to astronomical and plate-speed data, we must consider where his measurements really lay with respect to the true curve. From a number of our drift curves of August 21, 1920, I find the bolographic width from zero to zero of ordinates to have been 131.0 millimeters,[4] but the visually observed width, as indicated by the instantaneous cutting off of the record-light at ingress and emergence from the slit, was 128.4 millimeters.[4] The excess, 2.6 millimeters, was, we may suppose, symmetrically divided in Mr. Fowle's placement of the curve for measuring. Hence, he measured $\frac{2.6}{2 \times 131}$ of the solar diameter away from the orientation of the true curve. Therefore, in terms of the orientation of the true curve, he measured for the place 95 per cent out on the solar radius, for example, at 93 on the preceding and 97 on the following limb, or at places very close thereto, depending on accidental differences of individual curves. These very slight accidental second order shiftings might, of course, lead to changes of the order of a per cent or so between the advancing and following limbs in his measurements of individual curves, but, since the lowering of the one must produce the lifting of the other, these slight changes would be closely eliminated in his mean values. In what follows we shall assume that Mr. Fowle exactly bisected the bolographic curve by his zero setting of the plate for measurement.

[4] Comparable because measured on same plotting paper. Fowle's computed value, 129.02, on slightly different scale.

I then assume that at 95 per cent out on the radius, as indicated by the large scale plot above described, the preceding limb was measured at −2 per cent or −12 cm., and the following limb at +12 cm. from the 95 per cent place on the true curve. At the first named place it will be obvious that the bolographic trace was not only as high as the true curve had been at 20 cm. nearer the limb of the sun, but higher. For 20 cm. corresponds to the interval of time required for full response, and during all that interval the stimulus had equalled or exceeded the stimulus of the true curve at the place just mentioned. Similarly, for Mr. Fowle's place of measurement on the following limb, the ordinate must be inferior to the ordinate upon the true curve at 20 cm. nearer the sun's center. Thus we have a first approximation.

Our next inquiry is to find the effect of the excess of radiation persisting over the aforesaid 20 cm. interval for the preceding limb, and the defect of radiation persisting over the equal interval for the following limb. For this purpose I read the ordinates of the true curve at places 2, 4, 6, etc., to 20 centimeters towards the limb, counting from Mr. Fowle's place upon the preceding limb, and correspondingly towards the center upon the following limb. The differences of readings of ordinates corresponding to these 2-centimeter intervals were then obtained. We are now ready to proceed. For instance, the stimulus at Mr. Fowle's observed place on the preceding limb had exceeded that at 20 cm. back on the true curve by an amount corresponding to the first of the aforesaid differences active during 9/10 of a response interval, 8/10 for the second, etc.

Proceeding in this way, we find the following numerical values at place 95, preceding limb:

Places, cm	−20	−18	−16	−14	−12	−10	−8	−6	−4	−2	0
Ordinates	493	505	516	526	536	545	554	563	571	580	588
Differences	12	11	10	10	9	9	9	8	9	8	—
Fractional Response	.92	.85	.74	.61	.44	.31	.20	.12	.05	00	—
Products	11.04	9.35	7.40	6.10	3.96	2.79	1.80	1.08	.40	00	—

The sum of these products is 43.9. Adding it to the value at −20 cm., we obtain, as the second approximation to Mr. Fowle's reading, 536.9. It would have been possible by dividing the interval into 100 parts instead of 10 to get a very slightly higher result, but the difference surely for our inquiry is negligible. Proceeding similarly for the following limb, the correction becomes −38.9 and the second approximation there is 532.1. The difference between Mr. Fowle's readings on the two limbs in the sense preceding minus following would therefore be expected to be 4.8, or 0.9 per cent. Their mean is 534.5, which, as the Dutch authors claim it should be, is lower than 536.0, the true curve value at place 95, but by only 0.28 per cent. I have performed a similar analysis at place 92. It indicates a correction in the same sense of only 0.26 per cent there. Farther towards the center the correction sensibly vanishes.

Though we have no direct statement of the time required for complete response in the case of the Dutch authors, their figure 2 enables us to know that their instrument was far more sluggish than ours. For, as noted above, the sluggishness of a receiving instrument must cause the results on the sun's *following limb* to show *less contrast* than the *true curve* of solar drift. Hence, if we admit that P in their figure is the true curve, the following limb, had they published it, must have shown a continuation of the curve Q, higher than the continuation of the curve P. In other words, the difference between the two limbs indicated by a full curve Q taken with their apparatus would have been greater than that between P and Q in their figure 2. But this difference is actually no less than 8 per cent at 95 per cent out on the radius in their figure 2. In our work no systematic difference between the two limbs as great as this appears.

I suspect that the Dutch authors, being accustomed to the thermopile, have underestimated the quickness with which our bolometers respond. We have abundant evidence that our bolometers usually attained thermal equilibrium indefinitely sooner than our galvanometer could make its first swing, which usually occupied only 1.7 to 1.9 seconds.[5] But we have used bolometers of three different degrees of quickness of response. Prior to 1916, we used bolometers in air, which are quickest. In 1916, we used comparatively very coarse bolometer threads in vacuo, which made a far more sluggish instrument, almost indeed as sluggish as the most delicate of thermopiles. Since 1917, we have used finer threads again, but in vacuo, and therefore intermediate in quickness between those of years prior to 1916 and that of 1916 itself.

If, then, the Dutch authors were right in their criticism, our drift curves of 1916 ought to show lower " shoulders," or in other words greater contrast, than those of later years; and these, in turn, greater contrast than those of years prior to 1916. It needed only to have examined tables 68 and 74 of our Annals[6] to be convinced that no

---

[5] See also our Annals, Vol. II, p. 218.

[6] I draw attention here to a misprint throughout tables 73 and 74. For 1.0035 in the place headings read 0.92.

certain evidence of the kind appears. The accompanying figure 1 shows this. In order to save myself a future note, I admit that the *two* curves of *one* day of 1916 given in the table in this present paper do show lower values. But I hope readers will be fair enough to form their judgments from the *mean* results of many days and *many* wave-lengths given in the Annals.

As I have said, we did not take up drift-curve work for the sake of getting the most accurate distribution tables for the use of solar theorists. We were concerned only with relative measurements to compare distributions from day to day and from year to year. Hence, we

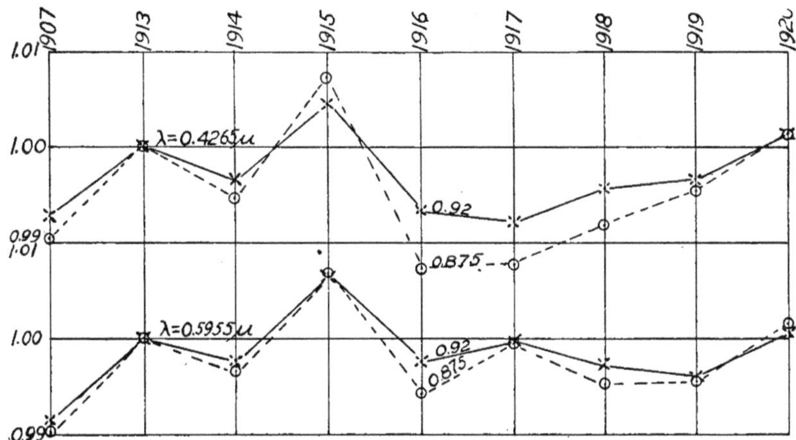

FIG. 1.—Comparison of drift results of different years for two different wave-lengths and two different places on the solar radius. Taken with four different bolometers and three different optical systems.

did not try to attack the difficult region near the sun's limbs, nor did we seek to produce absolute distribution curves of the highest attainable freedom from systematic errors. We were concerned with relative values. Nevertheless, I could not let pass without reply so hasty and unjust an attack on what, after all, was work of a pretty high order of accuracy.

From this investigation I see no ground for admitting that the defect of ordinates attributed by the Dutch authors to our results as a consequence of sluggishness of response is of much consequence. The main part of the difference between their results and ours must be due to other causes. Such may be:

1. Too hasty a conclusion. Further experiments proposed by the Dutch authors may not indicate such a discrepancy.

2. Too small a solar image. Possibly with a different optical outfit the results would differ. Perhaps, too, there is error on account of stray light from other spectral regions.

3. Error in wave-length. The change in form of distribution curve with wave-length is quite rapid.

4. Error in determining places of measurement. The ordinates of distribution curves vary rapidly along the radius.

5. Difference due to alteration in the distribution in the sun itself. See the accompanying curve where a range of over 1 per cent is shown independently by two wave-lengths.

6. Non-uniformity of galvanometer scale. We were accustomed to test this frequently and reduced it to negligible dimensions.

I am by no means prepared either to admit that our work is wrong or, on the other hand, to deny catagorically that it has appreciable error. I await with much interest, therefore, the further investigations which the Dutch observers promise.

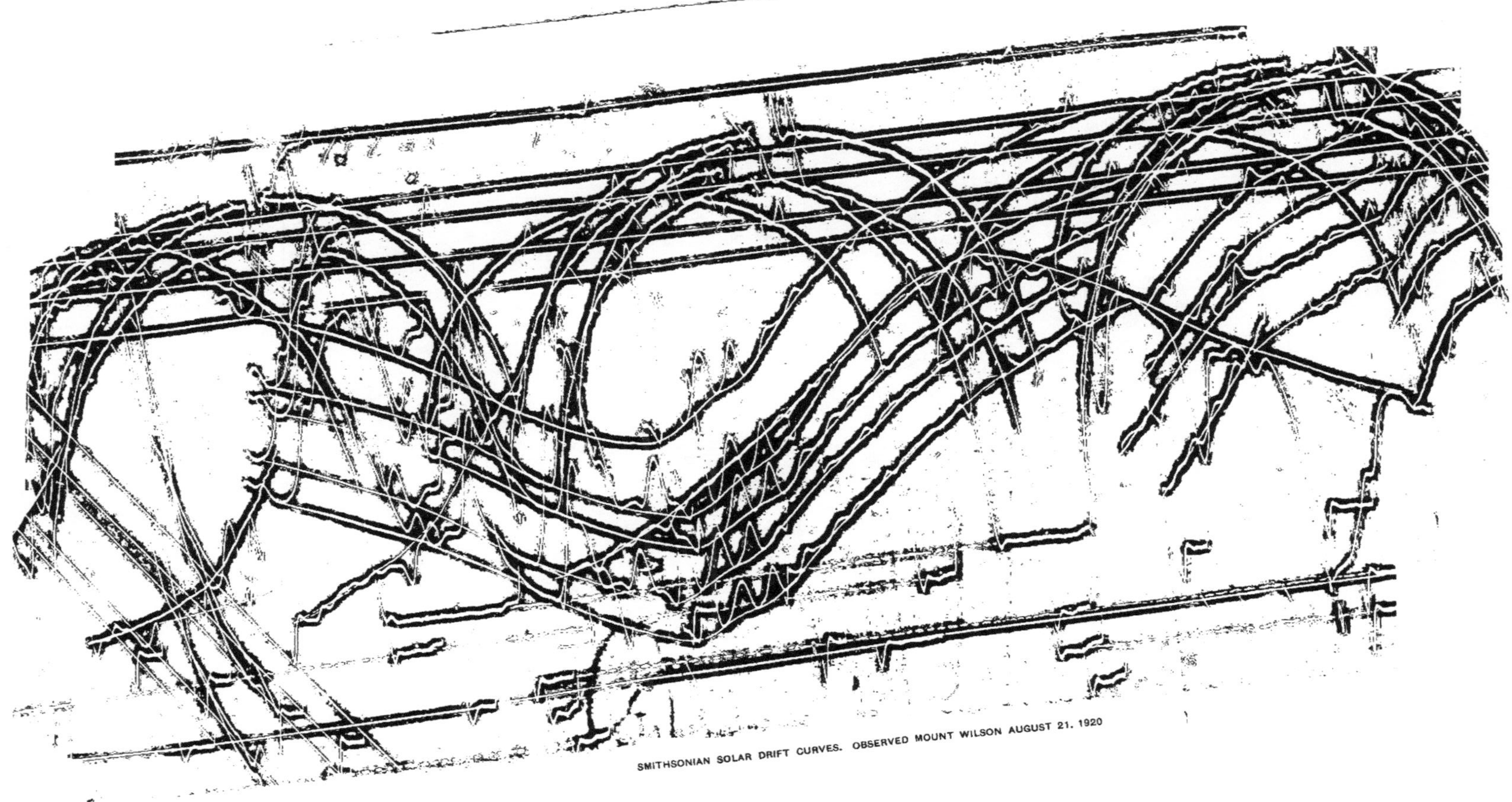

SMITHSONIAN SOLAR DRIFT CURVES. OBSERVED MOUNT WILSON AUGUST 21, 1920

SMITHSONIAN MISCELLANEOUS COLLECTIONS
VOLUME 78, NUMBER 6

THE LYELL AND FRESHFIELD GLACIERS, CANADIAN ROCKY MOUNTAINS, 1926

(WITH TWELVE PLATES)

BY
J. MONROE THORINGTON, M. D

(PUBLICATION 2911)

CITY OF WASHINGTON
PUBLISHED BY THE SMITHSONIAN INSTITUTION
FEBRUARY 5, 1927

The Lord Baltimore Press
BALTIMORE, MD., U. S. A.

THE LYELL AND FRESHFIELD GLACIERS, CANADIAN ROCKY MOUNTAINS, 1926

By J. MONROE THORINGTON, M. D.

(With 12 Plates)

LYELL GLACIER

The first glacier of the Canadian Rockies to be described with detailed accuracy was the Lyell, discovered by Dr. Hector, of the Palliser expedition, in 1858. His description enables one today to judge with a fair degree of certainty the condition of the ice at that time, its extent, and the alterations which it has undergone during the years intervening. It is the oldest record that we possess concerning a glacier of the North Saskatchewan system and, within limits, the deductions made from it may be applied to other ice-streams of that region.

Encamped in the valley of Glacier Lake, Dr. Hector wrote [1] of the present Southeast Lyell glacier as follows:

After crossing shingle flats for about a mile, we reached a high moraine of perfectly loose and unconsolidated materials, which completely occupies the breadth of the valley, about 100 yards in advance of the glacier. Scrambling to the top of this we found that to our left a narrow chasm, with perpendicular walls, brought down a stream from a glacier, descending by a lateral valley from the south,[2] but that the greater bulk of the water that formed the river issued from ice caves that were hollowed out beneath the great glacier of the main valley. By rough triangulation, I found that the width of the terminal portion of the glacier in view from this point was 550 yards . . . . we followed round the lower end of the glacier, having to wade through several streams issuing from below the ice, till we found the surface forming a uniform slope unbroken by crevasses. This was immediately beyond a point where a great longitudinal fissure seemed to divide the glacier into two halves up the centre of the valley; that portion to our left being pure ice much crevassed, but free from dirt on the surface; while to our right the surface we now ascended was less steep, smooth, and unbroken, but so discolored by foreign matters, that at a little distance it might have passed for a talus of rocky fragments. . . . . I now saw that the glacier I was upon was a mere extension of a great mass of ice that enveloped the higher mountains to the west, being supplied partly through a narrow spout-like ice cascade in the upper part of the valley, and

[1] Journals, Detailed Reports, and Observations relative to the Exploration of British North America, p. 110. Captain John Palliser. Folio. London, 1860.

[2] The present Mons glacier.

partly by the *resolidifying* of the fragments of the upper *Mer de Glace,* falling over a precipice several hundred feet in height, to the brink of which it is gradually pushed forward. A longitudinal crack divides the glacier throughout nearly its entire length,[1] sharply defining the ice that has squeezed through the narrow chasm, from that portion of the glacier that has been formed from the fallen fragments, the former being clear and pure, while the latter is fouled from much débris resting on its surface, and mixed in its substance. The more rapid melting of the dirty portion of the glacier[2] gives it a smooth undulating surface, which is much lower than the adjoining surface of the pure ice, which beside is much cut by crevasses and ice valleys, through which flow considerable streams, that often disappear into profound chasms. . . . . The ice was beautifully veined in some parts, and the streaks were often contorted in a manner exactly like the foliation in metamorphic rocks.[3] The precipice at the head of the valley stretches for more than two-thirds its width; the remainder is occupied by the ice cascade. The blue pinnacles of ice, tottering over the edge of the cliff, were very striking, and it was the noise of these falling which we had mistaken for thunder a few days before when many miles down the valley.

The present writer spent the period July 4-14, 1926, in making mountain ascents from the valley of Glacier Lake, devoting a portion of that time to observations of the glacier.

It has apparently changed but little since Dr. Hector's time. A huge rock promontory[4]—Gibraltar in miniature—rises from the gravel flats just below the ice terminus. Near its eastern timbered extremity it is split by the narrow cleft which Dr. Hector noticed, now containing a stream of clear water, but giving evidence that at no remote period it served as an outflow for the Mons glacier, lying in the adjacent southern valley. At one time, before Dr. Hector's day, the Mons and Lyell tongues were united, and swept over part of the great rock promontory, a portion of the stream following its present course directly into the main valley, while a smaller volume escaped through the narrow canyon at the eastern end of the promontory (pls. 1 and 2).

From Dr. Hector's description, and the size of the trees on the terminal moraine, one would judge that this moraine was formed at least

[1] This crack or fissure was not seen by us, although the pressure ridge forms a definite line of division between the clear and the dirt-covered ice.

[2] Dr. Hector fell into error on this point, probably because the dirt-covered ice was lower than the clear ice. Dirty ice normally melts more slowly than clear ice owing to the dirt cover absorbing the heat.

[3] Professor James D. Forbes had first pointed out this veined structure to Agassiz, on the Aar glacier, in 1841, seventeen years previously. It is not unlikely that Dr. Hector was familiar with the work of Forbes, as this was the period when glaciers were first studied intensively.

[4] Lake Moraine (5,116 feet), Station No. 65 of the Interprovincial Survey.

200 years ago. It forms a barrier, averaging 25 feet in height, across the main valley, and is cut through nearly at the center by the river. The Lyell tongue is now about 440 yards from the terminal moraine, a retreat of 340 yards since 1858, or 15 feet annually—an extremely slow recession rate, although one must not forget the possibility of short cycles of advance during the period elapsed.

At the present time, the Mons stream runs along the western margin of the rock promontory, the stream having been "captured" by the Lyell torrent, which it joins and deflects, the combined river flowing with great force transversely across the Lyell ice-front, eroding it as fast as it advances, well above the point where the ice terminus would normally be found.[1] The water swings in a great curve before starting down the valley, in a boiling flood, perfectly impossible to ford, carrying down blocks of ice weighing tons (pl. 3).

The ice facing the river rises above it in a cliff, 20 to 50 feet high. At the angle where the Mons stream joins the Lyell we found a bridge of broken seracs over which we could cross. This disappeared entirely within a few days, and one could never be certain of a route of approach. The tongue below the icefalls is flat, and about two miles long. The longitudinal crack splitting the glacier, which Dr. Hector recorded, is no longer present; but throughout its length there is a sharply defined midline, dividing the clear southern ice of the precipitous fall adjoining Division Mountain (9,843'), from the northern, débris-strewn segment derived from avalanches, pushed over the cliffs from the higher levels of the icefield. Dirt-bands are well formed in the area below the southern icefall; this ice is supplied under pressure, and in volume considerably exceeding that from the adjacent avalanche ice. Consequently the clear ice forms a huge pressure ridge, in the longitudinal direction of the dissipator tongue, rising in a bulge or fold along the junction with the northern, dirt-

---

[1] As recently as 1919, Dr. Charles D. Walcott, who visited the region, noted that: "All of the water from the Mons glacier passed out through the narrow canyon or cleft on the south side, and when the ice was melting on a warm day the stream from it spread out all over the flat. At that time the ice from the Lyell glacier abutted against the rock, forcing the Mons stream to pass out through the cleft." In the present writer's opinion, it is equally correct to state that the Lyell glacier abutted against the rock *because* the Mons stream found exit through the cleft. If the Mons stream in its present state be cut off (as it may be in winter), or if it should be diverted to its former bed, it is probable that the Lyell tongue would very soon re-advance and make contact with the rock promontory.

For the appearance of the terminal drainage in 1919, see Smithsonian Misc. Coll., Vol. 72, No. 1; Fig. 1, facing p. 1.

covered segment, maintaining a level of more than 20 feet above the latter (pls. 4, 5, 6, 7). The glacier, below the ice falls, presents the largest moulins we have seen in the Rockies: circular shafts of at least 20 feet in diameter and unfathomably deep.

The formation of dirt-bands in the dissipator tongue is unique. They are developed not only in the clear southern segment below the icefall, but also in the northern segment derived wholly from avalanches. The southern segment of clear ice moves downward with a faster rate of flow than the rock-covered northern segment, so that the bands do not connect throughout the width of the tongue. Furthermore, the northern and southern segments each have a motion faster at the center than at the sides—there is retardation along the central pressure ridge—with the result that an observer looking down on the tongue sees two series of concentric parabolas, one beside the other.

The southern third of the head-cirque of the glacier is occupied by icefall, with a narrow middle moraine near the southern margin, connecting the dissipator with the main icefield; the northern two-thirds is formed by the bare precipice, over which avalanches are pushed from the icefield, to be reconstructed and incorporated in the dissipator. During summer days the falls occur with great frequency, scarcely 15 minutes ever elapsing without one or more sizable avalanches. These occur in six distinct depressions on the cliff, almost evenly spaced across the face, several being occupied by waterfalls of considerable volume (pl. 8, fig. 1).

According to the Interprovincial Survey, the area of the Lyell icefield, with its outflowing glaciers, is 20 square miles. The Mons field, adjoining and loosely connected with it on the south, contains 10 square miles. The Lyell icefield is split into two nearly equal parts by the Continental Divide, the five peaks of Mt. Lyell rising on its northerly margin.

Into the lateral valley, immediately north of the dissipator tongue, a small glacier descends from the icefield, reaching a level of about 8,000 feet, terminating about a mile short of where it would join the main glacier (pl. 8, fig. 2). It is beyond the purpose of this paper to consider the remaining effluents of the Lyell icefield, although it should be remembered that on the west it drains to the north and south branches of Bush River; on the north to Alexandra River; and on the east to Arctomys Valley as well as to Glacier Lake.

There was neither time nor opportunity for making an instrumental survey of the southeastern tongue, draining to Glacier Lake, but the foregoing observations are of interest as being the first, in any detail, since the discovery of the glacier 68 years ago.

SUMMARY

The Southeastern Lyell glacier presents a dissipator tongue, partially reconstructed in type, which has undergone relatively slight recession during the period 1858-1926. It has receded at a rate slower than has been recorded in other measured glaciers of the Canadian Rockies. Its terminus affords a remarkable example of the effects of captured-stream erosion, with effacement of the ice above the normal balancing-point between forward motion and dissipation.

FRESHFIELD GLACIER

Arriving at the Freshfield Group on July 14, 1926, the afternoon and the following day were devoted to checking on some of the observations made in July, 1922,[1] in order to determine the advance or recession of the ice during the four years intervening. The results of this examination are as follows:

MEASUREMENT OF SURFACE VELOCITY

Station A, on the north lateral moraine, was occupied on July 15, and the vertical reference line on Station B, on the opposite side of the glacier, used for reestablishing the line A-B.

From this line direct measurements were made with steel tape upstream to the Great Boulder and the Glacier Erratic marked "1922." Their distances above the line were respectively 1,046 and 640 feet, as compared with 1,551 feet and 1,306 feet in 1922. This represents an advance of 505 feet and 666 feet in four years. In 1922, the two rocks were 350 feet apart; in 1926, they were found to be 440 feet apart.

	Motion 1,463 days Feet	Average daily motion Inches
Great Boulder	505	4.1
Erratic "1922"	666	5.44

On careful search we were able to locate all of the fourteen numbered stones, lined out 50 paces apart in 1922, except numbers 1, 9,

[1] The results were reported by Howard Palmer, whom the writer assisted in 1922, in two papers, "The Freshfield Glacier, Canadian Rockies," Smithsonian Misc. Coll., Vol. 76, No. 11; and "Observations on the Freshfield Glacier, Canadian Rockies," Journ. Geol., xxxii, 1924, p. 434. These papers should be consulted by anyone interested in making comparison with the present article. Designation of stations by name or number correspond throughout the papers.

During the observations of 1926, Mr. A. J. Ostheimer, III, assisted in the taking of measurements on the dissipator tongue.

and 14 which had no doubt fallen into crevasses. Numbers 12 and 13 were measured with steel tape and found to be respectively 390 feet and 381 feet below the line A-B. Numbers 10 and 11 were advanced several feet further, while the remainder had progressed a trifle less. Time did not permit of further measurements, but 380 feet fairly represents the average advance along the line A-B.

Observations on Movement of Stones in Relation to Line A-B Set Across the Freshfield Glacier July 13, 1922

Station	Distance from north margin of glacier Feet	Motion 1,463 days Feet	Average daily motion Inches
A	255	0	0
12	2285	390	3.2
13	2500	381	3.12

These figures check well with those of 1922, showing that the slightly slower winter activity brings down the average daily motion for the year as compared with the average daily motion during summer months.

It was surprising to find so many of the 1922 stations, which were left *in situ* and will be of use in further observations.

OBSERVATIONS ON THE ORIENTATION OF ERRATIC BOULDERS

The Great Boulder, the Glacial Erratic marked " 1922," and other sizable erratics, because of their large cubic content and consequent absorption of heat rays, are constantly rising on ice pedestals and forming glacier tables. The large surface area of the boulders shades the ice and consequently the pedestals are broader and taller than would be the case were the rocks smaller. The mass of the large erratics is so great that when they eventually fall from their pedestals there is considerably more displacement than with smaller rocks.

This emphasizes a singular phenomenon. The main axis of the Freshfield glacier is from southwest to northeast. The cutting action of the sun's heat upon the ice pedestals of glacier tables is chiefly from the south. Consequently the erratic is subject to the action of two forces applied from different angles; with the result that, in their rising and falling, the erratics, whose orientation in 1922 was determined and photographically recorded, are turning slowly in a counter-clockwise direction. In four years this rotation has been almost 90° (pl. 9, figs. 1 and 2; pl. 10, fig. 1). The little cairn erected by Edward Feuz on the downstream tip of the Great Boulder in 1922 has fallen over, and this point is now directed toward Coronation Mountain.

OBSERVATIONS ON THE TONGUE AND ITS RETREAT

The evergreen tree, its trunk painted with a white band, still stands at the Camp Station, now 2,379 feet from the forefoot of the glacier. The stream of clear water, formerly at the foot of the bank, has disappeared.

A photograph of the tongue from this point (elevation 5,300 feet) shows very well the vertical shrinkage since 1922, as well as the stream erosion at the extreme right (pl. 10, fig. 2). The moraine (M) has also suffered through erosion by the glacial stream.

Test photos of the tongue were taken from Station C, on the north lateral moraine, on July 19 (pl. 11, figs. 1 and 2). All of the three sizable stones near the edge of the ice in 1922 have now been left behind on the morainal flat. H, the most advanced, is 330 feet from the nearest ice, an average daily retreat of 2.72 inches during a four-year period[1] (pl. 12, fig. 1).

Two additional stones, now at the extremity of the ice, were marked, each with a letter T. The lateral abutment of the terminal ice against bed-rock was marked at the northwesterly angle by a vertical reference line and the numerals '26.

The forefoot drainage of the glacier has altered considerably, the streams issuing from the center of the snout and, to a greater extent, from the northwesterly angle where, at the extreme right, adjacent to the lateral moraine, a low broad ice arch is forming.

GENERAL FEATURES OF THE GLACIER

The dissipator tongue shows notable changes in the terminal portions due to retreat, vertical shrinkage, and lateral cutting from the stream descending the Garth-Coronation gully.

The main reservoir appears more broken than when previously examined, and the upper icefalls, especially between Mts. Gilgit and Pilkington, are more open. In the opinion of the guide, Edward Feuz, climbing routes followed in 1922 would now be more difficult.

In heavy showers, on the afternoon of July 15, the writer climbed to the Niverville meadow to obtain test photographs of the lateral-alcove tongue. This secondary tongue has followed the general retreat of the main tongue and, although measurements could not be made, appears to have receded at least 100 feet (pl. 12, fig. 2).

---

[1] This gives a figure of 85 feet per year, as against the 46 feet per year estimated by Palmer for the years 1902-22. Assuming the latter figure to be correct, it would indicate that the dissipator tongue has considerably increased its rate of retreat—a fact in agreement with what is known of the recession of other glaciers in the Canadian Alps.

SUMMARY

The Freshfield glacier is definitely in a cycle of retreat and, although the annual frontal recession during the period 1922-26 is less than has been observed in other glaciers of the Canadian Alps (excepting the Lyell), it is certain that the recession rate has increased during the past 20 years.

The topography of the areas discussed will be found in the detailed maps of the Interprovincial Boundary Commission, Sheets 18 and 19, which may be obtained from the Topographical Survey of Canada, Ottawa.

Southeast Lyell glacier and Division Mtn. from flats above Glacier Lake. Mons Pk. and glacier are seen at the left. Photograph by C. D. Walcott.

Division Mtn. and the rock promontory against which the Lyell tongue formerly abutted. Through the cleft is seen the Mons tongue, whose stream, formerly finding exit through this gap, now runs behind the promontory and joins the Lyell stream. Photograph by C. D. Walcott.

Southeast Lyell glacier, showing pressure ridge in tongue between clear ice supplied through the icefall and débris-covered ice reconstructed from avalanches. The old terminal moraine, through which the stream has cut a channel, is seen in the foreground. Photograph by C. D. Walcott.

Icefall of the Southeast Lyell glacier. Note junction of clear and débris-laden ice. Photograph by C. D. Walcott.

Mons glacier from the Lyell tongue, showing icefall by which it precipitates from the snow basin between Mt. Forbes and Mons Pk. Photograph by C. D. Walcott.

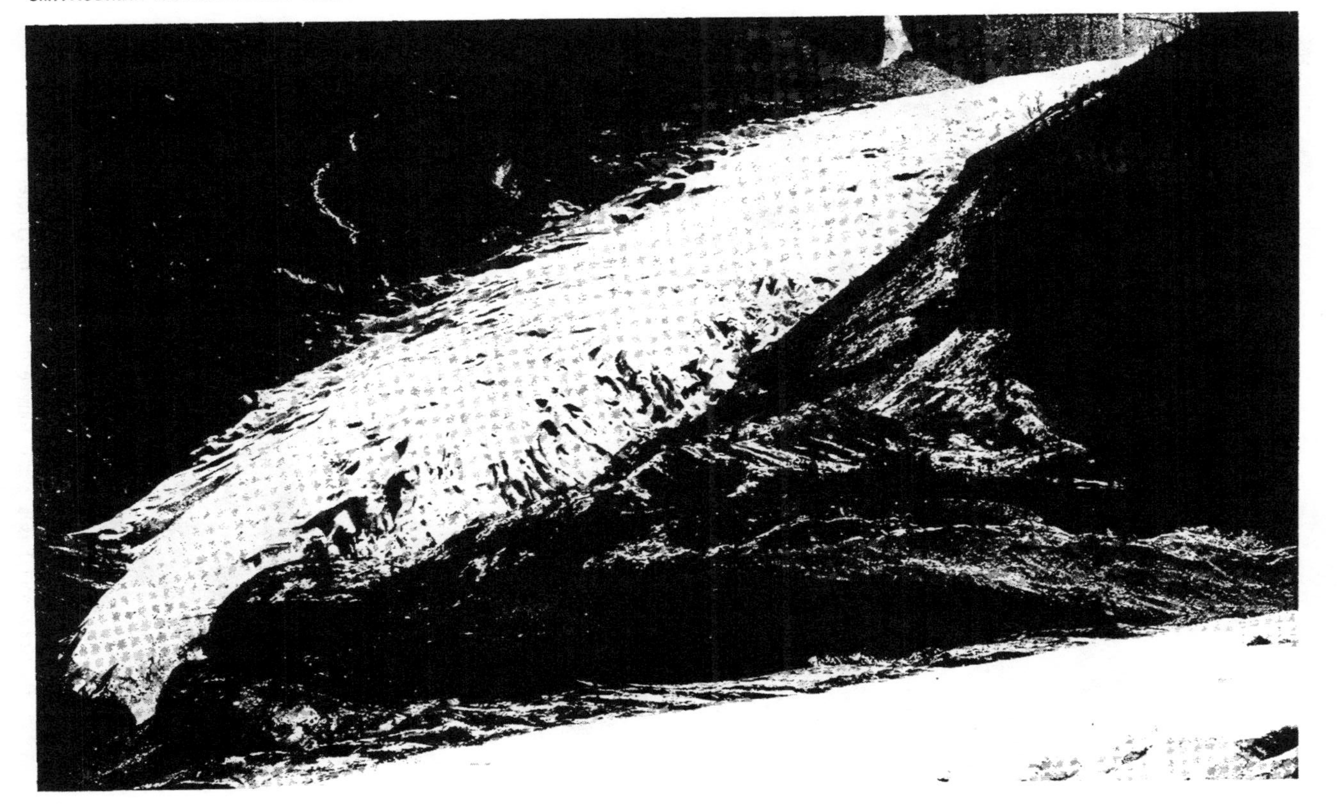

Mons glacier as it appeared in 1919. Due to retreat in succeeding years the course of its terminal stream has altered considerably. Photograph by C. D. Walcott.

1. Division Mtn. with icefalls above reconstructed portion of Lyell glacier. Bush Mtn. is seen across the Lyell icefield.

2. Central area of Lyell icefield, showing ridge of Continental Divide. In the foreground is a small tongue draining to Glacier Lake.

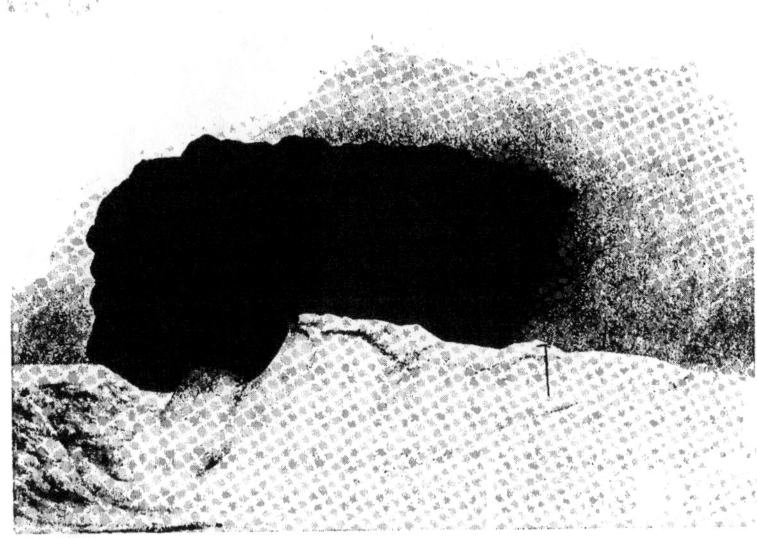

1. Great Boulder, showing ice pedestal formed since 1922. (Compare size with ice axe in right foreground.)

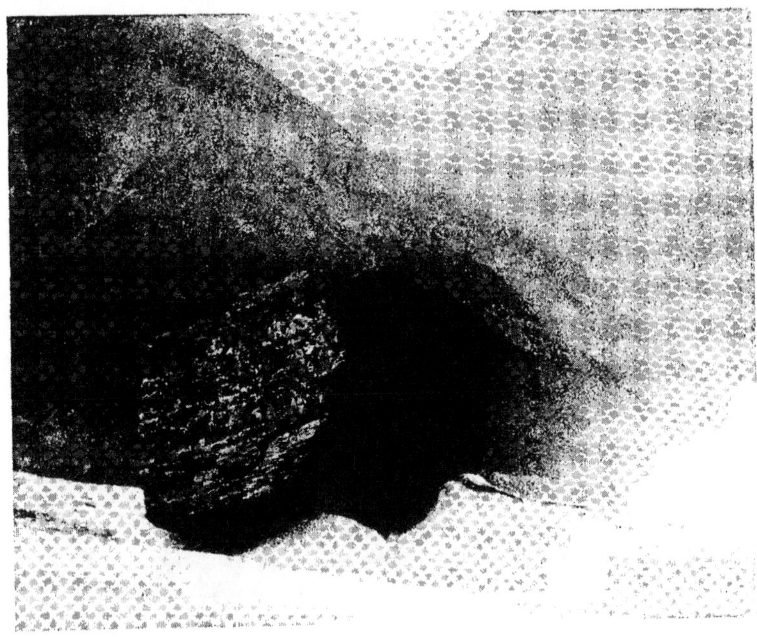

2. Great Boulder, from position of 1922 test photo, showing altered orientation.

1. Erratic Boulder marked "1922," showing ice pedestal and falling of block toward south.

2. Freshfield glacier from Camp Station in 1926. The old moraine (M) has undergone further erosion, while the ice-tongue shows marked vertical shrinkage and recession.

2.

Figs. 1 and 2 form a panorama of the Freshfield tongue from Station C. The boulder, H, was in contact with the ice in 1922. Boulders marked T were in contact in July 1926. The Great Boulder on the glacier is seen immediately below the arrow. The terminal stream issues mainly from the northwesterly angle (right).

1. Looking from the Freshfield tongue toward Camp Station. The emergence of the stream from the northwesterly angle of the glacier is seen. The boulder, H, in contact with the ice in 1922, is now 330 feet distant.

2. Lateral depression below Niverville meadow, looking toward the Freshfield reservoir and showing retreat of secondary tongue.

SMITHSONIAN MISCELLANEOUS COLLECTIONS
VOLUME 78, NUMBER 7

EXPLORATIONS AND FIELD-WORK OF THE SMITHSONIAN INSTITUTION IN 1926

(PUBLICATION 2912)

CITY OF WASHINGTON
PUBLISHED BY THE SMITHSONIAN INSTITUTION
1927

The Lord Baltimore Press
BALTIMORE, MD., U. S. A.

CONTENTS

FIG. 1.—Table Mountain Solar Observatory. The bolometer which measures $\frac{1}{1,000,000}$ degree temperature is in the tunnel, and sunlight is reflected into it.

EXPLORATIONS AND FIELD-WORK OF THE SMITHSONIAN INSTITUTION IN 1926

INTRODUCTION

Field-work is essential to the advance of nearly all branches of science, particularly those which the Smithsonian Institution is at present engaged in promoting, namely, geology, biology, anthropology, and astrophysics. The Institution therefore embraces every opportunity of putting expeditions in the field to obtain desired information or collections, either under its own auspices through financial assistance from its friends, or in cooperation with other agencies which will benefit equally from the work. During the past year more expeditions, in which the Smithsonian was represented, have gone out than ever before, and this in spite of the fact that the Institution has practically no unrestricted funds for field-work. Had it the unfettered income from an adequate endowment, much more extensive field-work in accordance with a definite plan would be accomplished each year, and the advance along the whole front of human knowledge would be greatly accelerated. The Smithsonian holds in abeyance a number of important projects in many branches of science, awaiting only funds to finance them. These include researches, nearly all of which involve work in the field, in astrophysics, meteorology, oceanography, entomology, zoology, botany, geology, archeology and ethnology, physical anthropology, mathematics, and chemistry.

The present pamphlet is intended as a preliminary announcement of the results of the year's field-work. The accounts, although written in the third person, were for the most part prepared by the participants in the various expeditions, and the photographs taken by them.

FIELD-WORK IN ASTROPHYSICS

Does the sun vary, and if so, what effects on our weather do the changes of solar heat produce? For eight years the field-work of the Astrophysical Observatory has been aimed to solve this question. With the generous aid of Mr. John A. Roebling, special observatories were erected on desert mountains in Chile, Arizona, and California.

Daily reports of the condition of the sun are received from the observers. who patiently carry on in these isolated deserts far from society, provisions, and even (at Mount Montezuma, Chile) from water.

Since January, 1926, these reports have been printed by the United States Weather Bureau on its daily weather map. Those who have followed them will have noticed that many days are missing and many rated unsatisfactory. It was to round out these records that the National Geographic Society on March 20, 1925, allotted $55,000 to Dr. C. G. Abbot, Assistant Secretary of the Smithsonian Institution, and Director of its Astrophysical Observatory, to enable him to select the best site in the Eastern Hemisphere, erect and equip there a solar radiation observatory and maintain it for several years under the title "The National Geographic Society Solar-Radiation Expedition Cooperating with the Smithsonian Institution."

FIG. 2.—Mt. Brukkaros. Looking towards the Observatory site near the top of the rim at the extreme right.

Accordingly, accompanied by Mrs. Abbot, he sailed from New York on October 31, 1925, examined the advantages of sites in Algeria, Egypt, Baluchistan, and South West Africa, and at length located the new observatory on Mount Brukkaros, South West Africa.

The mountain is 5,200 feet above sea-level and quite 2,000 feet above the plateau. The whole massif is composed of chocolate-colored rock with very little soil, though sparsely tufted with bunches of dry bush and grass, with here and there a queer cactus or dwarf tree. The cliffs are seamed into great cubes and the slopes are littered with fallen fragments.

The summit is like a cup with a flat bottom about half a mile in diameter and a steep rim 1,000 feet high. From a V-shaped

FIG. 3.—The precipice at Mt. Brukkaros leading up to the bottom of the cup. The water pools are just below the precipice.

break in the southeast side of the rim a precipice 60 feet high leaps to the bed of the dry stream, which leads down a 3-mile corridor to the plateau.

Since the observations require the use of the bolometer, that electrical thermometer sensitive to a millionth of a degree, they require very constant temperature surroundings. These are most easily obtained by making a horizontal shaft or cave, some 30 feet deep, right into the slope of the mountain near its summit.

The average yearly rainfall is $3\frac{1}{2}$ inches. Dr. Abbot was 12 days in the vicinity during March, which, equally with February, is

Fig. 4.—Observatory at Mt. Brukkaros.

Fig. 5.—Observer's quarters at Mt. Brukkaros.

the rainiest time of the year. On 11 days there were fine observing conditions each forenoon, though a bit of rain fell sometimes towards nightfall. That is surely a favorable record. There were absolutely no cirrus clouds—those wisps that slightly veil the sun and are fatal to our observing. There was very little wind—almost none in the forenoons—though in winter it sometimes blows hard.

Mount Brukkaros lies in a Hottentot reservation, and a vote of the tribe was necessary to permit us to locate there. This was

FIG. 6—Hottentot village near Mt. Brukkaros, South West Africa, which will be the headquarters of the expedition.

easily carried. Lying about 20 miles to the west of the railroad and 250 miles south of Windhoek, capital of South West Africa, Mount Brukkaros sticks out as the only peak of consequence in a circle at least 50 miles in diameter.

Thus the observers are exiled to a crater in the wilderness seven miles even from Hottentot neighbors (at Berseba) and 60 miles from a fair-sized town (Keetmanshoop). We hope their loneliness will be mitigated by the facts that they are both fine fellows, inured to camp life; that they will have interesting work; that there will be games, music, books, and radio to beguile leisure hours; and that the leopards and other wild game, so plentiful in the vicinity, will divert them. Perhaps they will raise a garden on the level bottom of the cup, if they can arrange for enough water.

The water supply is not so great a difficulty as it seems. By making a reservoir in the gulch in front of the observatory, the drainage of several square miles may be impounded. Even the few inches of yearly rainfall thus conserved will be abundantly sufficient.

For provisions and mail the observers must arrange with the Hottentots. With their own automobile, our men will make the

FIG. 7.—A Hottentot cattle-team near Mt. Brukkaros of the type which will be used to haul the expedition's instruments and supplies.

60-mile trip to Keetmanshoop frequently, where nearly everything needed can be obtained.

By permission of the Government of South West Africa, Mr. Dryden, of Keetmanshoop, Inspector of Public Works, constructed the observatory and approaches on Mount Brukkaros, last summer. Our expedition with its 60 boxes of delicate apparatus and supplies reached Keetmanshoop in September and was hauled to the foot of the mountain by two 12-ox teams in a 6-day journey. All the apparatus and accessories were at the top by mid-October. Observations began in early November.

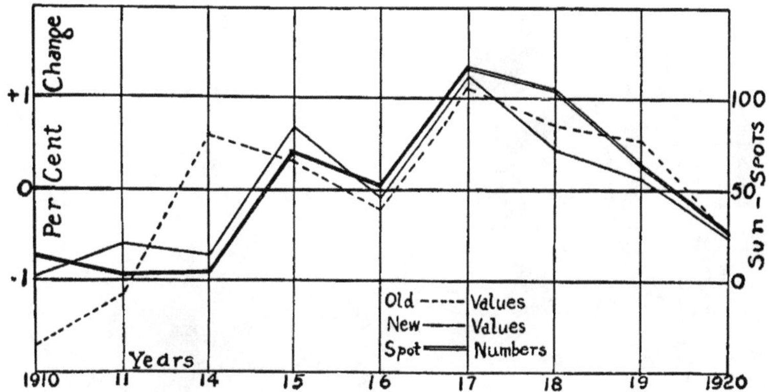

FIG. 8.—Selected pyrheliometry, solar constant and sun spot numbers compared. Mt. Wilson work, Julys 1910 to 1920.

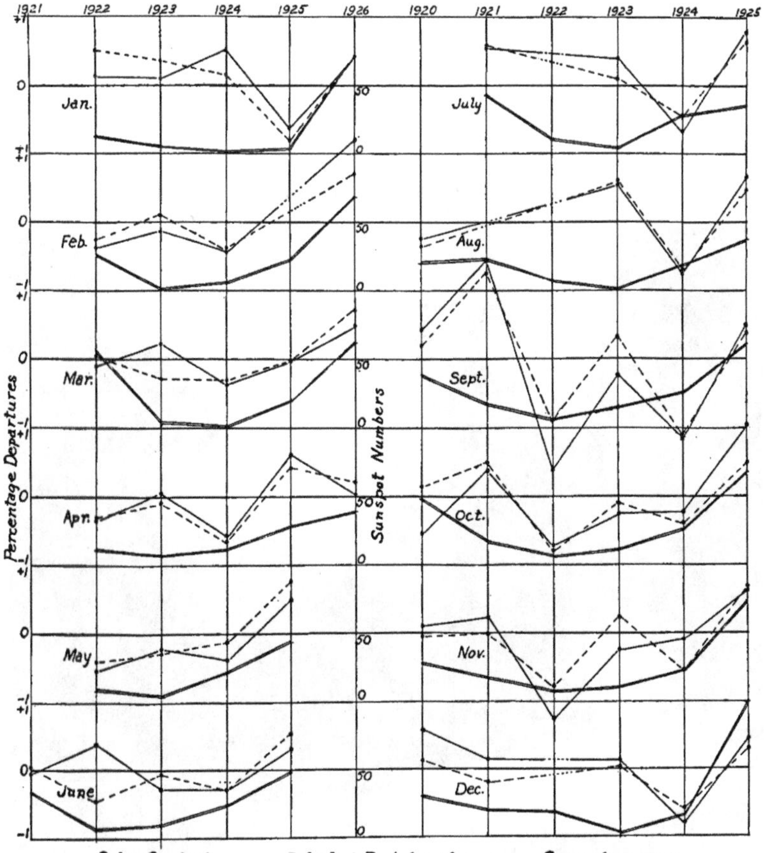

Solar Constant -----. Selected Pyrheliometry ——. Sunspots ——.

FIG. 9.—Montezuma observations, all months, 1920 to 1925.

FIG. 10.—Observer's quarters at the solar station on Table Mountain.

FIG. 11.—Looking toward the Mojave Desert from Table Mountain.

FIG. 12.—Reading the weather instruments at Table Mountain.

FIG. 13.—Reducing the observations at Table Mountain.

Returned from this expedition, Dr. Abbot, finding various critical papers published during his absence tending to express the view that solar variation is still doubtful, devised a new proof of it depending on a simple inspection, on selected days, of the pyrheliometric measurements of total solar radiation. By choosing for comparison only days of equal atmospheric transparency and humidity, the uncertainties which critics had stressed were largely eliminated. This new treatment supported closely the more elaborate and continuous indications of solar variation obtained by the complex process of spectrum analysis.

In figures 8 and 9 the new and old work is compared for separate months on Mount Wilson and Mount Montezuma, and the parallel variation of sun-spot numbers is also shown.

Having found from this new study, as well as from previous work at Mount Harqua Hala in Arizona, a source of error due to the bright rays of sky light immediately about the sun for very hazy days, an improvement of the silver disk pyrheliometer consisting of a very long vestibule has been made, designed to greatly limit the sky rays admitted to the field of view of the instrument.

In the frontispiece, figure 1, Field Director A. F. Moore is shown observing with the improved instrument at the new solar station on Table Mountain, California, built with the generous aid of Mr. John A. Roebling, and first occupied in October, 1925.

SMITHSONIAN-CHRYSLER EXPEDITION TO AFRICA TO COLLECT LIVING ANIMALS

The Smithsonian-Chrysler Expedition to collect living animals for the National Zoological Park sailed from New York on March 20 and arrived in Dar-es-Salaam, Tanganyika Territory, East Africa, on May 5. The expedition was financed by Mr. Walter P. Chrysler. Tanganyika was chosen as being one of the best localities in which to make a representative collection of the game animals of East Africa.

The party consisted of four members: Dr. W. M. Mann, Director of the Zoological Park and leader of the expedition; Mr. Stephen Haweis, artist and amateur naturalist; Mr. F. G. Carnochan, of New York City; and Mr. Arthur Loveridge, of the Museum of Comparative Zoology at Cambridge. The latter, having previously resided eight years at Tanganyika, some of which had been spent in the Game Department there, was conversant with conditions in the country, and in addition, had a good knowledge of Kiswahili language. The Pathe Review sent with the party Mr. Charlton,

Fig. 14.—Unloading crates at Dodoma. These were made at the Zoo and shipped taken down.

Fig. 15.—Wagogo natives at Dodoma.

one of their field men, to make a pictorial chronicle of the trip and the work of the expedition.

The United States Marine Corps supplied the expedition with cots, blankets, and certain other equipment, and the Freedmen's Hospital, of Washington, through the Chief Coordinator's Office, furnished the medicines.

At Tanga, the first port of call in Tanganyika, Mr. Fair, Assistant Chief of the Game Department of Tanganyika, joined the boat, and on the voyage between there and Dar-es-Salaam arranged a special permit for collecting, which was afterward signed by the Governor of the territory. There is in Tanganyika a well-organized department for the conservation of game. There is a chief, C. T. M. Swynnerton, and an assistant chief, several white rangers, and a considerable corps of native guards or scouts. The latter are much in evidence, and even in the remote parts of the territory would come into our camp, ask to see our license, take its number and our names, and find out from the natives with us just what we were doing. The very generous license that the Governor gave us proved invaluable, as it gave permission to capture specimens of practically all of the game in Tanganyika and, when necessary, to kill females in order to capture the young. This is not often necessary, and on the entire trip we did not take a single animal by the killing of the mother.

Headquarters were made at Dodoma, about 250 miles inland from the coast. The country about there is hot and dry, rolling and dotted with rocky kopjes, and reminds one strongly of parts of southern Arizona. The natives belong to the Wagogo tribe, an offshoot from the Masai, and are a pastoral and agricultural people, living on their flocks and herds and on the small amount of Kafir-corn that they cultivate. Not being a hunting tribe, they brought in to us very few large specimens but were very useful in collecting small things. Weaver birds do a vast amount of damage to the crops and the natives are in the habit of trapping these birds in quantities in small woven basket traps. So instead of destroying the birds, they brought them in to us, as well as anything else they happened on, and in this way were obtained a ratel, a fennec, a number of vervet monkeys, and a few reptiles. Mr. Loveridge and Mr. Haweis stayed at Dodoma a good part of the time, but made short trips out, and afterwards Mr. Haweis came down to Morogoro and stayed a month at Mhonde, 60 miles from Morogoro. In this locality we secured five blue monkeys, three golden baboons, and a pair of elephant shrews.

Mr. Carnochan worked in the Tabora district, where he lived among the Manumwezi tribe, experts in snake catching, and his division of the expedition succeeded in forming a considerable collection of snakes, several interesting mammals, including a female eland, and a rare caracal.

The party had been joined on the boat by a Mr. Lyman Hine, of New York, an American sportsman, and he and Dr. Mann, accompanied by two white hunters, George Runton and Guy Runton, left Dodoma by motor car and reached Umbugwe, about two hundred

FIG. 16.—Natives gathered for instruction on game drive.

miles to the north, where they made safari to Lake Manyara. Mr. Hine was able to stay in the field only a month when he returned, business calling him back to America. But Mr. George Runton, a professional hunter and guide, stayed with the expedition during its entire stay in Tanganyika and proved a most valuable addition to the party.

The Governor's license proved important at Umbugwe, for it enabled us to acquire 90 porters, men of the Wamboro and Wambugo tribes. The latter have the reputation of being swift runners, which they bear out from time to time by running down animals. The first catch on the lake was a water mongoose, which was run down on the lake shore and boxed.

FIG. 17.—Our first zebra colt.

FIG. 18.—Herd of white-bearded gnu.

FIG. 19.—White-bearded gnu calves.

FIG. 20.—White-bearded gnu calf.

On Lake Manyara we captured the three white-bearded gnu which we brought to Washington. As we marched along the shore of the lake, the herds of gnu would run in front of us, back and forth the whole day, and when evening came they were very tired and attempted to cut back between us and the lake. Our porters dropped their loads and headed them off. They got into the shallow lake and mixed together. When the splashing had died down, little groups of natives brought back the calves which they had seized. These became tame very quickly, fed readily, and were thoroughly satisfactory animals to catch and keep.

Mr. George Le Messurier joined the party to drive the motor car which the Chrysler factory in London had presented to the expedition. He kept the car busy between Umbugwe and Dodoma, carrying in specimens as fast as we had a ton of them accumulated. On the first trip, Dr. Mann accompanied the animals to Dodoma. sending Mr. Runton on to the Masai Steppe to look for young rhinos. We had been told that they were very abundant toward the Ngoro-Ngoro crater. Dr. Mann returned to Umbugwe in a week, climbed the escarpment to Umbulo, and started in the direction of the crater, but on the evening of the first day met Mr. Runton and his party coming down. In one week they had seen only four rhinos, no young nor signs of young, and since Mando, our best native guide, had told of a district, the Ja-aida swamp country, in which he said there were "Faro mingi sana" (very many rhinos) we went down into this region and found what he said to be true. Altogether we saw 22 rhinos. Our safari was charged once while on the march, and four times at night rhinos charged through our camp. But in all of these we failed to locate a single young specimen. Five different times we crawled into the scrub 30 or 40 feet from a rhino to see if it had young and were disappointed each time. One locates these rhinos, by the way, through the tick birds, which make a loud twittering at the approach of any suspicious object to the rhino on which they are clustered for the purpose of eating the ticks which are so abundant on its body. Theoretically they serve a useful purpose to the rhino by warning him of his enemies. Actually we found they were useful in leading us to where the rhino were lying, for we were attracted by the birds to each of the rhinos that we found.

The night charges are simply the result of the stupidity of the rhino. We camped usually in the vicinity of water holes, and when the near-sighted beast came to water late at night or early in the morning he would suddenly notice that there were fires and natives about. Whereupon he would put his head down and charge through in a straight line. On these occasions the natives have a frantic desire to get into the tents to be near the white men and the guns; the white men on the other hand have a frantic desire to get out of their tents, and the result is a collision at the entrance. Two rhinos came into our camp the same night. When the second one came, Le Messurier heard it snort and the sound of its tramp, and just then a native tripped over one of his tent ropes. He left the tent in a hurry, but was met by twelve boys entering it at the same speed,

FIG. 21.—A serval cub, one of our bottle babies.

and the result was that he was thrown and injured his knee so badly on a tent peg that we had to rush him back to Dodoma for medical attention. This, and the fact that we had spent 28 days without seeing a single young rhinoceros, made us decide to split up the expedition even more. So Messrs. Runton and Le Messurier stayed at Kondoa Irangi, while Dr. Mann secured the services of an additional hunter, Mr. C. B. Goss, and went down the railroad to Ngere-Ngere, and from there southward into the Tula and Kisaki districts to begin another search for rhino.

We saw, on the average, signs of five rhino each day, but again neither young nor fresh tracks of young. So after 10 days, time being short, we camped at Tula to collect whatever we could, but especially giraffes. Two native sultans, Chanzi and Chaduma, joined forces with us for a week, bringing with them about 500 natives. With the help of these we had the most successful trip of the expedition. Some of the boys from a mountain nearby had had some experience in netting game. They make a coarse seine of native rope in sections about five feet high and 15 feet long. These were placed in a row, until they made about 1,000 feet of native fence, one boy hiding behind each section. The two lots of natives would double over their ends and join in a circle about a mile in circumference, then closing in toward the net. The object was to drive animals into

FIG. 22.—Safari at rest.

the net, but nine times out of ten they would break through the line. Occasionally, however, they came straight on. One day a herd of over 50 impalla was surrounded. This is the most graceful antelope in Africa and a great leaper. Most of them sailed right over the net, but five fell short and we got them all. Fortune was with us as far as impalla were concerned, for it is one of the most delicate animals to handle, and yet all of ours reached Boston alive and in good condition.

Wart-hogs were captured in the same way and a troop of four were added to the collection.

Besides rhinoceros, giraffe was one of our important *desiderata*. They were abundant about Tula but not easy to catch because we had no horses. We tried time after time to run them into the nets, but a herd of giraffe runs in a file led by the biggest bull, and he

FIG. 23.—Hi-Boy, the male giraffe in crate.

FIG. 24.—Hi-Boy, the male giraffe in crate.

apparently enjoyed kicking the nets into the air as high as possible; whereupon they would all rush through and disappear in the scrub. But once we succeeded in separating from the herd a calf about eight and a half feet tall, and one of the natives grabbed it by the tail, another by the neck, and threw it. We got it into camp, carrying it on a native bed heaped high with grass, and put it in a room of the Kafir native-built house in which we were staying. It became quite tame in a short time, and fed readily on milk and mimosa leaves. Dr. Mann rushed to Dar-es-Salaam and had a crate built for

FIG. 25.—Handling stock at Dar-es-Salaam.

it, which was delivered to Ngere-Ngere. The animal was taken in a motor car from Tula, a distance of about 80 miles, and at last arrived safely in Dar-es-Salaam, where its crate was placed beneath a mimosa tree in the yard of the Government veterinarian. It was then that we telegraphed the Smithsonian that we had captured the giraffe. However, after 10 days, the animal was attacked by pneumonia and died very suddenly, leaving us with our homeward passage engaged on the last steamer to arrive in the States before cold weather and without the main object of the trip attained. We cabled the Sudan Government to see if they could let us have specimens and received word that they could let us have a pair of young giraffe. So all members of the expedition gathered with their respective cages in Dar-es-Salaam. For three days we maintained a menagerie visited by everybody in the city and finally embarked on the Crewe Hall, which took us to Colombo, where we transhipped on the City of Calcutta for a direct run to Boston. We landed with about 1,700 live animals, including the two giraffe (which were brought to us at Port Sudan by Mr. Skandar Armenius, assistant game warden of the Sudan, who brought also a shoebill stork), five impalla, a greater kudu, an eland, a blue duiker, red duiker, three white-bearded gnu, four wart-hogs, and quantities of birds, small mammals, and reptiles.

Fig. 26.—Aboard the steamship, Crewe Hall.

VISITS TO THE SERPENTINE DISTRICT OF SOUTHERN ENGLAND AND THE GEM-CUTTING TOWN OF OBERSTEIN, GERMANY

Dr. George P. Merrill, head curator of geology in the National Museum, availed himself of the opportunity offered by the Geological Congress in Madrid to visit some of the more important museums of England and on the continent, and also to visit sundry localities of geological and mineralogical interest. Among the more important, mention may be made of a visit to the historical quicksilver mines at Almaden, Spain. The party from the Congress was permitted to descend to the 280 meter level, where a fine massive body of ore (cinnabar) was exposed, embedded in quartzite. A series of typical specimens was collected and forwarded to the National Museum through the office of the United States Consul. At the close of the Congress a party was formed, of which Dr. Merrill was a member, visiting Barcelona, with side trips to the eminence Tibidabo and the monastery on Mount Serrat, and thence going to the Island of Majorka, one of the Balearic group. Here, from June 5 to 13, inclusive, a series of automobile trips took the party well over the Island, visiting many interesting points and examining in detail the evidence of overthrust faulting and folding which abounds. Later in the season, trips were made to the serpentine and tin mining districts

Fig. 27.—Kynance. The rocks are of the historic *Lizard Serpentine*.

of southern England and the noted gem-cutting town of Oberstein in Germany. A brief account of these two trips is given below.

THE LIZARD SERPENTINES

In the early literature relating to petrographic research are to be found many references to the serpentinous rocks of Cornwall in southern England, and their problematic origin. Concerning this last, it was long ago decided that they were the result of the combined forces of igneous intrusion and subsequent metamorphism. It is with the rocks as they are today that this note has to deal; the exact locality is Kynance, near Lizard Point and but a few miles west of England's "jumping off" place, or Land's End. It is a most fascinating place on a fair, warm day in July. The country is rough and the shore often precipitous. The rock where exposed to the action of the waves and windblown sand assumes an almost ebony-like hue and polish, but on close inspection it is found to be often filled with small streaks and gashes of a blood-red color. Broken fragments on the beach have given rise to a diversified series of oval pebbles of green, gray, purplish and red colors from which it is a delight to assort and select those which are most beautiful. Below the zone of pebbles the beach is of fine, clean sand over which the warm waters of the cove seeth with seductive softness, making it a bathers' paradise. As bathers are out in force one looks for bath

FIG. 28.—Pendeen. A glimpse of an old tin mining region in Cornwall.

houses, but not one of these ugly conveniences disfigures the landscape. A brief investigation reveals the fact that the numerous caves worn by the sea in the shattered serpentine are made to answer, and in their shallow and not very dim recesses may be seen groups of men and maids making the slight changes in raiment considered appropriate to the occasion. There is a primitive simplicity about it that is very pleasing, though it might have been a trifle shocking to a modest man had he witnessed it before his eyes had become accustomed to the scanty costumes of the present day.

Naturally the beauty of the stone, when polished, long ago attracted the artist and the artisan, and attempts were made to utilize it as a marble for interior decoration. The attempt failed for the same reason that has brought disaster to a large proportion of like attempts elsewhere, namely, the badly jointed condition of the stone—its dry seams and like defects that prevent the getting out of pieces of more than moderate dimensions. Nevertheless, in the museums and art galleries are frequently to be found turned bowls, vases, urns, and stands for busts, of good design and color, the work of these earlier years. A very considerable souvenir industry still exists. To be counted by the score are the little shops in Kynance, Penzance and elsewhere where the natives work up small pieces into objects more or less—usually less—artistic, but always interesting for the variegated veining and for their colors. One can but regret that an arrangement cannot be made whereby with larger means better work could be done.

For the tourist, the approach to Kynance is by bus from Penzance, a delightful little town, ancient but now becoming popular, though one may hope not fashionable.

A few miles eastward are the historic tin workings of Cornwall now largely discontinued owing to the increasing depth and low grade of the ores. The country is at best bleak, and the miles of stone-walled abandoned pits and ruined shaft-houses, often with little more than chimney standing, add to its picturesqueness but not to its beauty.

GEM CUTTING IN OBERSTEIN-IDAR, GERMANY

The towns of the Nahe Valley are not large. Indeed, there is not room in the narrow valley for large towns, and whatever attempts are made to attain greater dimensions must end only in increase in length. Doubtless the little stream has done and is doing its best, but geological processes are slow. The work that has thus far been accomplished is more in the way of depth than breadth. So narrow is the resultant valley that there is, for the most part, space for but a single street, along which the houses are jammed back against the solid ledges of rock which rise abruptly behind them.

The geological history of the region is interesting, and inasmuch as it is this that has given the towns their industry and importance, it may be briefly touched upon.

Very many years ago, so long ago that it can be approximated only in the rough divisions of geological time, there occurred here an enormous outbreak of volcanic activity. Huge steaming masses of lava flowed out over the region and in time became consolidated into the solid rock now forming the high hills on either side and down through which the patiently persistent Nahe has cut its way. But the upper portion of the consolidated mass was full of steam cavities, large and small, like those in a well-raised loaf of bread. Into these the heated waters carried silica in solution and formed agates. In turn, as the rock decayed and the stream cut deeper, these agates being less destructible, accumulated in the soil and ravines, to be collected doubtless at first as mere curiosities. Interest in their banded structure was probably aroused from broken fragments which led to laborious grinding of flat surfaces on sharp-gritted sandstones. Later yet they were ground by holding them against huge revolving grind-

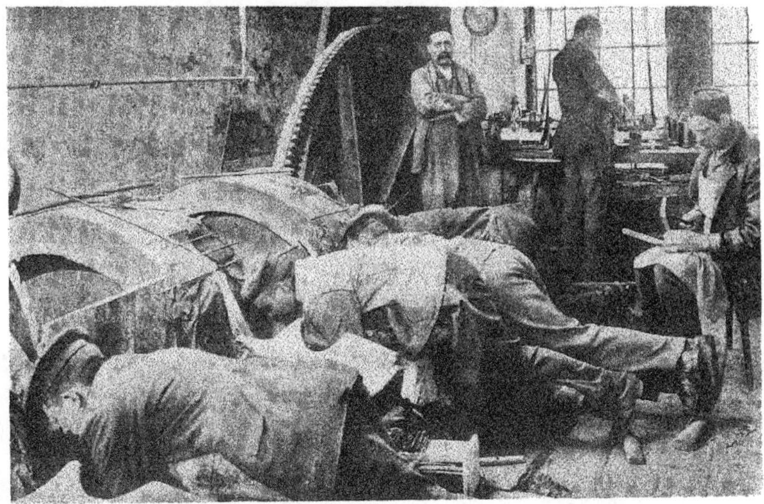

FIG. 29.—Grinding agates. Oberstein—Idar.

stones driven by slow-turning undershot water-wheels, the operator lying prone upon a roughly carved wooden form moulded to his chest and abdomen, and with feet braced against a wooden cleat nailed to the floor. This crude system is still followed in many of the smaller shops, but is laborious and unhealthy, the confined position of the lungs and constant dampness from the cold, wet wheel being productive of rheumatism and pulmonary troubles. In all the many "Schleiferei"—shops devoted to gem cutting, and there are some hundreds in the region—the slicing is now done on circular saws of thin metal charged with diamond dust, and the grinding on metal plates revolving horizontally, at which the workmen sit in comfort.

Exhaustion of the local supply it may be safely assumed resulted in the reaching out into world-wide sources for new materials, and

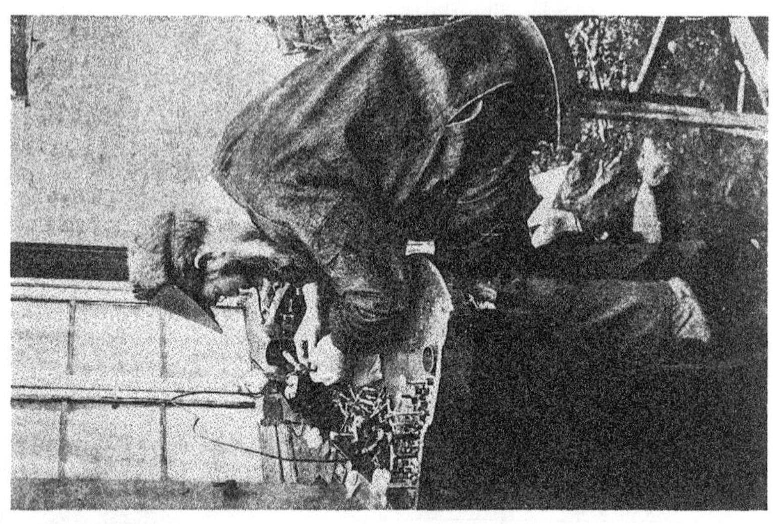

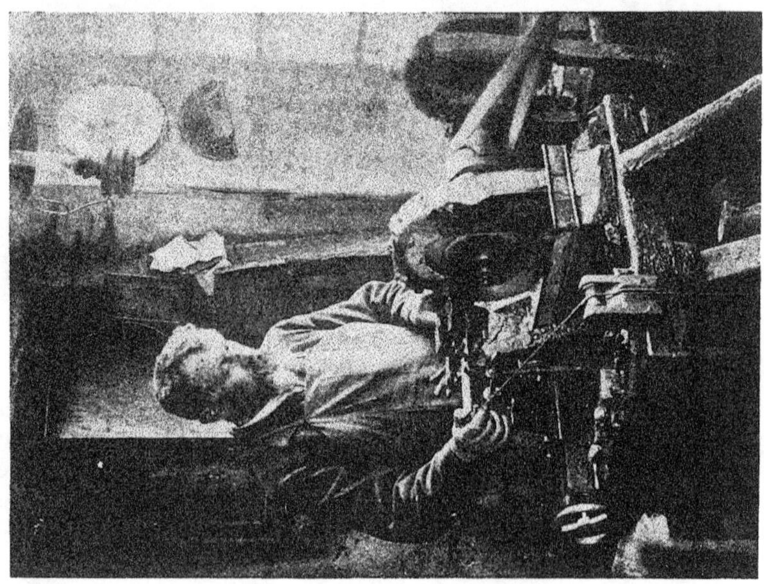

a contemporaneous development in the art of cutting as well. Agates, amethysts, and topazes from Brazil, kunzite from California or Madagascar, beryls including aquamarines, golden beryls and morganite from Madagascar and America, peridot from Oriental sources; amazon stone and rose quartz from America, are among the many varieties.

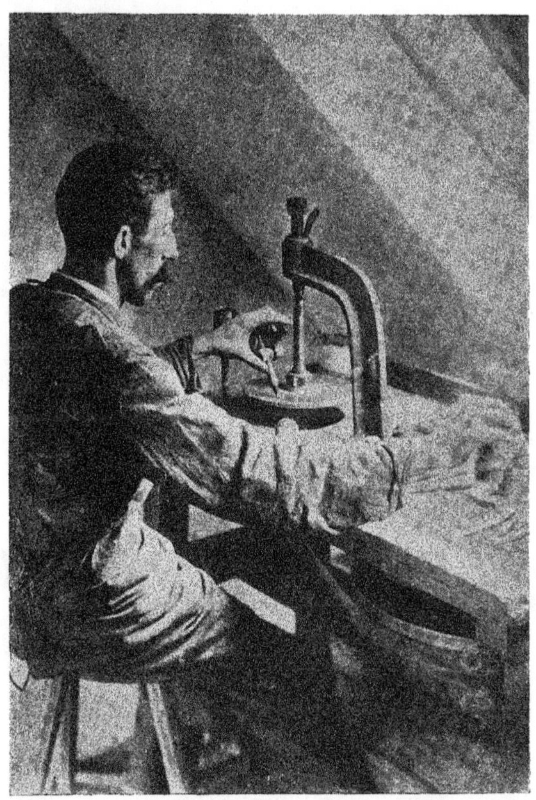

FIG. 32.—Grinding facets on gem stone.

Aside from the facetted forms, beads, pendants, etc., used in jewelry, a variety of ornamental bowls, trays, vases, paper knives, some of which are very beautiful, are cut from the agates, rose quartz, lapislazuli and other of the so-called semi-precious stones, many of which find their way into museums and mineral collections where they are highly prized. A visit to the little local museum is needed to convey a full idea of the variety and beauty of the work now carried on behind the walls of the unimpressive buildings with

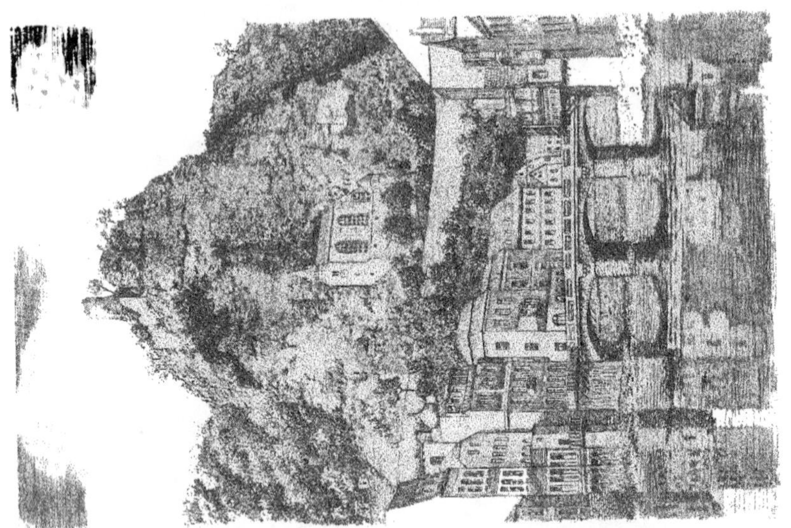

Fig. 34.—Oberstein.

Fig. 33.—Oberstein.

which the street is lined. Today the towns offer an interesting illustration of the continued development of an industry long after the conditions which gave it birth have ceased to exist.

Interest in the two towns is, however, not limited to the " edelstein " industry. High up on the ragged hills that dominate Idar are the ruins of two castles dating back to a very early period—it is said to about the middle of the eleventh century. Tradition has it that in one of these dwelt two brothers. Whether a cat or a lovely maiden was the exciting cause (the tradition varies on this point), a violent quarrel developed which culminated in the younger being thrown from the castle window and dashed to pieces upon the rocks below. Repentant for his hasty act, the elder wandered aimlessly forth, entering first the service of the prior of a Rhenish monastery and afterward becoming a crusader, ultimately receiving absolution on condition that he return and with his own hands erect a sanctuary upon the exact spot where his brother fell. The penalty was carried out, and the now existing church, erected it is said in 1482, occupies the original site. It is an interesting little relic, its steeple alone projecting beyond the face of the cliff, and is well worthy the climb of the 163 irregular stone steps to give it a visit.

EXPLORATIONS FOR MICROFOSSILS IN FRANCE AND GERMANY

Probably no branch of natural history has received more intensive study during the past 10 years than the microscopic fossils which have proved of such great value in the determination of underground structure, particularly in oil geology. The paleontological collections of the U. S. National Museum are rich in species of fossil microorganisms from many American Mesozoic and Cenozoic formations. Although many of these species have been separated and photographed during the past 20 years, their descriptions have never been published because of uncertainty regarding their relationship to the many described European species. Numerous large monographs and thousands of smaller articles upon European microfossils, particularly the moss-animals or bryozoa, the bivalve crustacea or ostracoda and the foraminifera among the protozoa, have been published during the past century but, in most cases, the illustrations are either freehand sketches or diagrammatic drawings which make the recognition of the species uncertain unless specimens from the type locality are available for study. To secure such typical European material for comparison with the American faunas, Dr. R. S. Bassler, curator of paleontology, U. S. National Museum, spent August and September, 1926, in a study of various classic localities in France and Germany.

The first two weeks were occupied in a study of the Paris Basin in company with Dr. Ferdinand Canu of Versailles, France, the most eminent student of microfossils upon the Continent. The various publications upon American fossil bryozoa by Canu and Bassler were prepared entirely by correspondence, dating back to 1909, so that the actual meeting of these co-workers was a long anticipated pleasure. The result was that more time was spent in personal conference upon the past and future work than in actual field investi-

Fig. 35.—Die Pfalz, Rhine Gorge. (Photograph by Bassler.)

gations. However, the main formations of the Paris Basin were studied in a general way, but the most valuable collections of microfossils from this area were donated by Dr. Canu from material secured in his previous researches. In remembrance of this meeting and of his years of pleasant association with the paleontological work of the National Museum, Dr. Canu also presented to the National Museum his entire collection of French Cenozoic and Mesozoic fossils, numbering not less than a hundred thousand specimens. This gift is of particular interest to American paleontologists in that all the specimens are most carefully labelled as to exact horizon and locality, a most necessary item in present-day studies but often lacking in many collections from foreign countries.

Leaving Dr. Canu with much regret, Dr. Bassler proceeded to the Rhine valley where he studied, in succession, the broad plain around Strassburg, the valley to Mainz and the valley of the Main River from Mainz to Frankfort. Next the trip through the Rhine gorge was made, which was particularly interesting in that a first-hand knowledge was obtained of the Devonian stratigraphy of this classic area. Important collections of Devonian fossils were secured here, and the classification of them as well as of other collections from this area secured in the past, can now be made intelligently.

FIG. 36.--Bavarian Plateau at Munich with Deutsches Museum.

Although Rhine valley scenes are familiar to all, the photograph of the Pfalz in the middle of the river, the ruined castle at the top of the plateau, the modern town at the water level and the terraces for vineyards, in addition to the stratigraphy (fig. 35) make a combination of geological and historical interest hard to surpass.

The Early Tertiary deposits of southern Bavaria were next studied, and opportunity was taken here to visit the wonderful Deutsches Museum at Munich (fig. 36) where one can study the underground geology of the earth's crust in the basement halls and proceed from story to story through all phases of human activity until in the planetarium at the top of the building the movements of the heavenly bodies are exhibited. A chance to study the Mesozoic limestones of the Bavarian Alps quickly and without effort was afforded by a visit to the Zugspitze on the Bavarian-Austrian boundary where the recently completed cable system to the top of the mountain, over nine thousand feet high, conveys one in 20 minutes over this distance which the best of climbers cannot accomplish in less than two days. The tiny car, operated as shown in figure 37, brings one at times

Fig. 37.—Cable road to top of Zugspitze, Bavarian Alps.

close enough to the limestone strata to give a good idea of the geological structure.

Proceeding northward from Munich, various regions in Germany were studied with profit both in the amount of good study material secured and in the information regarding stratigraphic relationships. The classic Mesozoic region north of the Hartz Mountains was visited in company with Mr. Ehrhard Voigt, an enthusiastic student of microfossils at Dessau, Germany. Mr. Voigt also accompanied

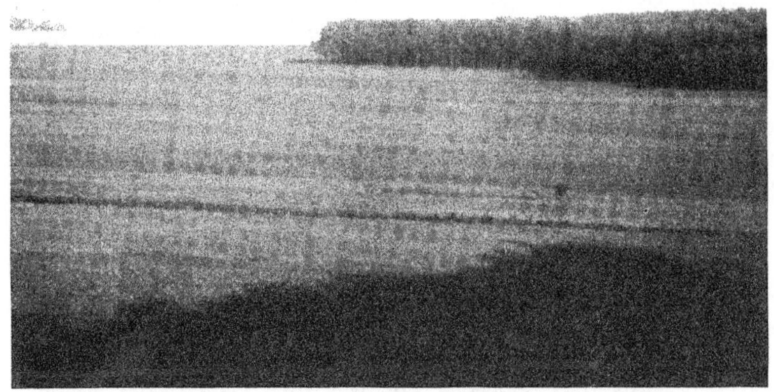

FIG. 38.—Baltic Plain, south of Berlin. Taken from train, the shadow of which shows in fore part of view. (Photograph by Bassler.)

FIG. 39.—Chalk cliff along coast, Island of Rügen.

Dr. Bassler to other regions celebrated in German stratigraphy, particularly the potash areas around Stassfurt, the drift region around Dessau and other regions to the north, and finally to the Island of Rügen on the Baltic. During these explorations, a short time was spent in the Berlin Basin where under the guidance of Dr. Kurt Hucke, of Templin, Germany, a good idea of the geology was obtained. This area and that to the north belongs to the Baltic Plain, a flat region covered with glacial deposits (fig. 38) in which the underground stratigraphy is made out with difficulty. However, the pebbles in these drift deposits afford such good clues to the geological strueture, from their contained fossils, that a new phase of research has been developed by Dr. Hucke and his associates and a special society for this study has been formed. *Der Zeitschrift für Geschiebeforschung,* the journal of this society, of which Dr. Hucke is President, contains discussions of drift problems which are quite new to the American investigator.

The classic Island of Rügen, off the north coast of Germany, with its chalk cliffs of Cretaceous age, was then visited, and although most beautiful from a scenic standpoint, it proved at first very disappointing paleontologically. Hundreds of species of Cretaceous microfossils have been described from this area and it was believed that the specimens would surely occur in great abundance. However, as shown in figure 39, the cliffs are almost inaccessible and only the chalk blocks which have fallen to the beach are available for material. Upon breaking up these blocks, few fossils were found. The disappointment was lessened by the fact that the shore is composed of undecomposed boulders of various igneous rocks, limestones, sandstones and other types deposited here as drift material during glacial and subsequent times. All the various formations to the north are represented and some of the formations are evidently from outcrops at the bottom of the Baltic Sea for they never have been found in place on the land. The latter formations, curiously enough, were particularly rich in microfossils. The disappointment over the few fossils in the chalk bed was dissipated entirely when the " Kreideschlemmerei " or chalk washing establishments at the town of Sassnitz on Rügen were located. It happens that an important industry has been developed around the use of chalk for various whitening purposes, but the chalk must be pure and free from fossils and flint fragments. To accomplish this, the chalk is passed through the washers and all the fine and coarse débris is sieved out and thrown aside leaving the water with its dissolved material to settle. The pile of débris resulting from such washing as shown in figure 40 explains why so many fossils have been discovered in this area. Not only were many excellent echinoids, brachiopods and other large fossils picked up in the dump heap, but literally billions of microfossils were obtained simply by shoveling up several boxes of the fine débris.

FIG. 40.—Kreide-schlemmerei at Sassnitz, Island of Rügen, showing chalk cliff in background and pit dump heap of fossils in foreground. (Photograph by Bassler.)

The last few days of the trip were spent in London where an idea of the stratigraphy of the London Basin was obtained although most of the available time was devoted to a study of Museum methods and to conferences with scientific friends. The material results of the summer's work have reached Washington safely and after two months' work are now unpacked and in proper form for study.

GEOLOGICAL FIELD-WORK IN THE ROCKY MOUNTAINS

In continuation of the geological field-work in Rocky Mountain stratigraphy, under the direction of Secretary Charles D. Walcott, Dr. Charles E. Resser, associate curator of paleontology, U. S. National Museum, accompanied by Mr. Erwin R. Pohl, of the paleontological staff, left Provo, Utah, with the regular field outfit late in July, 1926.

The first objective was the determination of the section in Shoshone Canyon west of Cody, Wyo. The strata were here found to be rather badly metamorphosed, and fossils were few, but enough information was obtained to determine the position of the exposed Cambrian beds. An attempt was next made to work out the difficult sections farther north in the Beartooth Range, with the result that the proper stratigraphic position of the rocks was determined although the minor details could not be ascertained because of the ruggedness of the country.

The major objective of the season's work was a restudy of the famous sections north of the Gallatin Valley in Montana. Camp was established at Logan and daily trips made into one or another part of the foothills north of the Gallatin River. Mr. Pohl here investigated the Devonian strata particularly, securing good study collections. The Cambrian beds yielded many fine trilobites and other fossils as they have in the past, but a new locality in Nixon's Gulch furnished a large collection representing the Middle Cambrian Stephen fauna of British Columbia.

After the extended stop at Logan, camp was moved to Pole Creek south of the Madison River in the foothills of the Madison Range. Here ample collections were obtained from Middle Cambrian beds from which the National Museum had previously but a few fragments.

Near the close of the season several days were occupied in the study of the exposures in the Wasatch Mountains north of Brigham City, Utah. Dr. Asa A. L. Mathews, of the University of Utah, joined the party and served as guide to the places where he had previously discovered outcrops of fossiliferous Cambrian and Canadian formations. Two very strenuous trips to the top of the range resulted in obtaining many instructive fossils.

The season as a whole yielded returns beyond the average for several seasons past, in quantity and quality of fossils and in new stratigraphic data.

FIG. 41.—Shoshone Canyon, west of Cody, Wyoming. Upper Cambrian beds directly beyond bridge. Massive limestone below bridge, Bighorn Formation. Wall of Canyon at right, Mississippian rocks. (Photograph by Milwaukee Public Museum.)

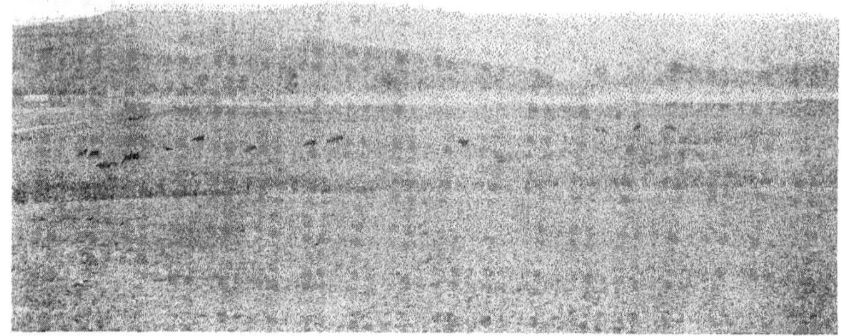

Fig. 42.—View northward across Gallatin River, north of Manhattan, Montana. Low hills of Paleozoic rocks. Nixon's Gulch on right side. (Photograph by E. R. Pohl.)

Fig. 43.—Gallatin Valley at Logan, Montana. Upper Cambrian strata at extreme right. Massive cliff, Jefferson limestone. Gully at left typical feature due to outcrop of the Three Forks Shale. (Photograph by E. R. Pohl.)

FIG. 44.—Gallatin Canyon near Cherry Creek Road. Cliffs formed by Paleozoic limestones. (Photograph by Chas. E. Resser.)

FIG. 45.—Details of Middle Cambrian strata, north side of Gallatin River, east of Logan, Montana. (Photograph by Chas. E. Resser.)

GEOLOGICAL FIELD-WORK IN NEW YORK AND ONTARIO

In continuation of work begun in 1925 by the division of paleontology of the U. S. National Museum, Mr. Erwin R. Pohl was detailed to spend four weeks during the summer of 1926 in the field in procuring the detailed data necessary to label correctly as to geological horizon the large collections of Devonian fossils from many classic localities in the eastern United States. Work was continued on the section at Kashong Creek in west-central New York where the beds of lower Hamilton age, including the Skaneateles and Ludlowville shales were made the subject of close stratigraphical

FIG. 46.—Falls of Kashong Creek over upper part of Ludlowville shale. Bellona, Seneca Lake, N. Y. (Photograph by E. R. Pohl.)

study. These strata contain many rare and excellently preserved invertebrate fossil remains, and not only were the collections increased but the information was also acquired to make those already in the Museum of real scientific value under the newer methods of study. Figure 46 illustrates one of the many charming waterfalls that abound in this section of the country. The strata here lie practically horizontal, and the falls are almost invariably due to the resistance of a harder bed of rock and the undercutting of the less resistant underlying layers.

At Ithaca, N. Y., due to the general southerly dip of the strata, excellent exposures of upper Devonian rocks are afforded, and the locality shown in figure 47 is a collecting ground that has long been favored by geological collectors. The beds exhibited are at the base of a considerable invasion of the Portage during the Upper Devonian and carry a distinct fauna of striking forms. Several other noted localities in central and western New York were also paid brief visits. The most important of these trips, both in purpose and result, was that to examine the Middle and Upper Devonian beds at the celebrated series of sections along 18 Mile Creek and the north and south shores of Lake Erie. In this vicinity the entire sequence of the Hamilton, which in central New York comprises some 400 feet of

Fig. 47.—Ithaca Falls over Ithaca beds of Portage age. Ithaca, N. Y. (Photograph by E. R. Pohl.)

rock and in eastern New York and Pennsylvania is as thick as 3000 feet, is represented in less than 90 feet of superimposed beds. The faunal sequence, however, is as well marked, and the zones even more distinct than those of their eastern equivalents, for here the succession lacks many of the invasions that are represented in the more easterly deposits of the same age. This portion of the expedition was undertaken in cooperation with the Department of Geology of the Milwaukee Public Museum, which institution was represented by Mr. Gilbert O. Raasch, assistant curator of geology.

Following the work at 18 Mile Creek, Mr. Pohl proceeded alone to Thedford and Arkona, Ontario, Canada, for the purpose of

Fig. 48.—Beds of Hamilton age at 18-Mile Creek. The protruding band is the Encrinal limestone, which separates the lower from the upper Hamilton. Section 7, 18-Mile Creek, N. Y. (Photograph by E. R. Pohl.)

Fig. 49.—Section of the middle Devonian shales at No. 4 Hill, Arkona, Ontario, Canada. The beds are divided into lower and upper Hamilton by the Encrinal limestone in the middle foreground. (Photograph by E. R. Pohl.)

determining the correlation of the beds exposed in that region with those of the New York deposits. The country intervening has been subjected to much erosion and is moreover covered under a heavy deposit of glacial material, so that the strata under study are nowhere displayed. This correlation was accomplished on the basis of a bed immediately above the Encrinal limestone, shown in figures 48 and 49, which contains the exact fauna that is carried in one of the zones of the section in western New York at East Bethany. The strata at Thedford and Arkona have been separated into a three-fold divi-

Fig. 50.—Rock Glen, Arkona, Ontario. The Encrinal limestone here forms a small falls due to relative resistance in contrast to that of the soft shales above and below. (Photograph by E. R. Pohl.)

sion according to their fossil content. The lower beds, or as they are locally called, the Olentangy shales, are composed of fine gray shales which easily weather into a soft mud used in making pottery, and carry thin lenses of hard crystalline limestone composed throughout of masses of beautifully sculptured fossil remains. Several large exhibition slabs, besides much other material, were returned to the Museum from these beds. The middle division includes the so-called Encrinal limestone which is made up of an innumerable mass of crinoid remains, and a three-foot bed of shale which is referred to as the "coral bed" on account of the large number of fossil corals that it contains. This is the bed whose fauna is also found in New York State. The Widder beds, or the upper division, contain an entirely different fauna and one that is not known elsewhere. The entire section is well exposed in several ravines such as those shown in figures 49 and 50, and is followed by black shales of Mississippian age which contain conodonts and plant remains. These beds are best exposed on the shores of Lake Huron at Kettle Point (fig. 51). Included in the shale, which has a high bituminous content, are many

FIG. 51.—Huron shales of Mississippian age, with included "cannon-ball" concretions. Kettle Point, Lake Huron, Ontario, Canada. (Photograph by E. R. Pohl.)

and often large peculiar spherical concretions with a radially crystalline structure. The nucleus of these concretions is usually a single plate of some armored fish of the Mississippian. In emerging from the shales under the action of erosion and frost, these concretions often split in half and lie about as the "kettles" of local terminology. The work in this district was aided by several amateur collectors including Messrs. Charles Southworth and J. R. Kearny, who have also been prevailed on to add to the Museum collections from their lifelong accumulations of local collecting.

COLLECTING FOSSIL FOOTPRINTS IN THE GRAND CANYON, ARIZONA

In continuation of the investigation of the fossil tracks found in the Grand Canyon, so well begun in 1924,[1] Mr. Charles W. Gilmore, curator of the division of vertebrate paleontology, U. S. National Museum, made a second visit to the Canyon in the early spring of 1926.

This investigation, made possible through an allotment granted by the Marsh Fund Committee of the National Academy of Sciences, had as its purpose the acquiring of additional specimens of fossil tracks from the Coconino and Hermit formations and the extension of the explorations into the older Supai formation in which the discovery of fossil tracks had been reported by Mr. J. R. Eakin. Superintendent of the Grand Canyon National Park.

Arriving at the Canyon on April 20, Mr. Gilmore with one assistant spent three weeks in active field-work collecting and exploring these formations as exposed on the Hermit and Yaki trails. The expedition was successful far beyond expectations, as the collection made consisted of a large series of slabs of some 2,700 pounds in weight on whose surfaces were preserved the foot impressions of a great variety of animal life.

The old Coconino locality on the Hermit Trail was revisited and a large series of beautifully preserved tracks and trails secured, which included many kinds that were new to this Ichnite fauna. In the Hermit shale some 1,400 feet below the level of the Canyon rim a large assemblage of fossil tracks and plants was collected. The presence of insect remains was made known for the first time by the discovery of the wing impression of a large dragonfly-like insect; and finally in the Supai formation 1,800 feet below the rim another footprint horizon was definitely located.

Fossil tracks were found in considerable abundance in all three of these formations and at several levels in the Hermit Trail section, and it was along this trail that most of the collecting was done. This later investigation shows that in the perfection of their preservation and in the great variety of footmarks found, there are few localities that outrank this one. It is further unique in probably being the only place in the world where fossil tracks of three successive faunas may be found in one nearly vertical geological section, separated by such great geological intervals. It is now known that these evidences of past life range through over 800 feet of rock strata.

[1] Smithsonian Misc. Coll., Vol. 78, No. 1, 1926, pp. 20-23.

Fig. 52.—Trackway of a small salamander-like animal showing the marks of a dragging tail between the footprints. More than natural size.

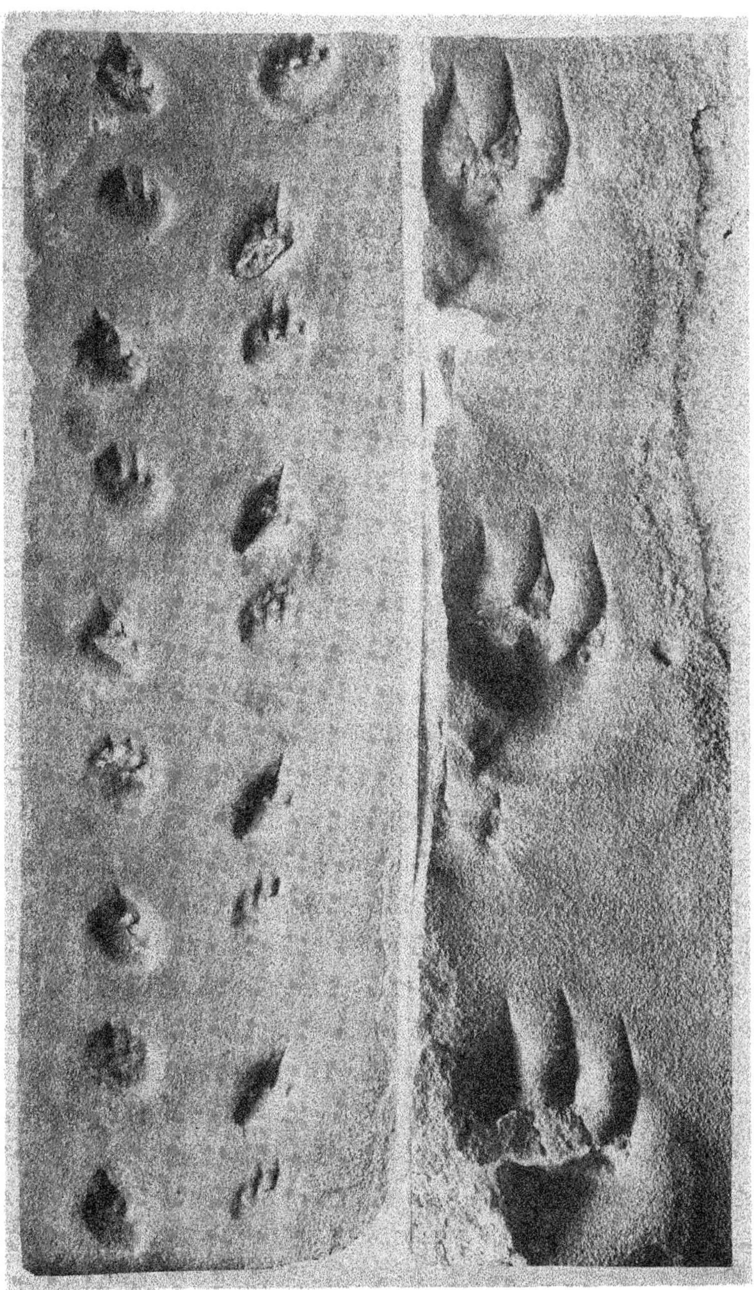

FIG. 53.—Trackway of *Laoporus noblei* Lull (on left.) New species of fossil footprints (on right). Both reduced.

At the present time fossil tracks are known in these formations only on the Hermit and Yaki trails, but doubtless their geographical range will be rapidly extended now that the precise levels of their occurrence has been determined.

Previous mention has been made of the curious fact that in the Coconino nearly all of the tracks and trails were going in one dircetion, that is, up the slopes of the crossbedded sandstones. Examination of many additional hundred square feet of track-covered surface all goes to confirm this original observation, for in the hundreds of trails seen in the field only three exceptions to this trend of direction were observed. While slabs of considerable size were collected, larger and more impressive trackways might have been secured but for the difficulty of transporting them to the top of the Canyon, as all of these specimens were brought out of the Canyon on mule back along a narrow and often precipitous trail. It will thus be seen that the size of the specimens collected was always determined by the carrying capacity of a mule, which is about 150 pounds on these trails.

Preliminary study of this collection of tracks indicates that the known Ichnite fauna of the Coconino will be doubled in the number of genera and species; that an adequate fauna will be established for the Hermit shale; and a beginning will be made in the development of a fauna for the Supai formation. The development of these Ichnite faunas may, in the absence of other fossil criteria, be of great assistance in the correlation of other track-bearing formations of distant localities.

AN ELEPHANT HUNT IN FLORIDA

A mountable skeleton of some species of American mammoth has been long desired for the National Museum exhibition collection. But while remains of these extinct elephants have been gathered from almost all parts of North America, they consist for the most part of isolated teeth and bones. Occasionally jaws and still less frequently skulls with parts of the skeleton have been located. The National Museum has two of these skulls. The smaller, representing the species *Elephas primigenius,* is from northeast Siberia; the larger one, *Elephas boreus,* was found in a Pleistocene deposit near Cincinnati, Ohio. The next in importance of remains of the mammoth in the National Museum is a specimen from Idaho, representing a third species, *Elephas columbi.* This specimen consists of part of a lower jaw and an upper tooth associated with several other parts of the skeleton. None of this material however, is sufficiently complete to form the basis of a successful skeleton restoration. It was therefore with renewed hope of realizing this desire for a mountable skeleton of one of these great elephants that a message was received, in the early autumn of the present year, from the Venice Company of Venice, Florida, reporting the discovery near that place of a mammoth skeleton. A cordial invitation to the Smithsonian to send and secure this specimen for the Institution's collections accompanied this report, and as this seemed sufficiently promising to warrant investigation, Dr. James W. Gidley, assistant curator, division of fossil vertebrates, was detailed to go to Venice for that purpose. Arriving there on the morning of the first day of November, Dr. Gidley was occupied for the next 10 days with collecting the fossil, and with studying the geological formations in its vicinity.

A preliminary examination revealed the fact that the bones all belonged to one individual but that the skeleton was by no means complete. However, the specimen was very far from valueless, and the portions remaining were of sufficient value to amply repay the time and expense required to collect and preserve them. Among the more important pieces obtained are the lower jaws including the teeth, both upper cheek-teeth, considerable portions of both tusks and many of the more important bones of the feet.

These pieces show that this mammoth, or extinct elephant, had been an animal of very large size. In life it would have stood at least 12 feet high at the shoulders and it carried a pair of great incurving tusks which were eight inches in diameter and, measured around the curves, were each 10 feet or more in length.

The location of this find is the north bank of the main drainage canal of the Venice Company at a point about four miles east of the Venice Hotel. The bones were first discovered by Mr. J. W. Parker, a farmer living in the vicinity, while walking along the water's edge after the subsidence of an unusually heavy freshet which had filled the canal to overflowing and had caused heavy caving of its banks. It was this caving that had brought the bones fully to view. It also broke up and destroyed many of the bones including the skull which evidently had been originally buried there, together with a considerable portion of the skeleton.

This partial skeleton was originally buried about five to six feet below the present surface of the ground in a stratum of sand belonging to the Pleistocene or so-called Ice Age. As revealed by the cut made by the canal, this sand deposit lies directly on an older Pleistocene stratum, several feet in thickness and of very wide extent, which is made up almost entirely of shells representing numerous species which can live only in brackish or salt water. This is evidence that the sea once covered this part of Florida and that in its shallow depth was formed this widespread layer of shells. Later, as it slowly emerged from the sea, the bone-bearing sand layer was formed upon it.

The mammoth bones themselves tell an interesting little story of conditions which prevailed here when Florida was in the process of being transformed into its present day appearance.

FIG. 54.—North bank of Venice Company's main drainage canal, about 4 miles east of Venice, Florida. Location of Venice mammoth bones. Exact spot of find is indicated by canvas shelter at middle-left. Arrow indicates uneven line marking contact plane between fossil-bearing layer and later deposits.

In excavating these bones it was observed that they had not originally lain in a natural position but had been somewhat scattered about and broken before their final covering many thousands of years ago. This together with the fact that they were buried in a sand which contained many broken sea shells, in fact a typical beach sand, suggested that although this spot is now more than four miles inland, it was in that earlier time a sea beach upon which the mammoth carcass had stranded and as the flesh decayed the bones had been considerably tossed about and broken by the waves. That this was in reality a sea beach at the time the bones were buried there is apparently fully proven by the fact that many of the bones including the lower jaws and teeth had a considerable covering of barnacles which can only live in salt or at least brackish water. The fact that barnacles had attached themselves to the broken surfaces is proof also that some of the bones had been broken before the final covering. The presence of the barnacles also indicates that the bones before their final covering had lain for some time just off shore in shallow water.

In connection with this story of the Venice mammoth, it seems in order here to express to the officers of the Venice Company appreciation for the courtesies extended to Dr. Gidley while engaged in collecting the specimen and to commend them for their public-spirited generosity in supplying all the material and labor needed for the work as well as assuming for the Smithsonian Institution practically all incidental expenses in connection with it.

COLLECTING MINERALS IN MEXICO

Regarding the mineralogy and geology of Mexico comparatively little is known. The country is rich in mineral wealth, producing over 40 per cent of the world's silver, yet very few Mexican mineralogical collections have been made. This present exploration was undertaken by Dr. F. W. Foshag of the Smithsonian Institution in collaboration with the Mineralogical Museum of Harvard University, for the purpose of collecting representative material from as many of the districts as possible. Field-work was confined to that portion of the plateau of Northern Mexico within the states of Chihuahua, Coahuila, and Durango.

The plateau region of Northern Mexico is characterized by a broad high plain from which have been thrust upward numerous monoclinal or anticlinal ranges of mountains. The plains and valleys between the ranges are clothed with grasses or scattered shrubs, and the slopes carry a sparse covering of brush characteristic of the Lower Sonoran life zone. This zone is characterized by a number of yucca and cactus species, the creosote bush and, in numerous places, the showy ocotillo. In a number of places the yuccas form extensive open groves. Some of these desert plants are put to economic uses; maguey and sotol for fermented drinks, lechuguilla for fiber, candelilla as a source of vegetable wax, and guayule for rubber.

The ore deposits of the area visited may be considered a mineralogical unit. The most important are found in limestone of Cretaceous age and are without apparent direct connection with intruded granitic rocks, although there probably is a connection. This class of ore is comparatively simple mineralogically, consisting of silver, lead and zinc minerals accompanied by only minor amounts of silicates. A number of the richest and most extensive ore deposits of Mexico belong to this type of ore occurrence.

After leaving El Paso, the first district visited was Los Lamentos in the northern part of the state of Chihuahua. This district con-

Fig. 55.—The mining camp of Los Lamentos, Chihuahua. The rocks are fossiliferous Cretaceous limestones. (Photograph by Foshag.)

tains one of the few great ore deposits of Mexico that was not known to the Spanish conquerors. The extremely dry and inhospitable nature of the region probably accounts for this. The ore deposits are of so simple a type and the relations of the ore to country rock so clear, that it has become a standard for this type of occurrence. The ore persistently follows the contact between a hard, dark gray limestone and a lighter fossiliferous one, and consists of the simple lead and zinc minerals without any development of silicate minerals but with a surrounding aureole of recrystallized limestone. Many other North Mexican ore deposits are similar, although not always so simple. The mine is rich in showy minerals of both lead and zinc.

The region surrounding the city of Chihuahua is a rich one. Sixteen miles away is the famous camp of Santa Eulalia, where since 1591 the mines have been in almost constant operation. The ore bodies are quite extensive, carrying silver, lead, and zinc, and occurring in

FIG. 56.—Large crystals of gypsum in a cave in the Maravilla Mine at Naica, Chihuahua. (Photograph by Foshag.)

"mantos," one ore body having been followed for over two miles. Some of the chambers left by the removal of the ore are enormous, one in the Potosi Mine being 600 feet high, 300 feet long, and 200 feet wide. Other nearby localities visited were the La Ceja district, producing lead and zinc; Placer de Guadalupe, a small camp well known for its gold placers and for the combination of gold and pitchblende in its lodes; and Cuchillo Parado, a lead deposit on the Conchos River.

About 50 miles south of Chihuahua is the camp of Naica, known to mineralogists for its two crystal caves, the larger being in the Maravilla Mine. Here are several chambers connected by passage

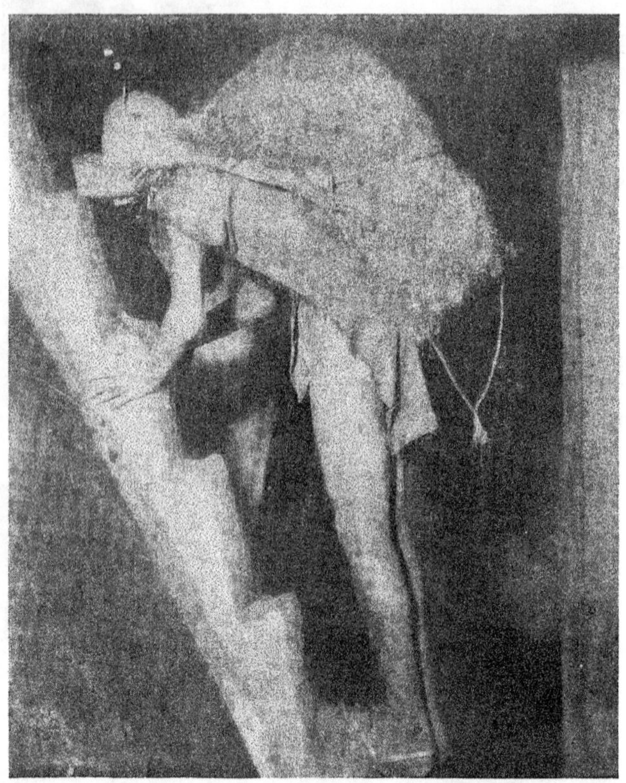

FIG. 57.—A peon carrying lead ore up a "chicken ladder." Encantata Mine, at Esmeralda, Coahuila. (Photograph by Foshag.)

ways, one chamber containing well-formed groups of long, radiating crystals of clear gypsum growing from the floor of the cave. Many of these crystals exceed three feet and some reach six feet in length. Another chamber, reached through a narrow opening studded with crystals, is completely lined with thousands of blade-like crystals one to three feet in length.

Sierra Mojada, the next district visited, owes its discovery to a band of smugglers attempting to elude pursuit. The ore bodies extend for a distance of six kilometers along the foot of a limestone cliff 2,500 feet high. The district is unusual in that lead, zinc, silver, copper, and sulphur have all been mined here. The great length but shallow depth of these mines makes it more economical to work them by the old Spanish methods than by modern ones. Much of the ore is brought to the surface on the backs of peons, often up ladders made of notched logs, popularly called "chicken ladders." It is said that a strong peon will carry loads in excess of

Fig. 58.—The mining camp of Ojuela at Mapimi, Durango. The rocks are Cretaceous limestone. (Photograph by Foshag.)

100 kilos (220 lbs.). Figure 57 shows a young lad carrying a sack of heavy lead ore by means of a tump line.

In the northeastern part of the state of Durango, near the village of Mapimi, is the Ojuela mine—one of the greatest lead mines of the world. Within this one mine are over 550 miles of tunnels driven to extract the ore. The camp itself is perched on a steep limestone mountain. Before the town, rises an almost vertical cliff of Cretaceous limestone 2,000 to 3,000 feet high. It is in the hills lying at the base of this cliff that the ore bodies lie.

Velardena, a district lying along the Cuencame River in Durango, has mines of copper, lead, and silver. The Ternares mine, located in the San Lorenzo Range, has been worked for over 200 years. The ore deposits are at the contact of granitic rocks with limestones, or in veins in diorite. At the Ternares Mine the limestone contact zone is characterized by great quantities of velardenite, spurrite, and hillebrandite, minerals first described as new from this locality.

About a mile from the railroad station in the city of Durango, rises a mass of volcanic rock and iron ore called Cerro Mercado.

FIG. 59.—Old church on Cerro Remedios, Durango. The rhyolite of this hill is a low grade tin ore. (Photograph by Foshag.)

The iron ore deposits of this hill constitute Mexico's chief domestic supply. The iron-bearing mineral is almost entirely hematite, the ferric oxide of iron, and it forms large masses in a rhyolitic rock.

Nearly five months were spent in the field and over two tons of specimens collected and shipped. The success of the field-work was largely due to the hearty cooperation of the Mexican government officials and of the American mining engineers in charge of the various properties.

THE CANFIELD MINERALOGICAL COLLECTION

Field-work of Mr. E. V. Shannon, assistant curator, division of physical and chemical geology, U. S. National Museum, was limited to a few trips to nearby Virginia and Maryland, and to nearly a month spent in Dover in northern New Jersey in packing the collection of minerals bequeathed to the National Museum by Dr. Frederick A. Canfield, who died during July of this year. This collection was begun by Mr. Mahlon Dickerson about 1808 and was

FIG. 60.—The collection of Franklin Furnace minerals, made by Frederick Canfield, Sr.

continued by Frederick Canfield, Sr., who secured many uncommonly fine specimens from old American localities no longer available and especially from Franklin Furnace, New Jersey, prior to 1870. Frederick A. Canfield, Jr., continued his father's collection and found in it his principal occupation and relaxation until his death. It now contains some 9,000 specimens, many of which are unique and all of them of exceptional quality. It was housed for over 70 years in the picturesque mansion at Dover, shown in figure 61, before being finally shipped to the National Museum. In order to insure its steady growth and relative standing, Mr. Canfield accompanied his gift with an endowment of $50,000, the income of which is to be utilized for this purpose. In the work at Dover, Mr. Shannon was ably assisted by Mr. James Benn of the Museum's staff and Geo. M. Hyland.

The opportunity was offered to visit several of the world-famous mineral localities of northern New Jersey. These included the various iron mines of Mine Hill, Wharton, and the vicinity of Dover. Of these the only one now in operation is the Hurd Mine owned by the Replogle Steel Company at Mount Hope.

Fig. 61.—Canfield house at Dover, N. J.

Through the courtesy of the officials of the New Jersey Zinc Company, it was possible to visit the underground workings and surface plant of the great zinc mine at Franklin Furnace, renowned as one of the most productive known localities for rare minerals. The mine and mill are models in efficiency, and the various methods which have been developed for handling the ore are of extreme interest. The quarries which furnish the beautiful zeolites at Great Notch and West Paterson were visited in company with the Newark Mineralogical Society, and a Sunday was spent at the well-known serpentine locality of Montville.

THE ROEBLING MINERALOGICAL COLLECTION

In July, 1926, a second of America's foremost collectors of minerals, Col. Washington A. Roebling, of Trenton, New Jersey, passed away. His noted collection was bequeathed to his son, Mr. John A. Roebling, of Bernardsville, New Jersey, who, in memory of his father, generously presented it to the Smithsonian Institution, with an endowment of $150,000 for the purpose of maintaining its present high standard.

On October 27, Dr. W. F. Foshag, Miss Margaret Moodey, and Mr. James Benn, of the staff, were detailed to go to Trenton to pack the collection for shipment to Washington. Mrs. Foshag also acted as a volunteer assistant for the greater part of the time. The work occupied a period of approximately six weeks, and the collection, contained in 258 boxes, was transported by truck and safely stored in the National Museum by December 7. It should be stated, incidentally, that Mr. Roebling defrayed all expenses connected with this work.

The career of Col. Washington A. Roebling, soldier, engineer, and mineralogist, is too well known to need narration here. The collecting of minerals was begun as a recreation and was energetically carried on from 1874 until the time of his death. With ample means to indulge this hobby, he spared no expense in securing rare minerals, both from American and foreign sources. In a large room in his beautiful home in Trenton, called by his family and friends "the museum," the collection was artistically arranged, and it was here that the Colonel spent most of his leisure time, delighting in showing his treasures to his friends, and being particularly happy if the caller chanced to be a mineralogist with whom he could discuss his rare finds. So methodical was the arrangement that he could place his hand almost immediately on any specimen desired.

Comprised of upward of 16,000 specimens, embracing almost the entire number of known mineral species, the collection contains much to delight both the eye of the seeker for mere beauty and the heart of the ardent student. Gem minerals are plentiful but there are only a few cut stones; showy examples of the more common minerals—malachite and azurite, quartz, calcites, and especially fine zeolites—are in abundance, but the chief value of the collection lies in the rare specimens which only the student can appreciate. Many of these represent the only example of a species in America. During the Colonel's lifetime, these were always freely loaned for study or comparison, and their usefulness will be increased when their final installation in the national collections makes them accessible to all accredited students.

BIOLOGICAL COLLECTING IN SUMATRA

Lieutenant Henry C. Kellers, Medical Corps, United States Navy, was detailed as representative of the Smithsonian Institution with the U. S. Naval Observatory Eclipse Expedition to Sumatra. Dr. Kellers reports the following:

The expedition left San Francisco September 1, and arrived at Sumatra October 26, stopping at Java for three days for visé of credentials of the members of the expedition by the Dutch officials. The Astronomical Station was located at Kepahiang, a small village on the Moesi River situated among the foothills of the western range, with an elevation of about 1,900 feet above sea level. Kepahiang is near Mt. Kaba, an active volcano, and about 40 kilometers from the Indian Ocean. This proved to be an excellent location for collecting trips. Most of the collecting was done around and on the slopes of Mt. Kaba.

Two ascents of this mountain were made to the rim of the crater, and on each occasion numerous specimens were collected. Snares and spring traps were set along the banks of the Moesi River and in the jungle, with good results. As the rainy season was on in force by the end of November, hunting in the jungle was rather disagreeable because of the countless leeches that attacked one, no matter what precautions were taken. They bite so softly that their presence cannot be detected except for the blood which trickles from the wound. After covering a few kilometers in the jungle, the legs of my native hunters would change in color from the natural brown to a bright red, when they would rest and scrape off the leeches. The terraced rice fields in the vicinity of the station were worked over with seine and dip net for fresh-water fish and amphibia, with good results, and many reptiles were captured.

Four trips were made to the village of Benkoelen on the coast of the Indian Ocean for marine invertebrates and tow net hauls. A double outrigger canoe was used for towing the plankton nets, and I was able by increasing or furling the sail area, to obtain all the speed needed with the afternoon trade wind. The canoe although rather small, was in no danger of turning over while sailing, as the outriggers used by the natives were the largest I have seen, either in the Solomon group of Islands or the Samoan Islands that I have visited. Two excursions were made to the Island of Poeloetikus in the Indian Ocean where large quantities of marine life were collected.

The advent of the automobile, with the whirr of the engine on the roads skirting the edge of the jungle, makes collecting more difficult, and one must now penetrate far into the jungle to hunt game which formerly might have been obtained comparatively near the borders of the wild territory.

FIG. 62.—Looking into the crater of Mt. Kaba, Sumatra. (Photograph by Kellers.)

FIG. 63.—Community rice mill on the Moesi River, Sumatra. (Photograph by Kellers.)

Fig. 64.—Hauling the seine on the beach at Benkoelen, Sumatra. (Photograph by Kellers.)

Fig. 65.—Sumatran outrigger canoe used for towing the plankton net. (Photograph by Kellers.)

Although malaria is very prevalent on the island, the members of the expedition enjoyed excellent health. Mosquitoes were a pest, and none of the houses was screened. The Europeans and natives alike protect themselves by means of mosquito nets only at night. No effort is made to rid the villages of breeding places for mosquitoes. I found that one of the main sources of mosquitoes is the large cement tub or tank that holds the bathing water at every dwelling, the daily bath being taken by dipping the water out of the tank with a small bucket and pouring it over the body. The water in these tanks becomes stagnant, as they seldom go dry, and so becomes a perfect incubator for mosquitoes. The water as used is replenished

Fig. 66.—Poeloetikus Island, Indian Ocean. (Photograph by Kellers.)

two or three times a week during the dry season from surface wells, and during the wet season, rain is caught by means of wooden troughs leading from the roofs.

It is remarkable what cleaning up and proper drainage will do to eradicate the mosquito. The Kaba Wetan Coffee Plantation, situated about four kilometers from the astronomical station, is clean and well drained. The bath house of the manager's residence is screened, but there are no screens in the living quarters. It was a pleasure to remain there and rest without the heavy nets at night. The manager stated that he seldom had a mosquito in or about the house.

All water used for drinking and domestic purposes comes from surface wells and the question uppermost in one's mind is: Has it been boiled? The day's drinking water is boiled in the morning and kept in covered containers until used. The natives are not particular in regard to the water they drink, and suffer considerably from dysentery. At hotels and boarding houses, to be on the safe side, the Europeans usually drink "Apollinaris" to such an extent that it is called by the natives, "Ayer Blanda," or white man's water.

In the department of natural history the natives are quite observant. Their nomenclature is fairly complete. They possess generic and specific names, the descriptive names being numerous and often a source of error. They class all the larger Felidæ under the generic name *harimau,* and the smaller under the name *kuching.* They then designate them by distinct descriptive names. They give the gen-

Fig. 67.—A typical coffee planter's bungalow, Sumatra. Kaba Wetan Coffee Plantation, where the mosquito is unknown. See text. (Photograph by Kellers.)

eric name *rusa* to the larger deer. They apply one name to the tapir and rhinoceros, namely, *bodak.* All of the squirrels are grouped under the name *tupai.* All rats and mice are known as *toekus,* except the bamboo rat, which is called *dkan.* The frogs and toads are classed together under one name, *kodok.* The names of snakes are confusing as all are called *ular,* the specific names being unsatisfactory from a scientific standpoint.

The tiger is the most feared of the mammals, although statistics show that crocodiles claim a larger total of victims. In the native legends and folk-lore, the men inhabiting the Korinchi district of Sumatra have the power of transforming themselves into tigers.

FIG. 68.—The covered wagons of Sumatra bringing in produce for market day. (Photograph by Kellers.)

FIG. 69.—Market day in Sumatra. Note the Durians in the foreground. (Photograph by Kellers.)

I was told that the natives observe a difference in the methods used by the tiger and the tigress in killing their prey, the former by breaking the neck and the latter by biting through it. The other mammals do not cause any terror among the natives, although many are dangerous when hunted.

The native Sumatran, whether an orthodox Moslem or not, believes in spirits both of animate and inanimate objects. On my second ascent of Mt. Kaba, I met a band of natives on the edge of the

Fig. 70.—Javanese coolies giving an ancient war dance New-year's day for the members of the expedition. Note that the horses are constructed of plaited bamboo and painted. (Photograph by Kellers.)

crater offering up as a sacrifice a white pigeon, sprinkling rice and water on the ground, and conducting religious ceremonies in order to invoke the favor of the spirits for bountiful crops and to the end that their wives would bear numerous offspring. The story of today of the semi-human ape that is supposed to inhabit the southern end of the island is nothing more than the old legend of the Sumatran orang-outang (*mawas*), a semi-human spirit of great strength and ferocity.

The present age is one of transition. How long are the old customs and beliefs to survive? The native Sumatran is being educated, he reads books, he is taking an interest in newspapers, and he has a good opinion of himself.

The Dutch officials and the managers of various coffee plantations in the vicinity of Kepahiang rendered all the assistance possible to make the expedition a success. Transportation for freight and personnel across and about the island was without cost, and the detailing of natives to accompany me on my daily collecting trips shows the interest taken by the Dutch in scientific pursuits. The two trips to Poeloetikus Island could not have been made except for the courtesy of the superintendent of the Dutch Mail Steamship Co., at Benkoelen,

FIG. 71.—A rainy day in Sumatra. Note the basket hats. (Photograph by Kellers.)

who placed at my disposal a sea-going motor boat and a native fisherman.

BIOLOGICAL FIELD-WORK IN FLORIDA

During March and April, 1926, I visited northern Florida for the purpose of making general collections of vertebrates and plants. Work was carried on at several localities in the region extending from Gainesville to Cedar Keys. It was mostly of such routine character as to call for no special comment.

To obtain skeletons of porpoises at Cedar Keys was the chief object in view. The animals were abundant and fearless, coming close to the wharves in pursuit of small fish. Here they were easily shot but less easily recovered owing to the dense opacity of

Fig. 72.—Preparing skeleton of porpoise on beach at Cedar Keys, Florida.

Fig. 73.—Black Skimmers on beach at Cedar Keys, Florida, showing good result of local bird protection.

FIG. 74.—Sponge boats and wharf at Cedar Keys, Florida. The sponge industry is carried on by Greek fishermen.

FIG. 75.—Sponge boat.

FIG. 76.—Building foundations dug in Indian shellheap, Cedar Keys, Florida.

FIG. 77.—Turpentine gatherer. A familiar figure in the pine regions of Florida.

the water, a condition locally attributed to the heavily flooded state of the Suwannee River. The process of skinning and skeletonizing a porpoise (*Tursiops*) is shown in one of the photographs.

Cedar Keys is one of the headquarters of the Gulf sponge fishery, an industry carried on by Greeks. Seaworthy motor boats are used, the divers working in the open waters of the Gulf of Mexico, and remaining at sea about a week at a time. It is said that good commercial sponges could formerly be obtained in the waters of Waccassassee Bay; but this supply was long ago exhausted.

The town of Cedar Keys occupies a site famous for its Indian mounds and shellheaps. Many of the houses rest on foundations dug into these remnants of a vanished culture.

GERRIT S. MILLER, JR.

VISIT TO A CALIFORNIA WHALING STATION

During the summer of 1926 a joint expedition was made to the coast region of northern California by the Smithsonian Institution and the San Diego Society of Natural History, A. Brazier Howell representing the former and Laurence M. Huey the latter. Opportunity was taken to make collections of the local mammals and some things of much interest were secured, including seven specimens of the rare rodent, *Phenacomys albipes,* of which but two specimens had been previously known from California. The main purpose of the expedition, however, was to secure data, rather than specimens, regarding the whales and whale fishery at Trinidad, Humboldt County. This station in summer, and one near Monterey during the winter, are both operated by the California Sea Products Company, a concern that has been active in this field for many years. It is now the only well established whale fishery anywhere on the coast of the United States, although at least one other has recently started work.

Whale hunting is a very old industry and hence is of great interest historically, as well as biologically and economically. The early type of fishery has disappeared never to return because of a number of reasons, among which are the fact that the more valuable species have been practically exterminated, leaving the speedier, less valuable ones which the old-time whalers not only did not want, but could not catch with the methods then in use; and the adoption of the present technique, including small, fast steamers and harpoon guns.

The catch at Trinidad consists almost exclusively of finback and humpback whales, the former reaching a length of about 75 feet

Fig. 78.—The "Hercules," largest of the three whalers, showing harpoon in position for firing and Captain Lane at the trigger. (Photograph by L. M. Huey.)

Fig. 79.—Captain Lane at the instant of firing a harpoon into a finback, shown in line of aim. (Photograph by L. M. Huey.)

FIG. 80.—The "Hercules" bringing a finback whale to the station, with the flukes made fast to the vessel. (Photograph by L. M. Huey.)

FIG. 81.—A large finback at the foot of the slip, ready to be hauled out. (Photograph by L. M. Huey.)

and the latter somewhat less. Each of the three small steamers used leaves the shore station and cruises usually to a distance from the coast of 50 miles, remaining out either until a whale is secured or until the fuel supply needs replenishing—a matter of some three days. If a whale indicates that it is wary, it is better to waste no time on it, for with the finback's maximum speed of better than 30 miles per hour it can play with its pursuer in an exasperating manner; but a close approach may usually be made, by cautious maneuvering, permitting a shot at less than 40 yards. If the bomb

FIG. 82.—The rare California gray whale being hauled up the slip. Note where sharks have bitten great pieces of blubber from in front of the flukes and also mutilated the latter after the death of the whale. (Photograph by L. M. Huey.)

fails to explode or the harpoon, weighing somewhat less than 150 pounds, is placed too far back, a fight of several hours may ensue. The cetacean may take out a mile of cable and must be as carefully played as a game fish, for although the line consists of a five-inch (in circumference) manila hawser with breaking strength of 18,500 pounds, the animal will snap this with ease if too much strain be applied or too much slack be given. When brought close alongside it is inflated with air and then may be marked with a flag and left to float while the vessel continues hunting for some time.

Upon reaching the station in tow of the vessel, a cable is attached to the tail and the carcass is hauled by steam power up the "slip" to the covered cutting platform, where the blubber near the head is first loosened by means of huge knives, and then torn off by cables. The larger muscle masses are similarly treated and the skeleton disarticulated. The great strips of blubber are cut by hand into pieces of convenient size and so fed into the blubber cutter, which consists of a number of sharp blades revolving at high speed. The resultant sludge is then led to giant, steam-heated vats and the oil tried out. The lean meat of better quality is cut into slabs and

FIG. 83.—Head of California gray whale, the animal lying on its back. Note eye (a), baleen or whalebone (b), and large tongue (c). (Photograph by L. M. Huey.)

placed within large iron retorts in tiers, with iron plates between each layer, where it is cooked for a number of hours, then dried and finally ground. The waste, including the blood and viscera, is dumped into open vats of boiling water and after being cooked. the oil is separated and the solids dried and ground for fertilizer. After most of the meat is cut from the bones the latter are also placed in retorts and cooked. Considerable oil is obtained from them and the dry bones are then dumped in a pile which grows to large proportions as the season advances. The baleen or "whalebone" of this group of whales is short and coarse and is not at present of commercial value, so it, too, is cooked and ground for fertilizer. The

Fig. 84.—A finback on the slip. Note corrugations in the skin of the throat which probably permit great expansion during feeding. The spots on the chin at (a) indicate the positions of short bristles, the last remnant of a hairy coat. (Photograph by L. M. Huey.)

Fig. 85.—Roof of the mouth, showing baleen plates, and bony part of lower jaw of a finback, after the corrugated portion of the throat, shown in figure 84, had been removed. (Photograph by L. M. Huey.)

Fig. 86.—Great mass of skeletal bones of whales—the accumulation from dozens of individuals. (Photograph by L. M. Huey.)

Fig. 87.—The western gulls gather by the scores and hundreds, ever hopeful of snatching bits of blubber and meat. (Photograph by L. M. Huey.)

blubber oil is produced in several grades, some being used by manufacturers of fine soap, some as lubricants, chiefly in combination with mineral oil, and still others in the manufacture of certain foods. The better quality of meat meal is sold as chicken feed, and the bones are shipped in bulk to Honolulu, where, after being made into a fine grade of bone black, they are used in the refining of sugars.

The overhead and other expenses of such a plant as that of the California Sea Products Company are considerable, as it is necessary to have much machinery of various sorts and large size, and a large pay roll, which continues alike through times of cetacean plenty or scarcity. Hence it is necessary that each whale be utilized to as complete a degree as possible, and in this the company is as successful as is the case with the proverbial slaughtered pig. And the work must be done expeditiously, before partial decomposition of the huge carcass has lowered the quality of blubber and meat. Hence, in about three hours after being landed the whale has disappeared to the last shred within vats and retorts.

The prepared products from one whale of average large size may be worth as much as $2,000 and from this it may be seen that our whale supply is a matter of economic importance, and not merely of aesthetic concern to those with emotional tendencies. When whales have become sufficiently numerous in any district a whaling station is at once started and invariably this continues operations as long as whales are to be had in sufficient numbers to offer a margin of profit, after which the station is abandoned. If whales existed in paying quantities along any of the remainder of our coasts, there would be whaling stations there for their capture. In some few cases the whales will regain a part of their former numerical status, but in others they can never recover, and in all cases the rarer species will have been reduced to the point of grave danger. Thus the gray whale was believed by scientists to be practically extinct on the west coast of the United States, but the whalers say there are a very few left, and by great good luck we were at Trinidad when one of these was secured—the first ever brought to that station during the six years of its existence. A number of extremely interesting observations, chiefly of an anatomical character, were made of this specimen, as well as of the commoner species, but it was saddening to realize that this might possibly be the last such opportunity afforded a mammalogist in the United States.

Few naturalists are now so situated as to be able to observe whales at first hand and it is felt that much of value will result from our stay at Trinidad, our work having been facilitated in innumerable ways by the cordial cooperation of the officials of the California Sea Products Company.

EXPLORATIONS IN SIAM

Many years ago the Smithsonian Institution received from the National Institute, in Washington, a collection of about 100 specimens of birds from Prince Momfanoi (also called Chawfanoi) of Siam, half-brother of the King of that period, and "in truth, rightful heir to the throne" according to Dr. Ruschenberger (Voyage around the World, 1838). He was reported to be an enlightened man, who loved pets, had many live animals and birds, and had a museum in which were many stuffed birds and animals, set up by himself. Due to faulty preservation, a large proportion of the specimens received from Prince Momfanoi have long ago been destroyed through the ravages of insects, but the record of his gift remains.

From 1896 to 1899, Dr. W. L. Abbott, the veteran explorer, spent several months of each year working through the province of Trang, in the peninsular part of Siam, giving the Institution its second collection of birds from that country. Later, in 1914-1916, he supported Mr. Cecil Boden Kloss in the latter's exploration of parts of southeastern Siam and the Franco-Siamese boundary, which resulted in over 130 additional birds. In 1916 and 1918, Mr. Kloss visited the southwestern and other parts of the country, contributing 496 more specimens. In 1924, over 200 birds were received from Dr. Abbott, collected by J. H. Chambai and K. G. Gairdner; these were from south and central Siam. In the same year, Dr. Hugh M. Smith, while engaged in fisheries investigations for the Siamese government, obtained a few specimens in the southern part of the country, especially from the island of Koh Chang, in the Gulf of Siam, including a new species of small "timaline" bird, *Pellorneum smithi.*

During the years 1925 and 1926, Dr. Smith continued his natural history work, with the aid of a Dyak collector, and has sent to the Smithsonian Institution 968 birds from southern and central Siam. In addition to birds, he has forwarded about 70 skins of mammals, several hundred mollusks, crustaceans, fishes, reptiles, and ethnological material. Dr. Paul Bartsch, curator of mollusks, says, "Dr. Hugh M. Smith's work in Siam has enabled him to make distinct contributions in the field of mollusks. I feel sure that quite a few new species and races will eventually be described when this material is worked over with a fine-tooth comb. All of it fills a decided gap in the collection, for prior to Dr. Smith's enterprise few mollusks from that region were found in our collection. The sendings in 1924 amounted to almost 2,000 specimens, while during the present year over a thousand specimens have been received, among them two new species of shipworms as well as other novelties not yet fully determined."

Count Nils Gyldenstolpe, a Swedish naturalist, has recently published a list of the birds of Siam, enumerating about 730 species and subspecies, of which 113 are water birds and waders, the remainder land birds. The list is confessedly only an incomplete one, for much additional ornithological work must be done before the birds of Siam may be regarded as well known.

Among the birds sent in by Dr. Smith is a specimen of the Openbill or "Shell ibis" (*Anastomus oscitans*), a remarkable species of stork, peculiar for the character of the bill, which is said to be "the result of wear, caused by the shells of the mollusca, on which the bird feeds." In the young bird the lines of the bill are straight, and show no gaping space as in the adult. Several species of fruit pigeons are represented in the collection, a group of birds usually of bright colors, green predominating in many of the species. They are native to the Eastern Hemisphere, a few species in Africa, an increasing number in southern Asia, with the majority in the Dutch East Indies and Polynesia. About 13 species have been recorded from Siam, of which five are restricted to the northern parts of that country.

BREEDING EXPERIMENTS WITH CERIONS

Dr. Paul Bartsch, curator of mollusks, U. S. National Museum, spent the period from August 10 to August 21 at the Tortugas and from August 21 to August 24 examining the Cerion colonies planted on keys between the Tortugas and Miami. An examination of the small island colonies yield the following results:

Island 1. Twenty-five *Cerion incanum* and 25 of a new species of *Cerion* with spiral sculpture were planted here 2 years ago. Last year, 22 of the *incanum* were found dead; this year no additional specimens of this species were noted. Of the *Cerion* new species, 18 were dead last year and 1 dead and 2 living this year, which accounts for 21 out of 25 of these specimens. The missing individuals may be buried in the sand, dead or alive, or they may have been carried away by some agency as, for example, the little Sparrow Hawk, which occasionally indulges in that pastime. In this island were

FIG. 88.—Dr. Bartsch going down with diving helmet.

FIG. 89.—Dr. Bartsch coming up with the diving helmet just removed. (Enlargement from movie film.)

FIG. 90.—A general undersea view showing a field of Gorgonians. (Enlargement from movie film.)

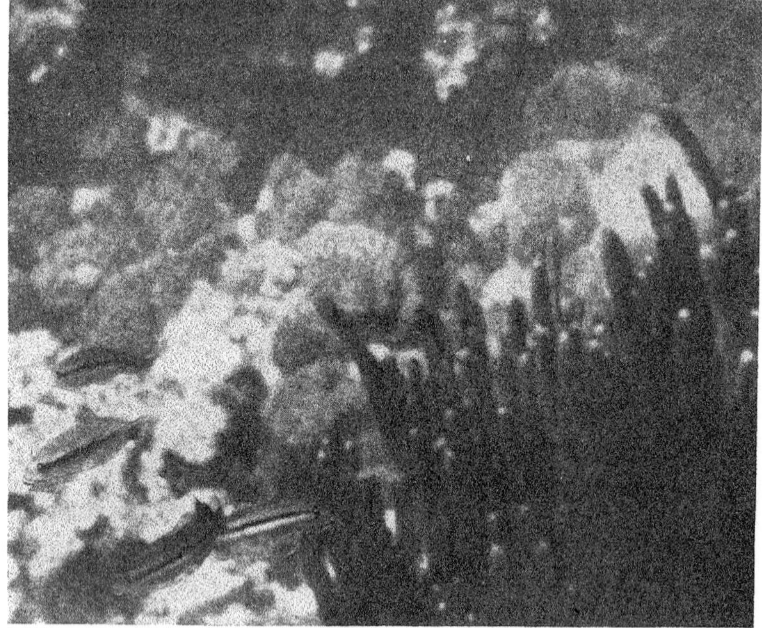

FIG. 91.—Gorgonians and Goat fishes. (Enlargement from movie film.)

also found a number of quite young shells, too young to determine their relationship at this time. Dr. Bartsch therefore refrained from seriously disturbing the sand for fear of exposing these small creatures to unfavorable stresses.

Island 2. In this were planted, 2 years ago, 25 *Cerion incanum* and 25 *Cerion chrysalis.* Last year 11 of the *Cerion incanum* were found dead and this year 4 dead and 1 living, accounting for 16 of the planted specimens. Of the *Cerion chrysalis,* 12 were found dead last year; to this were added 4 this year and 2 living individuals, which accounts for 18 of the 25 planted. Here, too, small young were in evidence.

Island 3. On this island were planted, 2 years ago, 25 *Cerion incanum* and 25 *Cerion mummia.* Last year 16 dead of *Cerion incanum* and this year 2 dead and 5 living were recovered. Of *Cerion mummia* 17 were found dead last year and 1 this year, and 1 living. All but 2 of the *Cerion incanum* are therefore accounted for in this island. Here, too, some young were present, and the interesting observation was made of finding a specimen of *Cerion incanum* mating with a *Cerion mummia.*

Island 4. Here were planted, 2 years ago, 25 *Cerion incanum* and 25 *Cerion tridentatum.* Of these, 13 *Cerion incanum* were found dead last year and 3 this year, and 2 living. Of the *Cerion tridentatum,* 10 were found dead last year and 3 this year, and 10 were living. No young were observed.

In the 8 small islands, all the adults seem to have died, but small young are present in some, which makes it wise to leave them undisturbed for another year as these may prove to be the desired hybrids. However an additional marked specimen of *Cerion incanum* and *Cerion viaregis* were added on each island.

Last year some of the old wooden cages were cut down to mere 4-inch base boards which were partly buried and fastened securely to the ground. To the free edge of each, which projects some 3 inches above the ground, a narrow strip of monel wire was tacked, bending inward to prevent the escape of the mollusks.

In each of these cages a specimen of *Cerion incanum* and *Cerion viaregis* was placed. Upon examining these inclosures this year, most of the shells were found dead and many were crushed, the damage probably having been done by rats. Each of these 75 cages were re-stocked with a marked specimen of *Cerion incanum* and *Cerion viaregis.*

Fig. 92.—Yellow-Tail and Slippery Dicks, and Sea Urchins. (Enlargement from movie film.)

Fig. 93.—Hogfish. (Enlargement from movie film.)

Fig. 94.—A school of Gray Snappers. (Enlargement from movie film.)

Fig. 95.—A school of Grunts and Gray Snappers. (Enlargement from movie film.)

The two original colonies of Cerions, namely, *Cerion viaregis* and *Cerion casablancae*, are simply swarming in their respective places and gradually spreading over adjacent territory.

The colony of *Cerion uva* from Curaçoa is gradually passing out, but it was a surprise to find that the colony of *Cerion crassilabre*, which was in bad shape a year ago, was not at all on the verge of extinction, but yielded sufficient material to give complete measurements of all the characters of enough specimens to complete the hundred series of both the first and second generations of Florida-grown individuals. This again shows that it is not safe to estimate the strength of a colony merely from the appearance of material above the surface of the sand, for Cerions under certain conditions have a way of disappearing beneath the surface and thus hiding themselves from view.

The huge colony of Cerions, brought from the Island of San Salvador and planted on the east side of the laboratory, does not appear to have found the habitat a suitable one, for there is an endless number of dead strewn about the ground, but not a single living specimen was seen.

The colonies planted on the rampart of Fort Jefferson 2 years ago are holding their own.

On the way north, September 21, a stop was made at Boca Grande Key where it was found that the part of the key bearing the plantings was again burnt over, and the same fate has overtaken the colonies which had been planted on Man Key, Boy Key, Woman Key, and Girl Key.

On September 22 a visit to the two plantings in the grounds of the U. S. Bureau of Fisheries Station at Key West showed both colonies doing fairly well.

On September 22, the *Anton Dohrn* was left at Newfound Harbor Key, and a visit made to the keys containing Cerion colonies between Key West and Miami. The hybrid colony on Newfound Harbor Key was doing splendidly. The last hurricane had not seriously damaged it. Some additional variant specimens were taken to Washington for dissections. This colony, as well as experiences on other islands, shows plainly the desirability of placing colonies not on unoccupied keys but, if possible, to select ground for them near some house where the people will take an interest in the experiment and see that the ground is protected against fire.

Bahia Honda, where the colony of *Cerion casablancae* was placed in 1912, was next visited. The hurricane some years later had cut the ground, where the original colony was planted, to such an extent that it was felt that it had been wiped out. Examination after the hurricane gave no indication of its survival, but this year's visit showed that it not only had survived, but was flourishing and spreading.

An examination of Duck Key, where a colony of *Cerion viaregis* had been planted in 1912, showed that the tall, dense grass and cactus, which had usurped the places of planting, still prevailed to such an extent that no specimens were visible.

FIG. 96.—Albino Noddy Tern, on Bird Key.

On September 23 Indian Key was visited and it was found that the colony of *Cerion casablancae,* which on our last visit seemed to be on the verge of extinction, was flourishing and doing well. There were enough specimens to warrant taking a series to Washington for measurement.

The colony of *Cerion viaregis,* planted on the adjacent Tea Table Key in 1912, failed to show a single specimen.

On September 25 a stop was made at Sands Key and Ragged Keys where plantings were made in 1912, but here, dredgings, fillings-in, human activity in general, and hurricanes have so overwhelmed the place that not a trace of any of these colonies was apparent. With the rapid spread of habitation in this neglected region of

Fig. 97.—A group of Black and Least Terns on Loggerhead Key.

Florida it will be possible to secure the sympathy of interested people to make further plantings among these keys and to have them secure from harm.

During the trip northward from Key West to Miami stops were made at a number of places not before visited and collections of the native *Cerion incanum* were obtained.

At the Tortugas Dr. Bartsch exposed 2,400 feet of moving-picture film among the coral reefs undersea, securing a series of pictures showing faunal associations of marine organisms *in situ*.

As in former years, Dr. Bartsch kept account of the birds observed at the Tortugas from day to day, as well as on the other keys visited.

STUDY OF THE CRUSTACEANS OF SOUTH AMERICA

As noted in the Smithsonian Exploration Pamphlet for 1925.[1] Dr. Waldo L. Schmitt, curator of the division of marine invertebrates, U. S. National Museum, the holder of the Walter Rathbone Bacon Travelling Scholarship, visited South America for the purpose of studying the crustacean fauna of the continent. His activities from August 1 until November 2, 1925, when he arrived at Itajahy, Brazil, were reported on last year. After a short sojourn at that port, he proceeded to Florianopolis, Brazil, where he made valuable collections. He next visited Montevideo, Uruguay, arriving November 10. Here Dr. Schmitt met Dr. Florentino Felippone, a correspondent of the U. S. National Museum, who showed him every courtesy and aided his work in many ways. Here also, through the assistance of Señor Tremolaras, arrangements were made for obtaining rheas for the National Zoological Park at Washington. He arrived at Buenos Aires, Argentina, December 14 and procured from the Buenos Aires Museum the loan of a valuable collection of crustacea. At this point it was found impracticable to proceed to Punta Arenas, Chile, as he had planned. On January 19 he returned to Santos, Brazil, and went by train to São Paulo where he visited the Instituto Butantan and obtained a collection of living Brazilian reptiles. On January 21 he embarked at Rio de Janeiro for New York, where he arrived February 1.

This year Dr. Schmitt is devoting his studies chiefly to the west coast. He left New York August 19, arriving at Cristobal, at the Caribbean entrance to the Panama Canal, August 25. At Panama Dr. Schmitt had the pleasure of meeting Mr. James Zetek, resident

[1] Smithsonian Misc. Coll., Vol. 78, No. 1, pp. 40-44.

custodian of the laboratory at Barro Colorado Island, who in the past has sent many valuable specimens to the National Museum. The ship left Balboa on August 26, and reached Salaverry, Peru, August 30. As hotel accommodations were not to be had there, Dr. Schmitt was obliged to move a short distance up the coast to Trujillo, where Mr. Sears, the agent for the Grace Lines, generously offered to take him into his own residence. The offer was accepted, and during his stay at Trujillo Dr. Schmitt enjoyed the most generous hospitality of a delightful private home. In regard to the climate he remarks, "You will not believe it, but it is like a cold California night. I am sitting here with my hat and overcoat on so as to keep warm enough to write. What will working in the water be like? I'm told the Humboldt current off the coast of Peru is way out of its course; that while it usually turns out to sea off Talara, it is now running way up into the Gulf of Panama; that it affects (kills) the supply of fish and has driven the pelicans up off the California coast in the vicinity of the Catalina Islands where as it is said here, they were never known to occur before. The heavier rains resulting from the shift of the Humboldt current have washed out a lot of land and killed off much of the marine life. The hotel at Salaverry was washed half full of sand by the last wet season's rains, thus the lack of accommodations."

Dr. Schmitt left Salaverry September 1 for Guayaquil, where he arrived September 4. He states that there is much more English spoken on this coast of South America than upon the east coast where he visited last year, and that there is a great deal of American machinery used everywhere. At three places along the route while the boat stopped to take on cargoes of sugar, he succeeded by means of his bottom sampler in getting mud samples. Guayaquil has but one fairly paved street, the main one running from the water front to the main Plaza. Dr. Schmitt finds that great quantities of shrimps and two species of crabs are sold in the markets of Guayaquil. A species of *Callinectes* much like our blue crab of the Chesapeake, and a red *Ucides,* a land crab, appear to be the only two species used as food. He says, "Oysters are very plentiful, but with all the hundreds of thousands that must have been shucked by the men with whom we had converse, none had ever seen a crab in an oyster."

After a week's unavoidable delay, Dr. Schmitt finally reached Santa Elena, which is on the coast directly west of Guayaquil. Colleeting here was excellent, but dredging was impossible as the only two motor boats of the place were laid up, and the only other available boats were the native dugouts. A number of species of crabs and shrimps were taken here, some of which Dr. Schmitt believes will prove to be new. A most interesting assemblage or association of marine animals was found to be inhabiting old worm galleries. Dr. Schmitt thought the coast around Santa Elena, if he could have given the time to it, would probably have proved to support the richest fauna of all the regions that he had visited in the past two years, but his time being limited, he was obliged to return to Guayaquil on September 19. He says, "We started early Sunday morning and between auto and train were under way from 6 a. m. to 7 p. m.

FIG. 98.—Paita, Peru, showing the rocky nature of the coast.

just covering 90 miles of country. It took us two hours to make the last 20 miles." Dr. Schmitt had hoped to visit Quito, but owing to the great loss of time in getting to and from Santa Elena he found it impossible. While at Guayaquil, Dr. Schmitt did some excellent collecting in the so-called Salada, an arm of the ocean which reaches up behind the city. There the black flies were a terrible torment, and he suffered severely from their attacks, but to offset in a measure this inconvenience he procured a species of *Upogebia*, the little shrimp that bores holes in rocks. This genus of shrimps had not until then been found on the west coast of South America. Here also were taken some pinnotherid crabs which are symbiotic in the shells of mollusks.

He left Guayaquil October 3 and arrived at Paita, Peru, late the same day. He says, "Here we drop into a holiday in memory of

San Francisco de Assisi. The holiday alone would not keep me out of work, but the Custom House was open only to let our stuff in. We arrived on the fourth, and the fiesta of San Francisco lasted two days, the 4th and 5th. And today the fisherman, families, etc., are resting up." Collecting proved to be very good here, and much fine material was procured from a point of rock jutting out into the water, and from a fine reef well north of Paita.

Dr. Schmitt left Paita October 15 and arrived at Salaverry October 17. Here the coast is just a bold surf-beaten headland with huge immovable boulders. At low tide the collecting was good, and the

FIG. 99.—Chorillos, a suburb of Lima, Peru, showing the cliffs at the water's edge which are in some places 200 feet high.

exposed beach yielded many invertebrates. Here also worm tubes were found to contain interesting and valuable specimens of crustacea and other forms. He found here a genus of shrimps not before reported from the Peruvian coast, and also procured a fine series of fresh-water shrimps.

On October 25 he left Salaverry, Peru, and reached Callao early the next morning. The Ambassador, Mr. Poindexter, procured him a pass to San Lorenzo Island, the naval base in Callao Bay. He made several collecting trips to the island and secured many amphipods and some remarkable porcellanid crabs, which live on the under side of a sea urchin. The fine material obtained here shows how much zoological work is yet to be done on this coast.

When last heard from, Dr. Schmitt was on his way to Valparaiso, Chile, where he expected to arrive on the 18th of November. From there he will proceed to Buenos Aires and then to his destination, Punta Arenas, Chile. Several cases of specimens have already arrived at the Museum.

INVESTIGATION OF THE AMPHIPODA OF THE TORTUGAS

Mr. Clarence R. Shoemaker, assistant curator of marine invertebrates, U. S. National Museum, spent the period from July 13 to

FIG. 100.—Fort Jefferson, Garden Key. Many amphipods and isopods were obtained from the rich growth of algae, hydroids and ascidians upon the wall of the old moat. (Photograph by Shoemaker.)

August 20, 1926, at the Carnegie Marine Laboratory, Dry Tortugas, investigating the Amphipoda of the keys. While studying the collections sent to the Museum by former investigators at the Laboratory, so many new and interesting forms were observed that it was thought advisable to spend some time at the Laboratory making as thorough a survey as possible of the amphipod fauna.

The waters surrounding most of the principal keys, and also the water of the moat at Fort Jefferson, were exhaustively examined. Pelagic species were scarce, but algae and old coral-rocks yielded many forms and almost countless numbers of individuals. Many of the species, as is usually the case in tropical waters, are very small, so that even sorting the specimens into the different kinds proves to be a long and somewhat tedious process, but enough has already been accomplished to show that many additional new forms have been discovered, and that the range of many species has been greatly extended. Some of the species were found to be very vividly and beautifully colored, while the coloration of others seemed to be of a protective nature, exactly matching their surroundings.

In order to escape their principal enemy, the fish, amphipods avail

FIG. 101.—Bush Key Reef. The abundant marine algae at the edge of this reef afford protection to innumerable amphipods and isopods. (Photograph by Shoemaker.)

themselves of the shelter of old coral-rocks which have been tunneled and bored by other marine organisms. The cavities in sponges and the interstices of algae also afford them the necessary protection. Each of these locations yielded many specimens which when finally identified will, with those already known to occur, probably be an almost complete list of the species inhabiting the waters of the Tortugas.

ENTOMOLOGICAL TRIP TO GUATEMALA

Dr. J. M. Aldrich, associate curator, division of insects, U. S. National Museum, went to Guatemala for the purpose of increasing

Fig. 102.—Native hut on sugar plantation, Guatemala.

Fig. 103.—Maya monument near Quirigua, Guatemala.

the Museum collection of tropical muscoid flies, while at the same time collecting such other insects as could be obtained. He left Washington on the last day of March for New Orleans, where he took the United Fruit Company's steamship for Puerto Barrios, arriving there April 7, and continuing by train to Guatemala City the same day.

He made his headquarters in that city for about a month, taking trips to the Mexican border at Ayutla, to the Pacific Coast at San José, to Antigua, to the interior lowland at Quirigua, and some nearer places. In the immediate vicinity at Guatemala City the collecting was poor in April, as it was in the latter part of the dry season.

FIG. 104.—Military headquarters on plaza in Coban, Guatemala.

Numerous courtesies were extended to Dr. Aldrich by the Agricultural Department of Guatemala, especially by the Director-General, Mr. J. G. Salas. The Department was at the time very much concerned with an outbreak of the migratory locust and had reared some parasites, which Dr. Aldrich was able to identify. This locust is a great pest of agriculture from Southern Mexico to Argentina, and there was much interest in its parasites. These, it happens, belong to a group of flies which Dr. Aldrich has studied extensively.

About May first the Minister of Agriculture proposed that Dr. Aldrich give up his plans for collecting in the western part of the country and join a government party in a trip to Coban for the purpose of making investigations on the locusts and their parasites and meeting a similar investigating party coming over from Mexico. As it seemed possible that some important information regarding

FIG. 105.—Morazan, Guatemala. Approaching the plaza.

FIG. 106.—Scene in market, Guatemala City.

the parasites might be obtained, Dr. Aldrich acceded to this request, and made a journey on muleback of about 100 miles to the interior town of Coban, where he remained 10 days. Unfortunately he was prevented by illness from doing as much work as anticipated, although continuing to collect flies in the vicinity of the town. The Mexican party was not heard from; and as soon as able Dr. Aldrich returned to the Atlantic Coast by way of the Polochic River, coming out at Livingston. He reached Washington June 6.

While the dry weather and change of plans interfered greatly with the expected results of the trip, a considerable number of specimens were collected and some discoveries of importance were made on grasshopper parasites.

The cordial relations which were established with the Guatemala Department of Agriculture, The International Health Board, the United Fruit Company, and with individuals, will no doubt be of value in future work.

COLLECTING AND REARING FRUIT FLIES IN PANAMA

On March 6, 1926, Mr. Chas. T. Greene, assistant custodian of diptera, U. S. National Museum, sailed from New York for Cristobal (Colon) Panama, on the Steamer Ancon. His primary object was to collect and rear the fruit flies of the genus *Anastrepha* in order to associate their larvae and pupae with the adults and to get all the information possible on their immature stages, as well as to rear any other flies which he might deem of economic importance. This work was carried on at the request of the Federal Horticultural Board. Ancon, Canal Zone, located at the Pacific end of the canal, was selected as Mr. Greene's headquarters because the United States government has a well equipped laboratory or experiment station there. He remained in Panama until May 27, sailing on that day for New York.

During his trip Mr. Greene visited several places on the Pacific and Atlantic coast of Panama, searching for fruit flies, and collected at several places along the Panama railroad which follows the canal across the Isthmus. All of the localities visited yielded valuable information on the fruit flies.

The three species of *Anastrepha* of economic importance were reared from native fruits and their immature stages are associated. The results will be published in the near future. Numerous other species of flies were reared and notes made on their immature stages in connection with this work. A large number of other species of

FIG. 107.—Boat landing and laboratory at Barro Colorado Island.

FIG. 108.—Entrance to one of the large trails on the island.

diptera were collected and many are new to our National Collection. Some insects of the other orders also were captured and brought back to the Museum.

One of the most interesting places which Mr. Greene visited in Panama was Barro Colorado Island. This large island of about six square miles is in the Panama Canal and located about half-way between the Atlantic and Pacific Oceans. Animal and bird life on this island is protected, and the abundant insect life makes it

FIG. 109.—Central Avenue, main business street in Panama City.

almost a paradise for the collector. Mr. Greene caught a large number of interesting species during his several visits to this island.

There is a large well-equipped laboratory on the island with facilities for studying and for lodging. Several large trails lead in various directions from the laboratory, but otherwise the jungle is undisturbed.

EXPLORING FOR FERNS IN THE BLUE MOUNTAINS OF JAMAICA

With the object of obtaining additional material needed in preparing a descriptive account of the fern flora of Jamaica, Dr. William R. Maxon, associate curator of plants in the U. S. National Museum, spent the months of June and July, 1926, in botanical exploration in that island, paying particular attention to certain areas in the difficult Blue Mountain region, which rises to nearly 7,500 feet.

The ferns of Jamaica were among the first to be described from the New World, but in many instances the names originally given them came later to be applied loosely to related but distinct kinds from other regions, with much resulting confusion. To afford a proper basis for studying the diverse fern floras of tropical America as a whole, it thus becomes of prime importance to know thoroughly that of Jamaica, an end that can be attained, naturally, only with the aid of adequate material.

Of the 500 species of ferns and fern allies described or known from Jamaica nearly all are found in recent large collections brought to American herbaria from that island; yet there are a few collected by Sir Hans Sloane in the latter part of the seventeenth century, and by Swartz about a hundred years later, that still are known only from the original specimens preserved in European museums. Present field-work is concerned therefore in the re-discovery of these "lost" species and of other very rare ones described more recently, but equally also in the discovery of new kinds, and in assembling data as to the distribution, characteristic habitats, habits of growth, and inter-relationship of those other species that are comparatively well known.

The equable temperatures and surpassing beauty of the Blue Mountains are proverbial. What is not appreciated by the traveler or the novice in botanical collecting is that the absence of springs and streams above 5,000 feet, the almost complete lack of trails and habitations, the steep, often precipitous character of the forested rocky inclines, and extremely fickle weather conditions make exploration difficult and laboriously slow. Approach must be through the drier coffee-producing regions on the south side of the range, except for two trails that cross at either end, nearly 25 miles apart; the wet northern slopes, from 1,500 feet to the top of the higher peaks, present an unbroken forest which can hardly be easier of penetration today than at the time of discovery.

It happens thus that our knowledge of the high mountain fern flora—which invariably in tropical regions is more luxuriant and diverse than that found at low elevations—is based mainly upon material collected on several peaks and in a few high passes near the former botanical station at Cinchona (5,000 feet) and on Blue Mountain Peak itself. The extremely rugged eastern end of the range and almost the whole heavily wooded north slope, with its

FIG. 110.—Looking westward from Blue Mountain Peak. The dense low forest is characteristic of the high peaks, several of which are shown in the background; Mossman's lies at the left. (Photograph by Duncan S. Johnson.)

deep, wet, hothouse valleys, have scarcely been touched; and this notwithstanding the work done by a host of collectors.

Accordingly, except for a few days of collecting in the denuded Port Royal Mountains, the region of Mount Diablo, the famous Fern Gully, and several localities reached readily from Kingston, effort was centered by Dr. Maxon at two points in the Blue Mountain range: Cuna Cuna, and the high peaks east and west of Portland Gap.

The Cuna Cuna region, which takes its name from a low gap (about 2,400 feet), lies at the eastern end of the Blue Mountain range

FIG. 111.—The fern collection about to leave House Hill, at the edge of the little-known Cuna Cuna region, facetiously dubbed "Back of Beyond."

and is traversed by a passable trail, upon which, however, there is no shelter. By good fortune it was possible to establish headquarters at House Hill, an estate lying southwest of the Gap at about 1,500 feet, in the lee of the range; and from this point many trips were taken to the Gap itself and to several of the neighboring peaks at 3,000 to 3,500 feet elevation—Gossamer Peak, Stone Hole Bump, Maccasucker Bump—all of them previously unexplored. The region is even wetter than the high peaks, an annual rainfall of 200 inches being by no means uncommon, and the luxuriance and variety of the fern flora, especially along the wet culminating crests and in the deep forested ravines, rich beyond all expectation, for so low an altitude. A most interesting feature here was the occurrence,

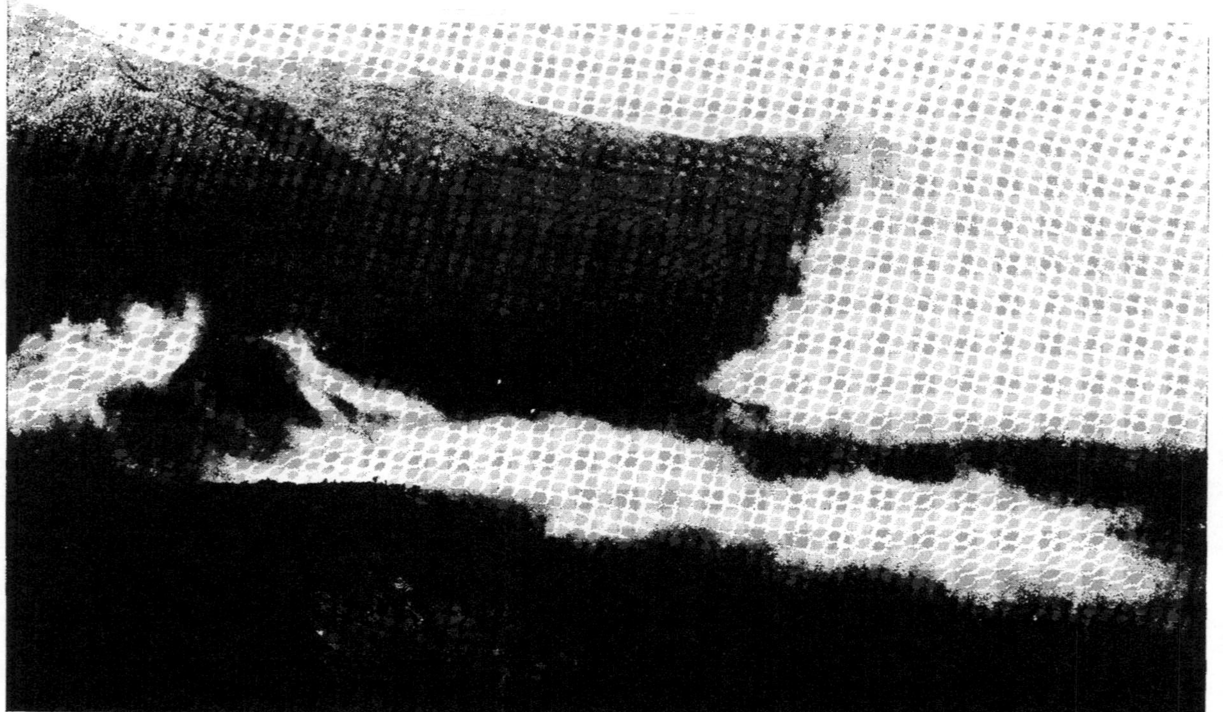

Fig. 112.—The main, heavily forested Blue Mountain range looking northeast from Cinchona. Mossman's Peak lies at the right. The bare patches in the foreground are newly cleared areas planted to coffee. (Photograph by Duncan S. Johnson.)

often in abundance, of a considerable number of ferns known previously only from the higher peaks to the westward at nearly twice the altitude. One day was given also to collecting in the isolated John Crow range, at a point opposite Mill Bank. The period of three weeks spent in exploring from House Hill was exceedingly productive, and thanks are gratefully tendered to the owner, Mr. C. E. Randall, and his family, for much cordial assistance, without which the work could scarcely have been carried out.

In the latter part of June Dr. Maxon joined a party of botanists at Abbey Green, a famous coffee property at about 4,000 feet, which

Fig. 113.—A "close-up interior" of tropical rain forest, Portland Gap. The ferns shown (*Elaphoglossum*) have simple stiff leaves, in strong contrast to the more usual lacy kinds.

had been leased by Prof. Duncan S. Johnson, of The Johns Hopkins University, and from this point, which lies 1,500 feet below Portland Gap, carried out intensive exploration of the higher peaks and ridges. Of especial interest was the ascent of Mossman's Peak (about 6,400 feet), lying just west of Portland Gap and hitherto unexplored, and the collection of large series of specimens upon the eastern and southwestern slopes of the mountain and at a high pass to the westward, known locally as "Main Ridge Gap." This whole upper region is densely forested and extremely humid. Fortunately a comparative drought prevailed at the time, allowing almost continuous work, and many rare and little known ferns were collected in quantity. In a deep pocket on the northeastern slope of Mossman's, filmy ferns

Fig. 114.—An early morning view of the coffee-producing region south of Portland Gap, with the steep denuded slopes of the Port Royal Mountains in the distance, southwest. (Photograph by Duncan S. Johnson.)

FIG. 115.—Entrance to a deep forest ravine in the higher Blue Mountains; the tree ferns are characteristic. (Photograph by Duncan S. Johnson.)

Fig. 116.—Dense epiphytic growth, chiefly bromeliads, ferns, and orchids, in wet forest near Morce's Gap, Blue Mountains of Jamaica. (Photograph by Duncan S. Johnson.)

(Hymenophyllaceae) were found in extraordinary abundance not elsewhere equaled.

To the east of Portland Gap similar exploration was carried out all the way along the trail to Blue Mountain Peak, the most extended trip being one of three days to the Peak itself (7,428 feet) and the

FIG. 117.—One of the rare tree ferns (*Cyathea Harrisii*), known only from the summit of Blue Mountain Peak. (Photograph by Alexander F. Skutch.)

deep forest sink between that and Sugarloaf Peak. Here a number of locally endemic ferns were collected, and observations made that will be of decided value in fúture exploration of the heavily forested northern slopes. At the Peak the panorama in all directions unfolds superbly, and one better realizes from the extremely broken topography why progress in exploration has been so slow. From this great height the sea, to north and south, lies only 12 or 13 miles away.

Fig. 118.—A filmy fern (*Hymenophyllum sericeum*), common in the higher Blue Mountains. The tawny-gray woolly fronds, about a foot long, hang limply in masses from tree bases in forest openings. (Photograph by Duncan S. Johnson.)

The present field-work was made possible through the allotment of a grant by the American Association for the Advancement of Science, a contribution from the New York Botanical Garden, and the hearty cooperation of the United Frúit Company. In all, about 15,000 botanical specimens were collected, representing 2,050 collecting numbers, and in addition a few other miscellaneous natural history specimens. The fern material will assist greatly in the projected manuscript, which is planned for publication by the British Museum. It is to be hoped that means will be found for continuing the Jamaican exploration, which, in spite of work already done, offers as fascinating an opportunity of botanical field-work as may be imagined.

BOTANICAL FIELD-WORK IN PANAMA AND COSTA RICA

In November, 1925, Mr. Paul C. Standley, associate curator, division of plants, U. S. National Museum, spent three weeks in the Canal Zone, in continuance of his previous studies upon the flora, as a result of which there is now ready for publication an account of the plant life of the Isthmus. At the invitation of Dr. Thomas Barbour and Mr. James Zetek a week was spent on Barro Colorado Island in Gatún Lake at the Laboratory for Tropical Research directed by the National Research Council.

This island, about six square miles in extent and for the greater part heavily forested, was set aside four years ago as a reservation for the wild life of the Canal Zone. The heavy rainfall, sometimes as much as 30 inches per month, is favorable to a luxuriant plant growth. The huge trees, of great variety, are loaded with aroids, bromeliads, orchids, and other epiphytes, affording a rich field for the study of these groups. Mr. Standley collected about 500 specimens, and has prepared for publication a list of the plants known from the island, amounting to about 600 species.

From the first of December until early April Mr. Standley was engaged in botanical work in Costa Rica. Nearly a month was devoted to exploration of the region of Santa Maria de Dota in south-central Costa Rica. Santa Maria, situated at an elevation of 5,000 feet, is surrounded on all sides by high mountains. Originally they were covered with dense oak forest, but much of this has been cut to permit cultivation. Inasmuch as the region was practically unknown botanically, it yielded a rich harvest of new or otherwise interesting plants. Of special interest were several sphagnum bogs lying at a considerable elevation in the dense forest, in which grow

Fig. 119.—An epiphytic cactus (*Epiphyllum ackermanii*) grown in Costa Rica. Flowers red. About half natural size. (Photograph by M. Gomez Miralles.)

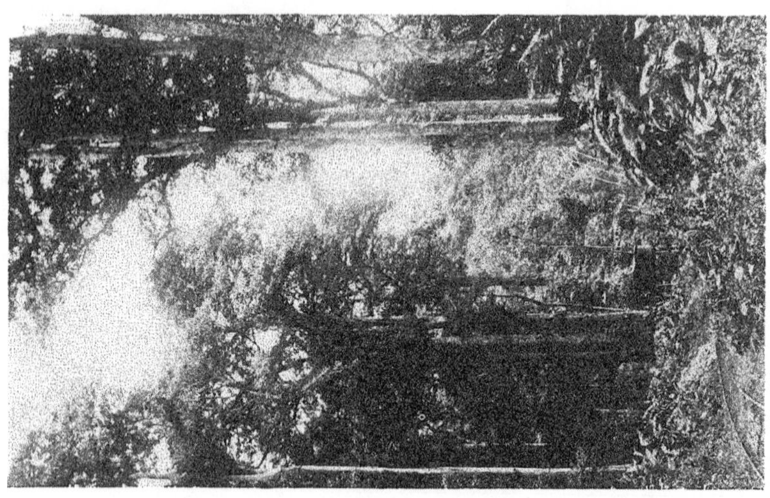

Fig. 120.—Scene in the wet lowlands of the Atlantic coast of Costa Rica. Heliconias ("wild bananas") are the most conspicuous herbaceous plants. (Photograph by H. Wimmer.)

many plants of South American types, nearly all of them new to Central America. New Year's eve was spent in a little cabin on the Cerro de las Vueltas, at an altitude of 10,000 feet. This mountain and the near-by Cerro de la Muerte (Mountain of Death) are unique in Central America, but are similar to many regions of great extent existing in the Andes of northern South America. The top of the Cerro de las Vueltas is an extensive tableland, partly forested but consisting chiefly of the type of grassland known in South America as paramo. In these paramos the vegetation consists principally of a velvety sward of fine grass one to two inches high, in which are

FIG. 121.—The volcano of Irazú, Costa Rica, in eruption. (Photograph by H. Wimmer.)

scattered many small plants which are essentially alpine, such as buttercups, violets, gentians, and other groups well represented in the mountains of the United States. Barberry bushes are common in some places, but the most remarkable plant is a giant dock (*Rumex*) 10 to 15 feet high. At this altitude the nights are very cold, owing to the combination of rain, fog, and wind. Ice a quarter of an inch thick forms frequently.

After leaving Santa Maria, a month was spent in company with Prof. Juvenal Valerio at Tilarán, in the Province of Guanacaste near the Nicaraguan frontier, a part of Costa Rica in which no plants had been collected previously. Tilarán lies at the foot of the cordillera bearing the same name. This range of mountains, one of the most reduced sections of the great backbone of the American continent, is only 2,500 feet high, but since it is the continental divide, separating the wet Atlantic slope from the comparatively arid Pacific slope, it influences to a remarkable degree the local distribution of the plants. Ordinarily a locality of Pacific Central America with such slight elevation as Tilarán (1,800 feet) would have a hot dry climate; but the rain-laden clouds of the Atlantic plains force their way westward over the crest of the cordillera and reach as far as Tilarán before they are dissipated. At Tilarán it is nearly always raining or misting, even when the sun is shining brightly. This might

FIG. 122.—A characteristic view along one of the rivers of the Pacific coast of Costa Rica. (Photograph by M. Gomez Miralles.)

well be called the land of rainbows, for with clouds, mist, and sunshine it is seldom indeed that a rainbow is not to be seen.

The rain and wind make the climate a cool one, much like that of the central tableland of Costa Rica, whose elevation is twice as great. The flora also is similar, and we find here, at only 1,800 to 2,500 feet, many species that grow at altitudes of 5,000 to 8,000 feet in central Costa Rica.

Westward from Tilarán, beyond the influence of the clouds, the climate changes quickly, and at a distance of one or two miles the soil is parched during the winter months, and the heat excessive. Some collecting was done in this arid region, particularly near the gold mines at Libano; nearly all the plants were different from those growing but a few miles away in the humid vicinity of Tilarán.

Fig. 123.—A typical view in the high mountains of central Costa Rica. Such forest is almost impenetrable. It is extremely rich in orchids and other plants. (Photograph by M. Gomez Miralles.)

A visit was made to Lake Arenal, across the cordillera on the Atlantic watershed. This lake, several miles long, is curious because of the fact that although the water is deep, its surface, at least during the drier winter months, is completely covered with vegetation. Viewed from one of the overhanging slopes one would never guess the presence of a body of water; it appears like a great savanna. At the end of the dry season the long grass covering the water becomes so dry that it may be set afire, thus affording the anomalous spectacle of a burning lake.

FIG. 124.—Páramo of the Cerro de las Vueltas, Costa Rica, at an elevation of 10,000 feet. The low clumps of vegetation are composed chiefly of a dwarf bamboo.

Field-work in Guanacaste revealed a large number of unusual plants, most of them, however, of technical rather than general interest. Much lumber is cut in this province, especially Spanish cedar (*Cedrela*) and guanacaste (*Enterolobium cyclocarpum*).

After leaving Guanacaste Mr. Standley had headquarters at San José for several weeks, making excursions from the capital and from Cartago to the surrounding high mountains, the richest region for plants in all Central America. In company with Prof. Rubén Torres an excursion was made to El Muñeco, south of Cartago, where there was discovered a new genus of trees closely related to the walnuts. Visits were made also to the lowlands of the Atlantic coast, which likewise have a varied and interesting flora.

In Costa Rica about 11,500 numbers of plants were collected, mostly phanerogams and ferns. As upon an earlier visit, much attention was directed to the collection of orchids, of which there were obtained 2,000 numbers, representing many species. In orchids no other part of the North American continent can compare with Costa Rica, and it is probably not excelled by any area of equal size in South America. About 1,000 species are known from Costa Rica. They range from the beautiful Cattleyas, with flowers equal to

FIG. 125.—Cerro de la Carpintera, near Tres Rios, central Costa Rica. This mountain, now nearly denuded of its forest, is a classic locality for plants. Black howler monkeys still live in the patches of forest about the summit. (Photograph by M. Gomez Miralles.)

those of any hothouse orchid, to diminutive plants less than an inch high, whose blossoms are so minute that they must be studied under a strong lens.

Although Costa Rica is a small country, about as large as West Virginia, its flora is so extraordinarily varied that it is still imperfectly known, in spite of the fact that this has been the favorite collecting ground of a large number of botanists. It may seem strange that so small a country should not have been exhausted long ago, but to one familiar with Costa Rican geography the explanation is simple. Some of the provinces are so difficult of access that they are practically unknown even to the national government. The Republic consists of a great mass of high mountains separated by steep-sided forested valleys, across which travel is nearly or quite impossible. Each valley and mountain possesses plant species all its own, hence many decades of exploration will be needed to exhaust this narrowly limited portion of the American tropics.

BOTANICAL EXPLORATION IN NORTHERN HAITI

Further field-work in Haiti was undertaken in November, 1925, by Mr. E. C. Leonard of the division of plants, U. S. National Museum. This work was fostered by Dr. W. L. Abbott, who in recent years has made a number of visits to Hispaniola, procuring much valuable material in several branches of natural history. The botanical specimens collected on these expeditions will be studied in the preparation of a flora of Hispaniola which is now under way.

Mr. Leonard commenced the winter's work in the vicinity of St. Michel, using as headquarters, with the permission of Mr. G. G. Burlingame, a house belonging to the United West Indies Corporation. With the invaluable assistance of Mr. F. C. Baker of the United States Department of Agriculture and Mr. E. J. Sieger, Manager of the Atalaye Plantation at St. Michel, it was possible to reach this region almost immediately after arrival at Port au Prince. The plantation is situated three miles east of St. Michel on the northern edge of a large savanna. During the winter season these plains are covered by a dense growth of a tall reddish-yellow grass, *Themeda quadrivalvis,* a native of Africa. It is of no value as a forage crop and is a serious menace to the plantation buildings, since fires started by Haitian farmers are constantly breaking out, and sweep rapidly over considerable areas. The principal crop raised on the plantation is tobacco; both cotton and sugar cane were tried but did not thrive in the black mucky clay of the savanna. A large number of Haitians, four hundred to three thousand, according to the season, are employed to assist with the crops, and, judging from the melodious cadences drifting in from the drying barns and fields, they find happiness as well as a congenial occupation.

The comparatively meager flora of the plains being soon exhausted, excursions were made to the neighboring mountains, the nearest of which, Mt. La Mine, lies a short distance to the north. Its arid, thicket-covered slopes produce many unusual plants. Of great interest in this region are several caves whose floors are covered by a thick layer of dust, débris, and fragmentary bones of an extinct fauna.

Other mountainous regions, more distant but accessible by horseback, were Mt. La Cidre, Mt. Platanna, both southwest of St.

FIG. 126.—The northern slope of Mt. Platanna, showing the xerophytic nature of the vegetation.

FIG. 127.—Logwood ready for export, at Gonaives, Haiti.

FIG. 128.—Cactus plains east of Gros Morne, Haiti. The tall tree-like cactus is *Cephalocereus polygonus*, the lower columnar cactus at its base is *Lemaireocereus hystrix*, the prickly-pear in the foreground is *Opuntia antillana*.

FIG. 129.—One of the interior towers of the Citadel, northern Haiti. The Citadel was constructed by the emperor Christophe as a last retreat in case of a French attack.

Michel; Kalacroix, south of Atalaye Plantation; and St. Raphael, Dondon, and Marmelade, to the north. In the regions of Kalacroix and Marmelade the altitude was great enough to permit an abundant growth of ferns. In climate and character of vegetation the region of Kalacroix resembles in many respects that of Furcy, south of Port au Prince, visited by Dr. Abbott and Mr. Leonard several years before. In the rich valleys of both Kalacroix and Marmelade a great deal of coffee is grown. On the low mountains, between the occasional streams, one usually finds xerophilous grasses or thickets. In some places high cliffs and steep inclines of jagged lime rock, partially covered by tangled vegetation, are the predominating features.

Fig. 130.—Three donkey loads of guinea grass, Gros Morne, Haiti.

Dondon, a picturesque but somewhat inaccessible village, was next chosen as a collecting base. It is near the famous Citadel, that stupendous monument to futile effort erected with so much suffering and loss of life as a final stronghold of Christophe, once Emperor of the North.

Lumbering possibilities in Haiti have long been a subject for investigation. In the regions visited on this trip, timber did not seem plentiful enough for profit. The common tree of the higher regions is *Pinus occidentalis;* yet even it is scattered and of poor quality. On the plains and foothills mahogany is fairly abundant, the trees growing a considerable distance apart and owned, as a rule, by individual landholders. Logwood is found mostly in dense thickets. When the native finds a tree which has produced sufficient heartwood, it is hewn into convenient lengths with a machete and the useless sapwood removed. The pieces of heartwood, which contain the dye, are then carried by head or on pack animals to the nearest clearing house, a shelter and rude balance erected by the roadside near or in some village, and there offered for sale.

On January 11, 1926, a new headquarters was established at Ennery, a village midway between St. Michel and Gonaives. The high mountains, rising abruptly on the south, resemble Mt. La Cidre and Mt. Platanna in both physiographic features and flora. On the north the Puilboreau or Cape road, winding upward in a continuous grade to an elevation of about 2,700 feet at the Pass, gave excellent oppor-

FIG. 131.—Market at Gros Morne, Haiti.

tunity for collecting the plants of the surrounding region. A dense thicket at the summit of the Pass, wet on its north slope and dry on its south, was especially rich in strange plants. The flora along Trois Rivières in the vicinity of Gros Morne, Pilate, and Plaisance, typical of most river vegetations throughout Haiti, was rather uninteresting, but the arid thickets between Gros Morne and Gonaives were extremely productive. These undoubtedly would furnish excellent botanical material if visited during the wet season.

After completing his field-work in the northern mountain region of Haiti, Mr. Leonard returned to Port au Prince, where several days were spent in packing specimens and in exploring the sources of Thor, Bizoton, and Mariani with Dr. Erik L. Ekman and Dr. H. D. Barker.

As a result of this expedition, a collection of 3,143 numbers (about 9,000 specimens) was procured, some of the species being new or rare ones not previously represented in the National Herbarium. In addition to the herbarium material, a number of land shells and insects were collected.

ARCHEOLOGICAL SURVEY OF THE FÊNG RIVER VALLEY, SOUTHERN SHANSI, CHINA[1]

In the latter part of December, 1925, Mr. Bishop suggested that I undertake some work in the field. The idea immediately occurred to me that before the spade be brought out, a preliminary survey should be made, so it was agreed that I should go to the southern part of Shansi and investigate the archeological possibilities along the Fêng River valley. The president of Tsing Hua College, Mr. Y. S. Tsao, kindly consented to cooperate. In his official capacity, he wrote to Governor Yen Hsi-san and successfully arranged for a permit for me to travel in southern Shansi. By a lucky coincidence also, the Geological Survey of China was on the point of sending Mr. P. L. Yüan to the same region for some field-work in geology. Mr. Yüan is a geologist of much experience, having travelled with Andersson in Kansu for two years and acquired a great deal of interest in prehistoric archeology. So we arranged to travel together. Our start was somewhat delayed, and it was not until February 5, 1926, that we left Peking, just a week before the Chinese New Year.

We arrived at T'ai-yüan on the 7th, and spent the whole day of February 8th in making calls and purchasing equipment. When I left Peking, I brought a large number of letters of introduction to the Governor and the various officers of influence in Shansi. Among these was one written by Mr. Liang Ch'i-ch'ao, now senior professor of the Tsing Hua Research Institute. Similar to many other letters, this one explained the purpose of my visit and the necessity of archeological work at present. The Governor, however, was too much occupied with matters of greater importance to see us; but we succeeded in having an interview with his secretary, who, on behalf of the Governor, promised us all the help we asked. These promises were well fulfilled later on.

[1] EDITOR'S NOTE: It is greatly to be regretted that, owing to lack of space available in the present publication, Dr. Li's excellent report, his illustrations and his interesting conclusions cannot be printed complete at this time. The following excerpts will, however, convey some idea, at least, of what he has accomplished by his preliminary search for archeological sites in southern Shansi.

We left T'ai-yüan on the 9th (fig. 132) and, traversing what is geologically a *loess* area, arrived at Chieh-hsiu three days later. I was much impressed by the extensive use of arches that is noticeable from T'ai-yüan southward. It seems to be quite a peculiarity of Shansi architecture. The first series I saw was along the Chêng-t'ai road. All the way, in houses that were built on a grand scale, as well as in the small inns, we found such arches employed. Buildings of this type are known as *yao-fang*. Native scholars told me that they are warm in winter and cool in summer, the style being derived from the early cave-dwellings. At present, we still find all the transitional

FIG. 132.—Our cart in front of a roadside temple.

stages from the *loess* cave type to the most complicated *yao-fang* represented in this region.

We took advantage of the Chinese New Year to see the city of Chieh-hsiu (fig. 133) and also made some measurements of the natives, who seem to be quite a heterogeneous group. I saw bearded men who can be compared with the average Armenian; I saw also men with 100 per cent yellow moustaches. One of the commonest physical types found in this district is the round headed individual with a long face, a disharmonious type according to physical anthropology. Such an occurrence did not, however, surprise me at all, as both dolichocephalic and brachycephalic people are found in this region. Very likely it is the mixing of these two fundamental types that has given rise to this disharmony.

On the 15th, we started our first trip to the mountain (Mien-shan) in the southwestern part of the district where some of the ancient temples are found. Of two of these, especially, I made some detailed study. Both temples are Buddhist, one located at the foot of the mountain, the other near the top of one of the peaks. The one at the foot was first built in the T'ang Dynasty, but has been destroyed and rebuilt many times since, only the bell and drum-tower still retaining a pre-Yüan style. The temple on the peak, also, was first built in the T'ang period, but burned during the Ming. The stratification of the three different layers of culture is here plainly visible.

FIG. 133.—Chieh-hsiu: in front of Kuan-yao Miao.

One of the halls is located very near to a cave. The images that are worshipped at present are evidently of recent origin. Behind them is another row of images cast in iron; and finally in the cave I found two broken statues of stone, one with a head and one without, carved in simple, bold style, showing T'ang workmanship; but they are cast away in the rear of the cave and covered by dust. Perhaps, however, the most interesting thing we saw on this trip is a stone ox washed down by a mountain torrent some time ago, from where no one knows.

The river Fêng has its source at Kuan-ch'in-shan and flows in a straight southerly direction till it reaches Chiang Chou. From I-t'ang southward, it cuts through the Ho-shan range for about 40 miles. after which it flows through the plain again. For a whole day, the 23d, after leaving Chieh-hsiu, we wormed our way through the Ho-shan along the river bed. About noon we reached Hsia-mên Ts'un which is one of the most beautiful villages I have seen in this province: the buildings of brick and limestone, the windows and doors arched.

From there southward, the mountains on both sides rise steeply, and in them are many limestone caves, some of which I explored, but found only traces of modern habitation. We stopped at Ho-chou, between which place and Lin-fêng Hsien there are many historic places; but as I did not intend to make any intensive study north of the latter, we passed this region rather hurriedly and reached Lin-fêng Hsien on the 25th.

Lin-fêng Hsien (or P'ing-yang Fu), a city that has aroused imagination in the past—the ancient capital of the Emperor Yao! What Chinese scholars are not acquainted with the list of virtues of this august monarch? Did he, however, also create a model city? Since he was, perhaps, the most self-denying emperor that the world has ever known, it would not be in harmony with his ethical principles if he should have used the national wealth to build luxurious palaces like those found in Troy and Knossos; but whatever he might have done in this respect, it is a fact that there is not even a tradition as to the exact location of his capital. The modern city of Lin-fêng Hsien is, like every other city in inland China, surrounded by machicolated walls. About a mile west of the city flows the Fêng River, and west of the Fêng is the famous Ku-i-shan where, according to the mythical tales of Chuang-tzŭ, resided many fairies.

We rested in Lin-fêng Hsien for a day, and on the morning of the 27th started to explore Ku-i-shan in the western part of the district, the place being popularly known as Hsien-tung (Fairies' Caves). A mile out of the western gate of the city, we crossed the Fêng River. Between the river and the foot of the mountain there is a series of *loess* terraces ascending higher and higher towards the mountain and dotted here and there by villages, some of which are merely a collection of cave dwellings. It took us more than half a day to reach the Northern Fairy Cave, where we were received by a monk—an old man, widely travelled, and evidently knowing something about his profession.

The temples in this mountain are for the Buddhists. They were originally built in the early part of the T'ang Dynasty; but in later periods they were repeatedly ruined and rebuilt. For some time in the

Mongol Dynasty, the Taoists took hold of them and converted them into Taoist temples. They were, however, soon restored to the Buddhists again.

The purpose of my own visit to this place was to explore the limestone caves; that of Mr. Yüan, to investigate the coal region still further west; so on the morning of the next day, each of us pursued his own task. The whole region here is limestone formation, divisible into many different strata. A deep ravine cuts the ground into two perpendicular walls, north and south, in which are several rows of caves, most of them inaccessible, while some are well fitted for early human habitation. Of these I visited five, in the hope of unearthing some paleolithic remains, but the search proved fruitless. We left the mountain on the next day by a different route and made a further search at the foot of the mountain—only to be disappointed again. I had a long discussion with Mr. Yüan in the evening as to the exact route we should follow, and finally came to the conclusion that, so far as my personal work was concerned, I should follow partly the historical sites and partly the probably pre-historic settlements as my guides, so at 10 a. m. on March 2d we left for Yao-ling.

The exact location of the tomb of the Emperor Yao is a long debated question. Previous to the Mongol Dynasty, this tomb was usually located in Shantung. The tomb in P'ing-yang Fu was not so well known at that time. The argument for its location in Shansi is that as Yao retired at quite an advanced age, it is improbable that he should subsequently have inspected his distant domains and died so far away from home. As the tomb in P'ing-yang Fu is near his supposed capital, it is probably the true one—if, indeed, there be a tomb of this Emperor; but since the very existence of such a person is doubtful, we can only consider both tombs as variations, merely, of the same myth. All these considerations passed through my mind while we were riding towards the tomb of the Emperor. When we inquired about the way to Yao-ling, the natives simply stared at us, and it was a long time before we made out that the local name for the place is Shên-lin (Spirit Forest). There we arrived late in the evening, and found it surrounded by a wall enclosing a building of modest size, a solitary temple in the midst of mountains, where we stayed that night.

The tomb is quite high, pyramidal in shape and half encircled by a rivulet (fig. 134). It was officially lost for a long time, but was rediscovered in the Ming Dynasty, according to the inscription in the temple. The arguments as to whether this is the real tomb are difficult to follow; but in spite of them the question remains a question today, and undoubtedly will remain so until the spade of an archeologist shall clear it up.

We worked about two hours in the morning and left this reputed resting place of the august Emperor at 10 a. m., arriving at Fou-shan Hsien late in the afternoon. On the fourth, we made very little progress. The ground was wet, and there were many steep ascents and descents. We covered about six miles in all and stopped at Hsiang-shui-ho. On this day I picked up my first piece of red pottery of an archaic character on a descending *loess* slope.

Fig. 134.—Supposed tomb of the Emperor Yao. From the southeast.

The next day we started early, while the ground was still wet. The road we followed lies deep down between *loess* cliffs. Such roads make it convenient to observe the exposed surface of the *loess*. The finding of the red pottery sherd was very encouraging and made me look carefully all along the way. Not long after we started from Hsiang-shui-ho, I began to see gray pottery sherds of the Chou and Han periods. All of a sudden I discerned a piece of red pottery decorated with black lying among the withered, wet grass. Then one after another came into view as we traced them to their source. It is a heap of earth about 10 feet in height and cut down vertically on one side to the public road. The upper surface is a long and narrow strip. This piece of land is owned by the Li brothers, who most politely received me and helped me with their spades to gather samples of the painted sherds from the exposed surface of their mound. When

I left them, they were very willing to pose for a picture (fig. 135) by the side of the heap where these pottery sherds were discovered—the first Yang-shao site to be found in southern Shansi. The rest of the day's journey was a cheerful one, and at four in the afternoon we arrived at the city of I-ch'eng Hsien.

It was a whole day's journey from I-ch'eng to Chü-wo which, in turn, is about 60 *li* east of Chiang Chou—one of the most important cities in southern Shansi and a center for curio-dealers. At this city, the Fêng River turns westward. While we were in Chü-wo, we decided to pay a visit to Chiang Chou to have a look at the various

FIG. 135.—Site at Chiao-t'ou-ho where prehistoric pottery of the Yang-shao type was found.

curio shops in that city. This, I thought, might perhaps serve to help us in forming a notion as to the kinds of antiquities which are unearthed in this region. But this idea proved to be a delusion. All the curio-dealers have a common secret: if one inquires about the exact location of the place whence the things they exhibit come, the unanimous and invariable answer is that they do not know. In vain one may tell them that the curios would increase in value if their sources were known. Thus a whole day in Chiang Chou only convinced me that so far as real archeological work is concerned, very little help can be derived from such people. Having gone to Chiang Chou from Chü-wo by the northern route through Hou-ma, we returned over a bypath through mountains in the south which has been gradually elevated from the Fêng River valley by *loess* deposit. A day and a half was spent in exploring the terraces, and although no particular archeological results were obtained on this trip, we had a fine chance to study the *loess* formation (fig. 136), a certain knowledge of which, I think, is necessary, if the archeology of southern Shansi is to be properly understood.

After we returned to Chü-wo, our next trip was to dash across Chung-t'iao-shan. This range, according to the local estimate, extends for about 800 *li* from east to west, parallel to the Yellow River on the south and the Fêng on the north, and inasmuch as early traditions about the Emperor Shun and the Hsia Dynasty are centered

Fig. 136.—*Loess* terraces south of Chiang Chou. (Photograph by P. L. Yüan. Courtesy of the Geological Survey of China.)

about these mountains, I decided to spend some time here. The next four days, therefore, were devoted to crossing and recrossing Chung-t'iao-shan; but as we found no archeological prospects here, we turned immediately northward to An-i Hsien and Yün-ch'êng.

We arrived at Yün-ch'êng in the evening of the 17th and entered the city on the 18th. On the 19th we set out to visit the supposed tomb of the Emperor Shun, and on the way stopped at certain temples in Yün-ch'êng. In *Shansi-t'ung-chih* (Vol. 52, p. 2), it is recorded that the stone pillars of these temples were formerly the palace pillars of Wei Hui-wang (335-370 A. D.), recovered from the ruined city south of An-i Hsien. Some of them are now used as the entrance pillars in Ch'ên-huang Miao and Hou-t'u Miao, and those of Ch'ên-huang Miao certainly show peculiar features which are worth record-

Fig. 137.—Details of the left hand stone pillar at the entrance to Ch'ên-huang Miao, Yün-ch'êng.

Fig. 138.—Supposed tomb of the Emperor Shun.

Fig. 139.—Type of Buddhist *stelae* collected by the Magistrate of An-i Hsien.

ing. Two pillars, hexagonal in section and carved with dragons coiled around them, are found at the entrance. The left one (fig. 137) is especially interesting, because in the claws of the dragon are grasped two human heads with perfect Grecian features: curly hair, aquiline and finely chiselled nose, small mouth and receding cheeks. One head with the tongue sticking out is held at the mouth of the dragon, while the other is held in the talons of one hind leg. It is an unusually fine piece of sculpture in limestone, wonderfully spaced and with the most graceful lines. The right one is inferior in its workmanship: evidently the two were not executed by the same hand. I saw 28

FIG. 140.—Supposed tombs of the Hsia Emperors. (Photograph by P. L. Yüan. Courtesy of the Geological Survey of China.)

of this kind of pillar in the succeeding two days; but most of them were crude imitations. It is possible, however, that some are of the ancient type and were made earlier than others. The whole subject is well worth more detailed study.

The tomb of Shun (fig. 138) has a very different appearance from that of Yao. It is located about 30 *li* northwest of the city of An-i Hsien in the midst of a vast plain with apparently no natural barrier on any side to shelter it from "the wind and the water." Half of the early references to the tomb of this emperor put its location at Ch'ang-wu. Yet Chang Chin-chün quite convincingly argued that it must be in An-i (*Shansi-t'ung-chih,* Vol. 56, pp. 20-23). The problem is similar to that concerning the tomb of Yao, and consequently the solution must be sought in the same way.

On the 20th, we were invited to dinner by the magistrate of An-i Hsien, Mr. Chêng, who is a student and collector of antiquities. He introduced us to a small museum in his *yamên* where he has gathered together a large number of Buddhist stelæ and ancient tablets (fig. 139) which were originally scattered all over the district of An-i. Only a part of all he found has been moved to his *yamên;* but his catalogue includes those which still remain in the different villages. He draws rather a sharp line of demarcation and leaves all the post-Sung sculptures unrecorded. It is an unusual work that he is doing, and gives one a ray of hope that some of the ancient monuments in inland China may yet be preserved. After the dinner we were taken about the city,

FIG. 141.—Site at Hsi-yin Ts'un where prehistoric pottery was found.

where we saw some more of the dragon pillars, mostly in Taoist temples. The ancient city itself, where some of these pillars are supposed to have been found, is less than a mile from the southern gate of An-i. The remains of the old wall are still visible, but, if the place was a city at all, it was indeed a very small one, measuring about 400 by 250 yards. It may, however, be the site of an important ancient building.

We left Yün-ch'êng on the 21st, and on the 22d we arrived at Hsia Hsien—center of the traditions of the ancient Hsia Dynasty. The temple of the Great Yü and the tombs of his descendants as well as many of the famous ministers of that dynasty are said to be located here, and all these I visited (fig. 140); but I must confess that I am not at all able to determine whether, judged by their appearance, they are the real tombs or not. They all look like the ordinary burial mounds, except that they are larger. However, while on our way to visit these tombs the unexpected happened. It was after riding through the village of Hsi-yin that, suddenly, a large field of prehistoric potteries was discovered! Mr. Yüan was the first to see it. The site (fig. 141), extending to several *mou* of land, is apparently larger than the one we found at Chiao-t'u-ho, and the pottery is somewhat different, too. While we were picking up at random the sherds exposed on the surface, the villagers gathered in large number, so we did not stay very long lest there be too much excitement created.

FIG. 142.—Votive *stela* found at Hsin-hua-shih.

When we left An-i Hsien, the magistrate had given us a copied list of the names and locations of the various votive stones which are still scattered in the different villages of his district, and as we were on our way northward again, we determined to stop and see some of them. Three places were visited on the 25th and the 26th, where besides *stelæ*, we saw also a number of individual Buddhist figures which are preserved in good condition, the most perfect piece being in San-lu-li Ts'un. Unfortunately, it is preserved in a dark room, where a picture could not be taken, as I had no flash light with me.

On the 26th, while Mr. Yüan went away on some special geological mission, I started for Chi-san Hsien to have a look at the so-called T'ang wall paintings at Hsiao-ning Ts'un, some of these paintings having been recently sold to curio-dealers who sent them to Peking for sale. Hsin-hua-shih is a Buddhist temple built in the 12th year of K'ai-huang of the Sui Dynasty near Ch'a-tien-chieh. It has been destroyed and repaired many times; but the front hall still retains some reminiscence of T'ang architecture. It is in the central and the third hall that the walls are painted on three sides. The paintings of the side walls of the third hall and those of the southern wall of the central hall (opening towards the north) have been taken away by curio-dealers. The rest is still intact; and the date is to be found on the northern wall of the third hall, being the *wu-hsü* year of Yüan (A. D. 1298 or 1358). The courtyard between the front hall and the central hall was locked, and my guide assured me that there was nothing worth seeing inside the yard. Nevertheless, I had him open it for me, in spite of his assurance, and by sheer accident I found the protruding corner of a stone which lay buried in the ground. I asked the villagers to dig it out for me, and it proved to be a votive stela (fig. 142) carved at the time when the temple was first built. This little discovery rounded out my trip, and from this day on I marched directly northward and returned to Peking.

NOTES

1. Anthropometrical measurements at Chieh-hsiu.

Through the kind arrangements of Mr. Huang Tzŭ-sen, Magistrate of Chieh-hsiu Hsien, I was able to measure 86 of the natives of this District who are serving in the Army Training Camp and the Police Court. With the exception of my series of Huang-p'i and Huang-kang, this is the largest of any series of anthropometrical measurements made in any one district. The following 13 measurements were taken of each individual: stature, auditory height, sitting height, head length, head breadth, horizontal circumference, minimum frontal diameter, bizygomatic diameter, nasion-menton height, nasion-prosthion height, nasal height, and nasal breadth. Observations were also made of the following descriptive characters: hair color, eye, brow, chin, shape of nose, malar bone, shape of face, prognathism, teeth.

2. Pottery Sherds from Chiao-t'ou-ho.

The total number of sherds is 127. Of these, 42 are painted. Of the painted, 20 are with rims: 6 with bent rim, 1 with thickened rim, and 13 with plain rim. The paint used is black. The ground color varies from light brown to dark brown. The patterns consist of triangles with three sides concave, or with two sides concave and one side straight or convex; straight lines; crescent moon; big round dot; cross-hatched lines; shape of X with horns elongated into straight lines; and parallel lines with big dots between. Of the 85 unpainted, 21 are gray in color and are decorated with incised lines in parallel, cross-hatched or mixed directions; 2 are black and undecorated; 2 are dark violet and undecorated; 60 have the same ground color as the painted, and of these 17 are decorated with incised lines, 1 with ornamentation in relief, and 42 are plain. In addition to these potteries, there are also two broken pieces of finely made black stone rings, and a small piece of the shaft of a human ulna.

3. Pottery Sherds from Hsi-yin Ts'un.

The total number is 86. Of these, 14 are painted. Of the painted, 7 have rims—3 bent and 4 plain. Triangles, straight lines and big dots are the chief decorative patterns. They are often combined. Of the unpainted, 15 are gray or black in color and 57 are red or dark brown. Of the gray or black, 11 have incised lines and 4 are plain. Of the red, 34 are incised and 23 are plain. One complete, rather ill-shaped cup is found in this collection. It is dark gray in color and not uniformly fired. The diameter of the rim varies from 5.5 cm. to 5.9 cm.; the height is 5.3 cm. There are 7 ridges at the bottom with 7 finger depressions between the ridges.

4. Votive *stelæ* seen in An-i Hsien.

In the list of ancient votive *stelæ* given to me by the magistrate, there are 41 dated previous to the T'ang Dynasty. Of these, 28 have been moved to Fang-kung-tz'ŭ, the museum in the magistrate's *yamên*. Twenty others are dated in T'ang and Sung, the latest date corresponding with A. D. 1101.

CHI LI,
Freer Gallery of Art Expedition to China.

ANTHROPOLOGICAL WORK IN ALASKA

Under the auspices of the Bureau of American Ethnology, Dr. Aleš Hrdlička, curator of physical anthropology, U. S. National Museum, made, during the spring and summer of 1926, a comprehensive survey of anthropological and archeological matters in Alaska.[1] The

[1] During the extended trip briefly outlined above the writer has received many courtesies and much help for which he wishes hereby to offer once more his grateful and hearty acknowledgments. It will be impossible to specially mention all who aided in the work, but in the first place thanks are due to Governor George A. Parks of Alaska; Mr. Harry G. Watson, his Secretary; Mr. Karl Thiele, Secretary for Alaska; Judge James Wickersham, formerly Delegate from Alaska; to Father A. F. Kasheroff, Curator of the Territorial Museum and Library of Juneau; Dr. Wm. Chase of Cordova; Mr. Noel W. Smith, General Manager, Government Railroad of Alaska; Mr. B. B. Mozee, Indian Supervisor, and Dr. J. A. Romig, of Anchorage; to Professor C. E. Bunnell, President Alaska Agriculture College at Fairbanks; to Mr. and Mrs. Fullerton, Missionaries at Tananá; to the Rev. J. W. Chapman at Anvik; to Father Jetté, at Holy Cross; to Mr. C. Betsch at Russian Mission and to Messrs. Frank Tucker and E. C. Gurtler, near the Mission; to Mr. Frank P. Williams of St. Michaels; to Judge G. J. Lomen and his very good sons and daughter at Nome; to the Rev. Doctor Baldwin, Fathers La Fortune and Post, and Capt. Ross, U. S. Coast Guard, at Nome; to Mr. Elmer Rydeem, merchant at Nome; to C. S. Cochran, the Captain of the "Bear," and his officers, particularly Mr. H. Berg, his excellent boatswain; to the Rev. F. W. Goodman and Mr. La Voy at

trip was in many respects a noteworthy one and rich in results, which, if followed up, promise to lead to valuable additions to our knowledge regarding the American aborigines.

Since American anthropologists started to study native man, there has always been in the background of the work the question of the origin and antiquity of the American Indian and Eskimo. There were always the questions: Who are they?; What are their true affiliations?; Where did they come from?; and When did they come to the New World? One man after another under the Smithsonian Institution, and more especially under the Bureau of American Ethnology, besides those elsewhere, has given his life to the study of these problems, and research was and still is carried on in all parts of the continent on these basic questions, without the final answers having as yet been reached.

Throughout this work there has always been felt a need of more definite knowledge of those parts of America and Asia that come closest together. In the studies on the origin of the American aborigines in particular, indications invariably point to the furthermost American northwest and thence to the Asiatic continent. Time and again, even to the present day, ideas or opinions have been advanced that the American man, or at least some of the American aborigines, may have reached this continent from other parts of the world—from Europe, Polynesia, Melanesia, Central or Southern Asia, and even Australia. The men who are advancing these ideas generally forget that when we are dealing with the peopling of America we are not dealing with the people of the continent within anything like historic times, but thousands of years back when man was by no means as civilized or apt in any part of the world as he later became, and when he did not control as yet sufficient means of navigation or, especially, of provisioning for any extended journey such as could have brought him into this continent. The best students of the question agree that man, up to relatively late times, could only have come into America over small stretches of ocean; and so everything points in the one direction of Alaska, and beyond to the Asiatic continent. And here many people have assumed that there must have been up to recent

---

Pt. Hope; to the American teachers at Wales, Shismareff, Kotzebue, Pt. Hope and elsewhere; to Tom Berryman, Jim Allen and Mr. Chas. Brower, traders respectively at Kotzebue, Wainwright, and Barrow; to Mr. Sylvester Chance, Superintendent of Education, Kotzebue, Alaska; to the U. S. Marshals, Deputy Marshals and Postmasters along the route; and to the numerous traders, miners, settlers and others who were helpful with specimens, advice, guidance, and in other matters.

FIG. 143.—Harding Glacier, Resurrection Bay, Alaska.

FIG. 144.—Episcopalian Mission, Yukon Tananá Jct.

Fig. 145.—Chief Thomas, of the Tananá Indians, with his young wife. (At Nenána.)

times a land connection. But Dall and others have shown that we have no sufficient foundation for this assumption within the time that the American man may have come over; nor was such a connection at all necessary. There may have been a land connection preceding, and possibly even during some part of the glacial period, but those times are so far away that they play no part in the peopling of America, however large a rôle they may have had in the exchange of various animals between the two continents.

When one looks on the map of Alaska, it seems a relatively small portion of the world and it would seem that the exploration of it for traces of ancient man should be fairly simple. There are a few large rivers; almost no harbors; there are in general only a few favorable spots in Alaska where ancient man could have established himself, and it would appear easy to reach these points, to survey them, and to see what they promise or can give. As a matter of fact, Alaska is as large as one-third of the United States; the whole of it has less than two hundred miles of good roads; and the interior—and by interior is meant here anything away from the shores of the seas and the banks of the rivers—is practically impassable except for short stretches during the brief summer. In winter the country can be traversed more easily with dog sleds, but winter is not the season for archeological work. And when the explorer comes to the rivers or shores, he finds that transportation facilities by boat have, since the gold rush is over, become very limited, suitable boats being hard to obtain and very expensive. So that the student from the very start is presented with serious difficulties which at times seem almost insurmountable.

Under these conditions it was difficult to carry out systematic work planned ahead from beginning to end; but the writer soon learned that Alaska in general is peopled today by the most helpful, big-hearted and generous men and women, and their help counted for much. With this many of the difficulties were overcome until the Bering Sea was reached, where by good fortune was found the Revenue Cutter "Bear," which was ready to help. On it the writer went to St. Lawrence and other important islands; and with it he was enabled to visit point after point of anthropological interest along the Seward Peninsula, the Kotzebue Sound, and then through the Arctic Sea right up to Barrow. The "Bear" could not give all the time needed, but enough was given to make possible at least the most essential observations on each spot and fair collections.

The journey began, strictly speaking, at Vancouver, for at the several stops of the boat between Vancouver and Juneau an opportu-

Fig. 146.—Chief Joseph, Yukon Indians, Tananá. (Photographed at the Mission.)

nity was had of seeing the natives of the coast and the islands. The striking characteristics of these natives are their relatively large faces and big chests, combined with rather short legs and stature. There are large numbers of mixed-bloods among the younger generation.

A stop of several days was made at Juneau, to see the excellent local museum under the able directorship of Father Kashevaroff, and to obtain the needed official papers. The occasion was utilized also for a visit of the last remaining known site of the Auk tribe of Indians, and for some important collections. The chief among the latter is an old Shaman mask of the Yakutat tribe beaten from one piece of copper.

From Juneau, transportation was taken for Seward on a boat which stopped at several stations for a sufficient length of time to enable the writer to see such things as could be found in these places, and to make inquiries of principal local men about Indian remains. Some specimens were collected on this part of the trip, including a very typical skull of an Indian child donated by Dr. Chase of Cordova.

From Seward the journey was continued on the government railroad to Anchorage, with some further collecting and information at the latter place. Through the courtesy of the general manager of the railroad, Mr. Noel W. Smith, a stop was next made at the now small but formerly more important Indian village of Eklutney. Here exists a new and well conducted Indian school in which, thanks to the principal, it was possible to examine a large number of Indians, with a few Eskimo children.

From Eklutney a train was taken to Nenana, where a number of local Indians, headed by their old Chief Thomas, were waiting for a brief talk; and then to Fairbanks, where additional specimens of interest were obtained, and where the writer was enabled, thanks to the kindness of Professor C. E. Bunnell, President of the Alaska Agricultural College, to examine the ethnological, archeological and paleontological collections at the college. From Fairbanks, the writer returned to Nenana, where he boarded a small river steamer with which he proceeded down the Tananá River on which are several small Indian villages.

The main part of the inland journey began at Tananá, at the junction of the Tananá and Yukon Rivers, and extended down the Yukon for a stretch of over 900 miles. It was covered mostly in small boats, some owned by traders, some by Indians. We zigzagged from bank to bank, from settler to settler and camp to camp, making inquiries, observing natives, examining old sites, and collecting what it was possible to collect. In this way considerable light was finally gained on our problems along the Yukon, which grand stream must have been one of the principal arteries of the ancient movements of population; and the impressions increased until by the time the mouth of the river was reached it was possible to formulate the following conclusions:

1. The living Indian population along the Tananá, and the Yukon below the Tananá, is scarce. It is doubtful if the total number of the natives on both rivers, as far as covered on this trip, reaches 1,000. And many of the younger adults and especially the children, are mixed-bloods. Due to a lucky coincidence—a potlatch at the mission above Tananá, and other conditions—about 400 Indians were actually

FIG. 147.—Anvik, on the lower middle Yukon.

seen. They all belong to one type, of moderate stature and features, moderate pigmentation, and brachycephaly. They are identical with, or very near to, the Alaska Indians further south.

2. The boundaries between the Indian and the Eskimo both culturally and physically, are indefinite and vague. Moreover, in olden times the Eskimo, according to indications, extended somewhat farther up the river than he now does. But the Indian seemingly occupied always the middle two-thirds of the Yukon, and the Tananá. As the trip proceeded down the Yukon below Ruby, the more specimens obtained and especially the more skeletal material gathered, the more difficult it became to say just where ended the Eskimo and where began the Indian. There is no clear line of demarcation between the

two; they interdigitated and mixed together, culturally as well as physically.

3. From Tananá down stream to its mouth, the Yukon is now and was evidently in the past peopled almost exclusively on the right or northern side. This side is essentially that of the heights, the left being mainly that of flats. But even in the flat country towards the mouth of the river (north mouth) the settlements are all on the right bank. The Indians and the Eskimo behaved alike in this respect, as in a good many others.

4. The old banks of the Yukon, where preserved, show occasional sites of human occupation and yield stone tools, fragments of pottery, more or less fossilized animal bones, and now and then barbed points of ivory or other bone objects.

The habitations in general were of the partly subterranean "igloo," or pit-and-tunnel type. Traces of Russian presence are common. The occupation of some of the sites is still remembered. A few appear earlier than the coming of the white man. No trace was found or heard of any remains that would suggest geological antiquity.

So much for the Yukon. When the Bering Sea was reached, the first more important visit was to St. Lawrence Island. This island was formerly not much considered in archeology and anthropology. It was believed to be rather an out-of-the-way place with small if any connection with the American side, and on which there was not much of importance. This idea must now be given up, for as a result of what was seen and learned this large island should be one of the principal points of attack for future research and is one of the most promising. On the St. Lawrence and the little "Punuk" Islands southeast of it, there are now being recovered by the Eskimo quantities of objects made of ivory which through age has become more or less "fossilized," and these objects in some cases show remarkable and beautiful decoration. Definite information was obtained that on one of the Punuk Islands numerous such implements are actually visible frozen in old refuse heaps. A small party from the Revenue Cutter "Algonquin" reached these islands this year (1926) and obtained, according to reliable information, several bushels of such articles and of old ivory. These objects were brought away to be worked up into beads, pendants, etc., which find ready sale. A number of them, parts of which were still extant, showed human workmanship. These and other facts indicate that these islands are of much promise, and it is urgent to explore them before what they can give disappears.

From St. Lawrence the voyage led to the Diomedes, passing King Island. This island was at that time bare of population, all of the inhabitants having gone to Nome for the summer. At Nome, where they now come every year, these Eskimo work at various trades, and manufacture articles from the walrus ivory which they have gathered during the rest of the year. They do not return until late in the season, when they have sold all they have made. We also passed Sledge Island, with two interesting dead villages which ought to be explored.

The next stop was on the smaller or American Diomede. At this time the writer did not have permission from the Russian Government (received later) to visit the larger island, which belongs to Russia. We found, however, that both the larger and smaller Diomede were

FIG. 148.—June 26, 1926. Bird's-eye view of Cape Prince of Wales, Alaska. Picture taken at midnight on above date showing the midnight sun. The point on the left hand margin of the picture is the nearest point to Siberia.

Note: The village can be seen extending along the coast to a point about 300 ft. beyond Village Creek. Large body of water on the extreme right is Lopp Lagoon. (Photograph by Clark M. Garber.)

nearly deserted, the inhabitants having gone to Nome where they, too, stay through the summer making articles of fossil (mammoth, walrus) or fresh (walrus) ivory. If the student wants to see these islanders in summer he must go to Nome, where on one side of the town live the Diomede, on the other the King Island people. The majority of both were seen there before the start with the help of the very efficient and good Father La Fortune.

Although the small Diomede is difficult for exploration, some interesting features were found. They consisted of rock burials, old refuse heaps, and ruined habitations. The burials have been made among big boulders of granite that cover the steep slopes of the island. Beneath these boulders are many large crevices in which nest thousands of the little auks. One hears the birds constantly chattering deep beneath one's feet, but never sees them. The crevices are deep and spacious, and into them have fallen beyond redemption many of the skulls and bones of the people buried once among the rocks above. Only here and there a bone or a skull that escaped such fate remained to be collected.

As to the village on the little Diomede, it is a poor little hamlet of only five houses, and could never have been much larger. But it is built upon something that preceded it and the ruins, the tumbled

FIG. 149.—Portion of Native Burial Grounds from which specimens of skeletons and crania were collected for Smithsonian Institution. *Location:* Northern Talus slope of Cape Mountain about ½ mile from Village of Wales. (Photograph by Clark M. Garber.)

buildings of older times, have served for a foundation of the newer dwellings. Dr. Jennes of the Ottawa Museum came out with us from Cape Prince of Wales to do some excavation on the island.

The next stopping place after a brief visit to Wales, was the site of two old native settlements with a living village between them, called Shismareff, located near the middle of the northern coast of the Seward Peninsula. Along this part of the coast are several other old dead villages, some of them particularly promising for exploration. Of the two at Shismareff itself, the more important was unfortunately appropriated recently for a fox farm, the burial grounds were razed, the skeletal remains mostly dumped into holes, the surface of the old igloos levelled and cages for the foxes erected upon the flats thus made.

It seems that there is no protection for such sites and nobody cares sufficiently to save them, and so, in this case at least, what is probably one of the most interesting sites of that coast is lost to science. And there are others.

From Shismareff, the "Bear" proceeded to Cape Blossom, and the Kotzebue Sound, where, with the valued help of Mr. Sylvester Chance, Superintendent of Education for the North-West district, valuable information was obtained concerning a large number of dead villages in this region. Some of these are old and some fairly recent, but all deserve to be explored. Along the Buckland River there is also apparently much archeological as well as paleontological material.

The Kotzebue Sound is an especially important region to anthropology because, as long as either natives or whites remember, here have congregated every year natives from all parts of this region—from the Diomedes, from the East Cape in Asia, from the villages of the Seward Peninsula, and from those of the Arctic coast as far north as Barrow, as well as from the inland rivers. Here doubtless is much to be found and learned, though even here much has already been obliterated or scattered.

From Kotzebue the journey led along the coast to Point Hope. This coast is barren and unoccupied, or almost so; and, except at Cape Krusenstern there is apparently but little in the way of older remains. One of the most important and interesting points however of all these coasts, is Point Hope itself. There we meet with a dead village which was occupied up to 30 years ago but whose beginnings are very old. The people have abandoned the village because, it is said, of the encroachment of the sea. They are now occupying a site a little back of it, and are assiduously excavating the old remains and selling the proceeds to whoever comes along. We were there twice—going north and coming back, and had some interesting experiences. Going north the writer bought a good number of old implements, etc., from the natives, especially from one young woman. When they found that skeletal material also was desired, they brought willingly what they could find from the old village—three human skulls, skulls of dogs and a fox, and other specimens. When the "Bear" came back, the young woman from whom a skull was obtained before, came against all expectation to the boat with a bag on her back containing five very good skulls, which she had excavated from the old burial grounds with her own hands. It was an illustration of the helpful and matter-of-course nature of these natives, who in general are progressing rapidly in civilization.

The old village at Point Hope is yielding large numbers of specimens of great variety—of such variety that one stands astonished at the extent and refinement of the former culture of these far north people. Point Hope is at the extreme end of a spit of sand and gravel many miles long extending into the sea and exposed to all the winds and storms—scarcely a place where one would expect to find a people of such varied or advanced culture. Many old articles of native trade appear from the diggings, trade with the people of Kobuk River to those of the Bering Sea and perhaps even Asia. And in the upper layer occur occasional articles (metal, beads) of white man's introduction.

The further journey to Barrow was a series of difficulties, with not much of anthropological interest except at Barrow itself. Between Point Hope and Barrow there are but few settlements, old or present. There is however Kevaleena, a small new village, with two old and possibly important sites, one along the lagoon and one further inland, on the river. Then comes Wainwright, another small recent village with an older site nearby, followed by a few little camps of reindeer herders. Then, just outside of Barrow, is an interesting old village site, and again another, further on, known as the " Hunting Place," the latter yielding good archeological specimens. Still further north there is a village at Point Barrow; but from Point Barrow eastward the now northern coast is seemingly barren until one reaches Barter Island, where there is a large dead village, which however was still occupied in the time of the earliest sailings in these waters. A collection of material from this village, seen in the possession of Mr. Charles Brower, the intelligent trader and collector of Barrow, proved interesting, though of much the same nature as the material along the coast this side of Barrow and, except for a few objects not as refined or beautiful.

On the return trip, each of the villages along the Arctic coast was re-visited and some small places were seen in addition; and then the " Bear " endeavored once more to stop at Cape Prince of Wales, but was prevented by a storm; it next tried to stop at the Diomedes—prevented by storm; at St. Lawrence—prevented by storm; once more at the small Diomede where Dr. Jenness awaited us, but unable to approach. Then a landing was tried at Nunivak Islands further south, but the boat was again driven away by storms and had to turn towards the Pribyloffs and Unalaska.

The total experience among the Bering and Arctic coasts may be summed up as follows: We are confronted here with an extensive region that is but sparsely peopled—the total population of today probably not exceeding 5,000 individuals—and which evidently has not been much more peopled at any time during its existence. Here again, as along the Yukon, the only sites available for man and which were doubtless utilized in the past, were the beaches and especially the " spits " or low sand and gravel bars reaching into the sea, or separating the sea and inside lagoons, for these offered man the best facilities for getting at the animals and birds that he most needed for his food. All that was actually seen was, however, recent. The old beaches, the old flats, the old accumulations of rivers, the old lagoons, are filled up or washed away. In places there may be seen three, four, five—a whole series of beaches, and it is not known on which of these ancient man did settle. In some places these older beaches have been or are now being cut away. New lagoons have formed or are now forming, and old ones are filling up before one's eyes, to be converted into pools and marshes.

Still further, entire regions during a large part of the year, that is, during the open season when the ice goes north and again before the new ice forms, are subject occasionally at least to violent storms; and what these storms can do to human remains the writer himself saw. Eighteen miles to the east of Nome is a dead village, one of the largest on that coast. Near this village was an old burial ground, well known to some of the old white pioneers. One of these, who did not know what had happened, advised the writer to go there to collect, for he said he saw the ground covered with skeletal remains and various objects placed with the dead. The writer had, however, already visited the spot and this is what was found. In 1913 there had been a very violent south-wester—so bad that the cemetery at Nome itself was washed out and the bodies were scattered over the country. This storm absolutely washed off and left barren of human specimens the old burial ground east of Cape Nome, and had it not been for the depressions that still show where the ancient igloos of the village stood, no one could possibly guess today that an important burial ground had ever been in that vicinity.

Such storms doubtless happened repeatedly in the past, and they must have destroyed or covered many of the old sites. But many sites and remains of man of moderate antiquity still exist there. Many dead villages invite exploration and will repay excavation. And such explorations, judging from the experience acquired on this trip, will not be as expensive as might be feared.

A few words as to the problem of Asiatic migrations. The last summer's studies gave much definite light on this question. It so happened that upon reaching the upper parts of the Bering Sea we had the three clearest and most peaceful days of the whole journey; and all this time the sensation was that of floating on a big lake, all boundaries of which could be seen at one and the same time except in the southeasterly direction. There is no problem of migration here. It was no great effort for people to pass from Asia to St. Lawrence Island, or the Diomedes, or Wales, Nome, Teller, and even as far as the Kotzebue or Norton Sounds and along the Arctic coast to Point Hope and northward. The people today think nothing of such trips. They have excellent big skin boats, much like the wooden Haida or Tlinkit boats of the south, which are so seaworthy that in going to

FIG. 150.—East Cape of Asia.

Nome from the King or Diomede Islands the natives fill them to the gunwales with dogs, ivory, and all sorts of household articles, and on the return trip they pile in boxes and barrels of provisions. An example of how little the Eskimo think of these journeys was witnessed during the last call of the "Bear" at Nome. As the year before, the King Islanders at Nome were offered the facilities of the "Bear" for transporting them homeward; but they preferred to be left behind because they had yet some purchases to make and some few articles to sell. They preferred to make the return journey in their umiaks later, regardless of the storms and distance, which shows how seaworthy these boats are and how practical native navigation was and is in these parts of the world. And once they reached the northeasternmost parts of the American continent, it was natural for the Asiatics to pass on. They were not emigrating into a *new* world; they merely saw another land a bit ahead of them and went to it, and they had no reasons to return with the table better and better spread before them the farther they reached, and no opposition. The only questions can be, what were the exact routes the different contingents of these people took; and where are we to look and watch for their remains?

In this last respect the inquiries made along the Yukon and the coast and the islands were very instructive. There was certainly no single large migration. The people came over the Diomedes, through the Bering Sea, north and south of the Bering Strait; they came in small tribal groups, and this was doubtless repeated over a long period

Fig. 151.—Just south of East Cape of Asia. Native village to left.

of time. Then, judging from historic evidences, their movements were as follows:

It may be presumed that ancient man adhered to very much the same routes as those that are in use today, as the most practical, the most natural and sometimes the only available ones. Undoubtedly the greatest and most frequented route of farther spread was that along the coast down to the base of the Alaskan Peninsula, where with the inlets and lakes there is but a little portage over land to the Gulf of Alaska. Then there were the routes up the Yukon and Kuskokwim Rivers. On the Yukon, the earlier contingents probably did not go right up the whole stream, but branched off at the Tananá River and then went towards the Copper River which brought them into Prince of Wales Sound and the Gulf of Alaska. There were two other routes, one down the Koyokuk River, one of the largest tributaries of the Yukon, by which they would reach the latter; and the

FIG. 152.—Eskimo family of Wales on the "Bear."

FIG. 153.—Eskimo, St. Lawrence Id. Rounded, low vault.

FIG. 154.—Eskimo at Nome, Alaska. Indian-like in type.

other along the Noatak River. The latter route was probably used by but few people, but there are indications that some did follow it until they reached the Colville River, peopled its region and extended farther eastward.

To sum up, as to the routes of migration, outside of the Aleutian Islands, which are a problem of their own and point to Kanitchatka, the most plausible and doubtless most used route was that southward along the shores of the sea; the second, along the Yukon River and its

FIG. 155.—Eskimo mother and child, Indian-like in type, Reindeer Camp, n. Pastolik, Norton Sound.

tributaries; the third and fourth, along the Selavik, Koyokuk and Noatak Rivers in the north; the fifth along the Kuskokwim; and the sixth along the Arctic coast.

The questions of the culture of the older people and their physique are of much interest. So far as culture is concerned it was appreciated, more than ever before, that there existed in these parts of America, not so many hundreds of years ago, remarkable development, especially in what may best be characterized today perhaps by the term "fossil ivory culture." The people of this culture, whoever they were—doubtless ancestors of the Eskimo or Indian or both—reached a high degree of industrial differentiation and art—so high that we have nothing to compare with it in America except among the more highly civilized and developed tribes of the northwest coast, Mexico, Yucatan and Peru. They attained a high grade of native art, which was characterized especially by decoration in curves and soft lines. There seems to be a distinction between this and the Eskimo art of today, as if some other people were responsible for the older culture; but when one examines the skeletal remains there is no indication of any other people except Eskimo and Indian in this

FIG. 156.—Eskimo woman, Indian-like type, and child, Kevalina (Arctic).

region at any period thus far represented in the collections. Therefore it seems that for the present at least it must be accepted provisionally that this culture was connected with the ancestors of the present natives of these regions. This interesting old culture seems to reach away along the American coasts; but it is not certain that its arts were actually practiced everywhere along these coasts. One of the most striking phenomena in these parts of the world is the extensive trade that, according to many indications, was here carried on in implements and other cultural objects. There is on the Kobuk River a mountain which is called Jade Mountain. This mountain was early known to the Indians or the Eskimo of the river and its green stone was utilized by them for making adzes, drills, knives, lamps, and other objects. This particular stone is found only in that one place, yet objects and implements made of it occur scattered all the way from

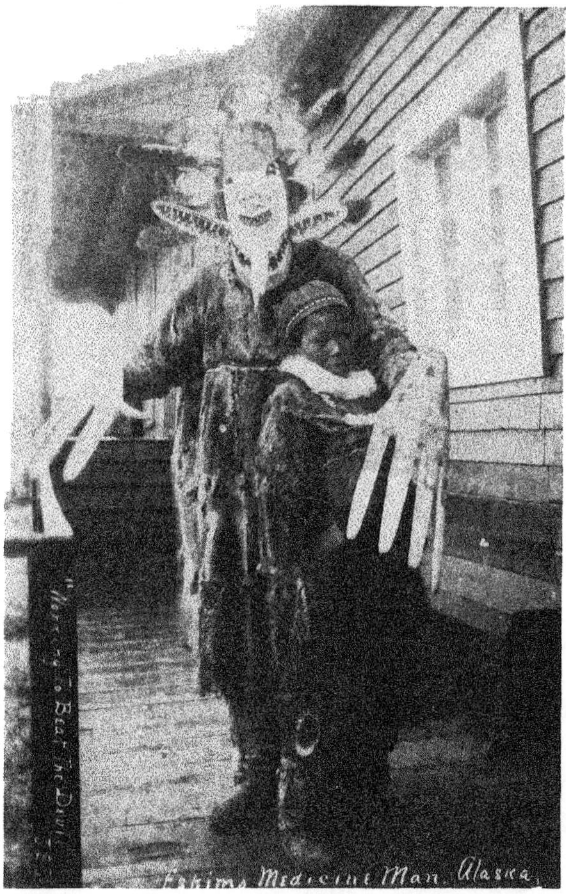

FIG. 157.—Eskimo medicine man, treating a boy.

Point Hope down to Nunivak Island, and probably even the Gulf of Alaska and the northwest coast. Similarly, one finds the highly decorated objects of now fossil ivory on the Diomedes, the St. Lawrence Island, the Asiatic coast, and from Barrow and Point Hope again down to, if not beyond, Nunivak Island. The indications would seem to point to the old ivory culture having been central on the Asiatic side, whence it spread by trading along the American coast.

The skeletal remains collected on this journey will probably prove to be of much importance. They represent skeletal material from Barrow down the coast, spot by spot, including the islands; and they comprise not only recent material but also some older. These remains show at first sight that the Eskimo of these regions are by no means the highly differentiated Eskimo of Labrador and Greenland, but that they approach, in some cases almost to an identity, on one hand the Asiatic and Mongoloid types of people, and on the other the American Indians, more particularly those of Alaska. The writer has no longer any hesitation in believing that the Eskimo and Indian originally were not any two distinct races nor even two widely distinct and far-away types, but that if we could go a little back in time they would be found to be like two neighboring fingers of one hand, both proceeding from the same palm or racial source.

ARCHEOLOGICAL INVESTIGATIONS IN CHACO CANYON, NEW MEXICO

The Pueblo Bonito Expeditions[1] of the National Geographic Society, under the direction of Neil M. Judd, curator of American archeology, U. S. National Museum, were concluded in October, 1926. Pueblo Bonito is a prehistoric Indian village in the Chaco Canyon National Monument, northwestern New Mexico; archeological evidence suggests its abandonment about one thousand years ago. Pueblo Bonito may justly be regarded as one of our finest aboriginal apartment houses for, in its hey-day, it stood four stories high, comprised approximately 800 rooms, and covered more than three acres of ground. Its occupants were sedentary Indians who had surpassed all their known contemporaries in civic organization, architectural development, and dexterity in the manual arts. The National Geographic Society's Pueblo Bonito Expeditions, inaugurated in 1921, have contributed a vast store of information concerning these primitive Pueblo folk and others who, in pre-Columbian times, tilled the desert soil not only in Chaco Canyon but elsewhere throughout the plateau regions of the southwestern United States. The spacious communal dwellings these ancients built and occupied form the major antiquities of their respective periods and cultures.

During the past six years Mr. Judd and his associates have investigated the archeological evidence of the domestic and ritualistic life of the Bonitians and their neighbors at Pueblo del Arroyo and have

[1] Smithsonian Misc. Coll., Vol. 72, Nos. 6 & 15; Vol. 74, No. 5; Vol. 76, No. 10; Vol. 77, No. 2; Vol. 78, No. 1.

Fig. 158.—Pueblo Bonito as seen from the north cliff of Chaco Canyon. In the middle distance, the National Geographic Society's camp and the present arroyo, formed within the past century. (Photograph by O. C. Havens. Courtesy of the National Geographic Society.)

Fig. 159.—Doorways connecting rooms of the fourth or last major period of construction in Pueblo Bonito. (Photograph by Neil M. Judd. Courtesy of the National Geographic Society.)

Fig. 160.—Huge, blocked, T-shaped door in a wall of characteristic third period masonry. (Photograph by Neil M. Judd. Courtesy of the National Geographic Society.)

Fig. 161.—From the north cliff of Chaco Canyon, towering close behind the walls of Pueblo Bonito, enemy warriors hurled defiance and more destructive missiles at the village folk below. (Photograph by Neil M. Judd. Courtesy of the National Geographic Society.)

also inquired into the probable geophysical changes brought about in Chaco Canyon since Pueblo Bonito was inhabited. It is now known, for example, that Pueblo Bonito was occupied by two distinct groups of Indians; that the type of architecture developed by each group was as unlike the other as were the household utensils employed in the corresponding sections of the village. Both peoples were farmers, however, and their fields of corn, beans, and squash may well have lain side by side. Hunting played no essential part in their means of livelihood despite the variety of mammal and bird bones found in

Fig. 162.—Bonitian cooking pots buried just beneath the floor of Room 350 and used for storage purposes. (Photograph by Neil M. Judd. Courtesy of the National Geographic Society.)

the rubbish piles. Agriculture was the main dependence of both groups, but there came a time when the harvests were no longer sufficient to support a population of from twelve to fifteen hundred. Arroyos formed, rain waters drained off quickly, helpful chemicals were leached out of the soil, the latter became impervious to water, crops failed to mature. This condition, it appears from the data at hand, was the indirect result of the prodigal manner in which the Bonitians utilized their available timber supply. Altered agricultural conditions, then, in addition to the harassment of nomadic, enemy tribes, unquestionably contributed to the disintegration and ultimate abandonment of both Pueblo Bonito and Pueblo del Arroyo.

Each successive period of constructional activity at Pueblo Bonito witnessed a marked intramural rearrangement of dwellings and a distinct effort to strengthen the outer walls, thus increasing their impregnability. External doorways were eliminated; ventilators were closed; the people drew closer and closer together. The single passageway which gave ready access to the inner courts was barred by a wall through which a narrow door opened; but this door was subsequently closed and thereafter entrance was had only by means of ladders which could, in time of attack, be drawn up to the housetops.

FIG. 163.—The overhang above the Indian's head shows the union of a new addition with older, partially razed walls in Pueblo Bonito. (Photograph by Neil M. Judd. Courtesy of the National Geographic Society.)

Sporadic warfare is even more vividly evidenced by discoveries made in individual rooms of the ruin.

Among other researches pursued by the 1926 expedition was that concerned with sub-court walls in Pueblo Bonito. Its closely-grouped houses were not hastily constructed from previously prepared plans. Throughout a considerable portion of the village deeply buried walls have been found. These pertain in each case to earlier periods of occupancy; they represent dwellings partially razed and over-built by later structures. Some of these demolished walls have been found as much as 12 feet (3.65 m.) below the last utilized court level. So great an accumulation of blown sand, débris of reconstruction and

Fig. 164.—Superimposed, sub-court walls in Pueblo Bonito, representing successive periods of construction. (Photograph by Neil M. Judd. Courtesy of the

Fig. 165.—Plastered and whitened masonry of the second period at Pueblo Bonito underlying the last utilized court level. (Photograph by Neil M. Judd.

floor sweepings furnishes mute evidence of the passing of many generations.

Mr. Karl Ruppert, one of Mr. Judd's associates in the Pueblo Bonito Expeditions, again supervised explorations at Pueblo del Arroyo. These, in 1926, were chiefly confined to excavation of a much-ruined, lesser structure close on the west side of the larger pueblo. The excavations were complicated and laborious; but upon their conclusion it was found that the site had originally been occupied by a circular tower, 73 feet (22.24 m.) in diameter, of a type

FIG. 166.—Outer south wall of Pueblo del Arroyo, showing varied stonework and later, abutting rooms. (Photograph by Neil M. Judd. Courtesy of the National Geographic Society.)

well known to the northward but not previously recorded so far south of the Rio San Juan. Interest in this structure was augmented by its apparent relationship to the super-kivas or huge ceremonial chambers of the Bonitians. Later walls joined the tower to Pueblo del Arroyo, but, subsequently, both tower and adjacent buildings were almost wholly razed by prehistoric masons who desired the dressed sandstone blocks for use elsewhere.

In addition to his study of the Expedition's ceramic collections, Mr. Frank H. H. Roberts, Jr., now of the Bureau of American Ethnology, undertook exploration of two small-house sites about nine miles east of Pueblo Bonito. Both settlements belong to an earlier horizon than that represented by the great communal dwellings of Chaco Canyon. One of the two sites had been occupied by pre-Pueblo peoples whose local form of habitation was a semi-subterranean, circular structure the lower wall of which was formed by upright sandstone slabs. Similar houses have been noted elsewhere in the Southwest; their characteristically flat or low, conical roofs were frequently supported by four upright posts, as in Chaco Canyon. The second village examined by Mr. Roberts was much later than the first, both in time and culture. But it was never inhabited; indeed, it was not even completed.

Fig. 167.—South half of a prototype kiva at a pre-Pueblo site 9 miles east of Pueblo Bonito. (Photograph by Neil M. Judd. Courtesy of the National Geographic Society.)

Excavation established the fact that although some of its rooms had approached completion the larger number was represented merely by foundations prepared but never built upon.

This latter site had been wholly concealed by sedimentary deposits washed down from the borders of the canyon and was only brought to light by a narrow, deep arroyo cut within the past 15 years. As further indication of the not inconsiderable length of time during which prehistoric folk inhabitated Chaco Canyon prior to construction of the great pueblos, it is interesting to note that the arroyo which disclosed the unfinished settlement last mentioned also exposed a "pit-house"

Fig. 168.—In the branches of Chaco Canyon infrequent logs and partially decayed stumps witness the former presence of limited pine forests, exhausted by Bonitian housebuilders. (Photograph by Neil M. Judd. Courtesy of the National Geographic Society.)

Fig. 169.—Nine miles east of Pueblo Bonito a deep arroyo has exposed the unfinished walls of an ancient communal house and a still older pit-dwelling (at right of Indian boy). (Photograph by Neil M. Judd. Courtesy of the National Geographic Society.)

whose floor level lay 13 feet 6 inches (4.1 m.) below the present valley surface. No one may say what chapters of human history still lie buried beneath the alluvial floor of Chaco Canyon.

Mr. Roberts also superintended stratigraphic studies, authorized by special permit from the Department of the Interior, in the rubbish piles at the two major ruins known as Pueblo Alto and Peñasco Blanco. These enormous heaps consist principally of floor sweepings containing ashes and fragments of pottery broken upon the hearth, intermixed with blown sand and débris of reconstruction. Potsherds are to the archeologist what fossils are to the geologist! The oldest fragments lie at the bottom of the pile; the latest, on top. The evolution or decadence of pottery technique at any one site is thus represented by a cross-section of its ash heap. In consequence of its studies at Peñasco Blanco and Pueblo Alto, the recent Pueblo Bonito Expedition has obtained data which illustrate not only the development of the art of pottery manufacture in both villages but which also indicate their probable positions in the chronology of Chaco Canyon. Mr. Judd finds no reason to alter his original impression that Pueblo Bonito was the earliest of the major Chaco Canyon villages and that it was inhabited for a longer period than any of the others.

The method by which it is hoped to ascertain the absolute age of Pueblo Bonito and Pueblo del Arroyo, namely, through study of annual growth rings in their ancient roof timbers, is already familiar to readers of the Smithsonian Explorations volume. These studies have gone forward during the past field season under the direct supervision of Dr. A. E. Douglass, of the University of Arizona. The data assembled by Dr. Douglass hold much of interest; although his conclusions may not yet be published, it is permissible to say that this phase of the National Geographic Society's explorations in Pueblo Bonito is also certain of contributing to the prehistory of the southwestern United States results not previously recorded.

INVESTIGATING EVIDENCE OF EARLY MAN IN FLORIDA

All research for the accumulation of knowledge pertaining to the early history of man is of human interest, and especially so in that part of it which reaches back into primitive time.

It has long been known that in Europe primitive tribes of the human race for long centuries lived contemporaneously and were well acquainted with many kinds of wild animals which are now extinct. Among these animals were the hairy mammoth, the woolly rhinoceros, the great cave bears, cave hyenas, wild horses, great oxen, etc. In

America also human bones and artifacts have been found associated with the remains of extinct animals of the Pleistocene Age. But usually either the associations or the circumstances of discovery have been of such a nature that doubt has been cast on the contemporaneity of deposition of objects found in the same stratum, it being assumed that the human remains and artifacts belonged to intrusion through later burials or other accident.

While the general problems relating to the first appearances and early development of man in America belong more properly to the ethnologist, this phase of it comes definitely within the province of mammalian paleontology and geology. Thus, members of these sciences from time to time have taken an active part in investigation of the evidences of early man in America. Perhaps the work in this line that has aroused most interest in recent years is that carried on by Dr. E. H. Sellards at Vero Beach, Florida. This work, which followed the digging of a big drainage canal at that place, resulted in the discovery of human remains and artifacts associated with fossil bones of a Pleistocene fauna. Although careful observation was made by Dr. Sellards and evidence was produced to show that the association was normal, this evidence was not accepted by some of the leading anthropologists as conclusive. After much discussion interest in this discovery lagged for a time, but was revived again a few years later by the discovery at Melbourne, about 30 miles north of Vero, of associated fossil bones and human remains in similar deposits and under similar conditions to those observed at Vero. Following this discovery in the summer of 1925, a joint expedition under the auspices of the Bureau of American Ethnology of the Smithsonian Institution and Amherst College was organized for the purpose of further research at Melbourne. I was detailed to cooperate with Prof. F. B. Loomis of Amherst in carrying out this work, and the six weeks' carefully directed field-work which followed revealed three additional localities where human remains or artifacts were found in direct association with bones of extinct species of animals. All these finds, however, were at or near the top of the fossil bone-bearing layer, and for this reason were not accepted by the anthropologists as undisputed evidence of normal association.

Time and funds did not permit completing the exploration of these important and extensive deposits at Melbourne, and in February of the present year (1926) I was again temporarily transferred to the Bureau of Ethnology and detailed to continue this work begun by the Amherst-Smithsonian Expedition. Again lack of funds prevented completing the investigation, but the six weeks' work accomplished yielded very satisfactory results. In the additional fossil material secured were specimens representing some important species new to this locality, and the geologic observations made at this time are proving to be valuable aids in the study of the general problems involved. For example, one of the most important specimens obtained on the previous expedition was a crushed human skull and jaws which had been found associated with fossil bones. But the exact position in the geological strata was not definitely determined, owing to the fact that much of the overburden or covering layer had been

FIG. 170.—Canal bank on golf course 2 miles west of Melbourne, Florida. Crushed fossil human skull *in situ* in upper layer of fossil-bone-bearing deposit. Skull is at right of trowel handle in middle foreground, as indicated by arrow.

removed before the skull was discovered. At the time it was made certain that the sedimentary layers above the specimen had previously not been disturbed since their deposit, but it was not quite clear whether the fossil skull belonged in the upper six inches of the Pleistocene fossil-bearing deposit, as it appeared, or whether it lay in the dividing zone between this bed and the overlying deposit. This doubt was caused mainly by the fact that the upper 10 inches of the lower bed was changed into a perceptibly darker hue than that observed in the main mass below. Going back to this locality last spring I extended the relatively small excavation which had been made the previous summer, but concluded to follow a different plan than

FIG. 171.—Excavation trench exposing vertical section near spot where was found human skull shown in figure 170. The undulating contact plane is here plainly seen.

FIG. 172.—Section of the deposits near spot where was found human skull shown in figure 170, showing the uneven bedding-plane between the lighter-colored fossil-bearing deposit below and the later deposits above.

FIG. 173.—Section on south bank of main canal about 3 miles west of Melbourne, Florida. Point of trowel indicates pieces of Indian pottery *in situ* in top layer of fossil-bone-bearing beds.

FIG. 174.—Typical flat country in the vicinity of Melbourne, Florida, showing scattered pines with heavy undergrowth of scrub-palmetto.

FIG. 175.—"Mulberry Mound." An Indian burial mound and kitchen-midden at the north end of Lake Poinsett, St. Johns River, west of Cocoa, Florida.

FIG. 176.—Remnant of great Indian shell mound at Grant, Florida. This mound before being excavated for road-building was about 1,000 ft. long, more than 200 ft. wide, and in places 14 ft. deep. Dr. Henry M. Ami, of Ottawa, Canada, is seen pointing to broken pottery *in situ*.

that before adopted. Instead of stripping from the top, the excavation was carried forward with nearly perpendicular walls so that the formation could be studied in diagrammatic cross-section at all points. In this way it was clearly observed: first, that the bone-bearing layer was everywhere plainly distinguishable from the overlying strata; second, that the separation plane was uneven, showing there had been an erosional interval between the two periods of deposition; and third, that the portion of the bone-bearing layer following the undulations of the separation plane presented a uniformly darker color on top which faded gradually into the light color of the formation below. This was exactly the condition observed at the spot where the human skull was taken, hence it may now be confidently stated that its original burial place while near the top of this layer was definitely within it. It is unfortunate that the investigation at Melbourne could not be continued. The field is a promising one and doubtless would amply repay further exploration.

James W. Gidley.

ARCHEOLOGICAL AND ETHNOLOGICAL STUDIES IN SOUTHEAST ALASKA

During April and May, 1926, H. W. Krieger, curator of ethnology in the U. S. National Museum was detailed to the Bureau of American Ethnology for the purpose of inspecting native houses and totem poles at the National Monument of Old Kasaan, with a view to their preservation. The National Monument of Old Kasaan was originally established by Executive order in 1907 amplified by the Presidential Proclamation of October 25, 1916. The monument thus established contains the abandoned Haida Indian village of Kas-a-an and the surounding forested area containing about 40 acres. It fronts on Skaul arm of Kasaan Bay, on the east coast of Prince of Wales Island, and is about 40 miles by motor boat from Ketchikan, the largest town and the first port of call in Alaska for American steamers out of Seattle.

Kasaan, like most of the native villages of southeast Alaska, is abandoned, its former occupants having moved to fish-cannery towns or to towns like Ketchikan where a number of occupations and industries await them. Indians of Alaska have adopted white man's ways, and have never been wards of the nation like the Indians assembled on reservations within the United States. In accepting the new, however, they have forgotten or learned to disregard their own culture with its splendid claim to distinction as possessing the most unique and realistic examples of plastic sculpture of all aboriginal America.

Fig. 177.—Eagle House "hut-nes," Kasaan. Built by Sanixat (Southeast), who assumed the name of the Southeast Wind in retaliation for the wrecking of his canoe in a southeast gale. The house is an old one, and originally there were no windows nor sawed boards on the front wall. Entrance was gained through a small opening reached by a stairway. The charred smoke hole frame may still be noted on the roof, which is covered with split cedar "shakes."

The abandoned village of Kasaan today consists of the ruins of houses and memorial columns. Many of the tall totem poles profusely decorated with carvings of animal and human figures representing

Fig. 178.—Frame and roof timbers of Eagle House "hut-nes" with "Eagle King's" totem pole at the front. This house is the sole remaining structure at Old Kasaan, and with the house of Jim Peel at New Kasaan, is the only surviving native house in southern southeast Alaska. Crests on pole are: eagle, at top; beaver, at bottom; bear, with protruding tongue, beaver, with large projecting upper incisor teeth, and figures illustrating the myth of "Raven Traveling," at center. See fig. 177.

the family crests are still standing. The region surrounding the abandoned village is virgin forest consisting of the giant cedar, spruce, hemlock, and a few other species. The village site itself was overgrown with alders and dense masses of the salmon berry. The salmon berry looks delicious, but is somewhat flat to the taste and is extremely perishable when picked. Several black bear were seen feeding on these berries and on the succulent grasses within the shadow of the towering memorial columns or totem poles. As the island is uninhabited for many miles in the vicinity of the National Monument of Old Kasaan,

FIG. 179.—Side view of "Eagle House" with totem pole in foreground erected by a wealthy woman "Big Smoke Hole," the first to encounter white men; hence the crest representing a white man at the top. Beneath this is Raven with the moon in his beak. At bottom is the carved figure of "Duchtut," the strong man, splitting open a sea lion with his bare hands. See figs. 177 and 178.

game is rather abundant. Deer come down from the hills out of the forest to feed on the grasses and wild celery near the water or wherever there is a clearing. As there are literally hundreds of similarly situated islands scattered all along the coast of southeast Alaska, it has been thought that fox farming might prove a profitable industry. Thus far, however, the industry is in an experimental stage.

The view from the village of Kasaan is beautiful, looking out towards the distant islands with their hills and occasional snow-capped mountains, and the intervening water channels and inlets. If the site of this ancient Indian village had originally been selected from

Fig. 180.—Interior of Eagle House. Two benches or floor levels rise on each side of the centrally excavated pit. The hearth is at the center of the pit, and the remainder of the floor including the two benches is covered with split cedar slabs. See figs. 177-179.

the standpoint of beauty of location, the choice could hardly have been excelled.

Kasaan was originally a village belonging to a rival tribe, the Tlingit, who were probably driven away by the Haida, according to their traditions, more than 100 years ago. The name Kasaan in the Tlingit language means the " village on the rock." The Haida came from the south, originally from the Queen Charlotte Islands far out to sea off the coast of British Columbia. The Haida Indians, like all the tribes along this island-studded coast, were great travellers, making journeys of hundreds of miles in their huge dugout war canoes, often for the mere love of adventure, but usually in search of new fishing grounds or to carry on trade with neighboring tribes. The Indian had to follow the salmon to its new spawning places whenever for unknown reasons it migrated to different spawning beds from those near the Indian's ancestral village. In his quest for food, the Alaska Indian was often forced to abandon his well established village with its large framed houses of split cedar slabs and decorative crested totem poles.

The reason for the coming of the Haida to Old Kasaan was of an entirely different order. Family life among them was communal and consequently rather complicated. It often led its members into difficulties. Each house was built large enough to accommodate two or three generations, together with their slaves and retainers. The house floor was arranged in platforms, each succeeding platform being built on a level two or three feet higher, beginning at the deeply excavated centrally located fireplace, until the outer platform or the one next to the walls of the house was reached. This platform was flush with the ground level on the outside. Each section of the house was assigned to different divisions of the large family. The head of the house, who was often the chief of the clan as well, together with his wife occupied the place of honor on the platform back of the carved house posts at the rear of the house. The slaves, strange to say, gathered and slept at the front of the house nearest the only exit.

The fire burned at the center on the lowest part of the excavated floor sections. In one house at old Fort Tongass there are nine different levels excavated so that the fireplace at the center appears from the front entrance to be at the bottom of a pit. The fireplace is a squared section of bare earth or stone; coals and ashes raise it a few inches above the floor level immediately surrounding, and it is enclosed with a frame of hewn logs or slabs. On this level, about the fire on the bare floor or on mats of woven cedar bark, with their feet toward the fire, slept the members of the family during cold and inclement weather.

The narrative of the coming of the Haida to Kasaan is an involved one and includes a story of family dissension culminating in the murder of one chief by his own brother, who was also his rival. The murder caused the villagers to take sides and led to the removal of the slayer and his adherents far to the north and to the ultimate settlement of Old Kasaan.

After an occupancy of considerably more than 100 years by the Kaigani family of the Haida, Kasaan was abandoned. About the year 1900 its entire population removed to the newly established village of New Kasaan some 40 miles distant on another arm of the same Kasaan Bay. This removal was due to the establishment of a salmon cannery there and to the offer of good wages during the canning season. This proved too much of an inducement to the Kasaan natives who were only too ready to adopt white man's ways and wages and who had already forgotten most of their ancestral lore in

Fig. 181.—Western end of Old Kasaan showing cemetery with carved memorial columns representing the family crests of killer-whale, sea lion, eagle, and bear. The dense forest growth in the background is mostly Douglas spruce, hemlock, and yellow cedar.

woodcraft and decorative art. No new totem pole has been built for more than 50 years and the art of totem pole carving is lost to the present generation.

As Kasaan appears today, after being abandoned for more than a quarter of a century, there is practically nothing remaining to remind one of its former glory but a row of tall totem poles facing the beach, still standing as erect as they were when placed there many years ago. The most recently erected pole is more than 50 years old and remains in a fair state of preservation. Many of the older poles still have sound heart wood, although the incised surface carvings crumble to the touch. The oldest poles have completely rotted away leaving mere traces. It was the duty of the writer to attempt to preserve these realistically carved representations of human and animal figures and the totem pole itself wherever possible.

The task of restoring the abandoned village of Kasaan is practically an impossibility owing to the deterioration and decay of many years. No repair work has ever before been done. Houses have with one exception completely fallen into ruins. The Indians never repainted their memorial columns, once the pole was erected. The rotting process is practically continuous throughout the year. This may be better realized when one considers the number of rainy days in a year,

FIG. 182.—Central section of Old Kasaan as it appeared in 1924. Eagle House may be seen on the left, while the front house posts and roof beams of another house, "More-back House," is visible in the background on the right.

reaching a total of 235 at Ketchikan, the nearest observing station. Winters are mild and are hardly severe enough to freeze the rank vegetation. Alders and salmon berry bushes grow in profusion and have completely hidden from view the fallen timbers of the house frames. Large cedars and spruces grow out of what was once a house interior. In one case a cedar sprang up at the base of the hollowed back of a totem pole, and it has now grown up and filled the hollowed cavity, splitting the totem pole which still adheres to the living tree, forming two decorative panels. All remaining poles which were worth preserving were scraped. The rotted wood was removed and the pole was then given a treatment of creosote. No attempt to paint the poles in their original colors was attempted, as such expense would hardly be justified unless tourist travel should increase materially and a caretaker could be maintained to keep down the alders and salmon berry brush so as to prevent any future danger of destruction by fire during the short dry season in the fall of the year. At Wrangell, at Ketchikan, and in the park at Sitka there are poles which have been repainted in their original colors. Attempts have been made to use the old native colors which consist of red ochre, chrome yellow, and a mordant composed of water and a gluey mass derived from crushed salmon eggs. These colors and several other native coloring agents produced a well blended effect and are not the glaring paints produced by modern white and red leads. Pigments derived from the mineral coloring matter in rocks and ores were formerly much used by the natives. The Indian was indeed an artist who was capable of carving the designs on the totem pole according to highly conventionalized patterns, and who also had the technical knowledge required to procure the proper paints and to properly apply them to the patterned designs in the conventional style.

The totemic memorial columns of tall cedars represent the highest achievement of the Northwest coast Indian. They were also his show property. If the Indian who wished to erect a memorial pole in honor of his maternal uncle was himself incapable of the task, he employed an expert or artist. These skilled wood carvers took as much as a year to complete the task. For his guidance, the native artist had a complicated set of pattern designs in wood. These panels were incised with sections of the conventional designs. It became thus the task of the artist to carefully fit these patterns to the pole to be carved, special effort being made to estimate the proper size of the pole, the entire front surface of which must be covered with the carved figures. No essential part of the animal figure crest must be omitted, although if cramped for space the figures were so highly conventionalized that what might appear to us as essential parts of the animal or human figure represented were often omitted, and no one was offended for all understood what was intended. It was high art with futuristic leanings. Then too the carver had to take into consideration the number of animal totems or crests to which the owner was entitled. If the one in whose honor the pole was being prepared was a great man in the village, he probably had so many totemic crests that the problem was one of overcrowding. If, on the other hand, there were not enough crests to fill the space of the front of the pole, the artist probably suggested that a smaller pole be selected. If this was not done, he could conveniently fill in the spaces between the animal crest carvings with representations of frogs and the ground worm. No one claimed these lowly creatures as their animal protector, nor represented them among their crests. It was, moreover, an unpardonable offense against the amenities to have on one's pole more totems or crests than one was entitled to.

It is customary among the Tlingit and Tsimshian Indians to erect their totem poles several feet in front of their dwellings or in line several feet to the right or left of the house. At old Tongass village the writer found in front of the ruins of one of the largest houses in the village a number of poles which had been erected in honor of several maiden aunts of the former occupant of the house. The Haida at Kasaan built their totem poles in contact with the front of the house itself, a hole two or three feet square cut through the base of the pole serving as entrance to the house, ingress being possible only by crawling.

Occasionally a niche about one foot square was cut at the hollowed back of the pole some 15 feet above the ground. Into this recessed niche was placed a carved or painted box containing the cremated remains of the former head of the house in whose honor the pole had been erected. Several of such grave boxes were seen by the writer in the poles at Kasaan and at Village Island, another abandoned village.

The houses at Kasaan were placed in an irregular long row facing the shelving beach. The totem poles at the front of the houses are almost at the water's edge at high tide. The action of the salt water on the base of the pole undoubtedly served as a preservative of the wood as all the poles thus situated are still free from decay at their base while those farther removed from the beach are much rotted and decayed at the base.

The frame work of the house and the roof usually rest upon four posts commonly hollowed out at the back. Upon these house posts rest two enormous unhewn log plates sometimes each more than 50 feet long. These unhewn plates extend horizontally the entire length of the building without any other support than that of the end posts. These huge plates, the purlines, the hewn cedar planks for the side walls, "shakes" for the roof, and the logs for the posts and carved columns must be gathered from the forest with great labor, sometimes being brought from a considerable distance. They were towed to the village site where they were hauled up on skids. Forests of southeast Alaska are quite irregular, the large cedar and the spruces growing only in certain favored places where soil deposits are thick enough above the rock substratum to support their growth. Southeast Alaska will never prove a satisfactory place for farming operations as nowhere is the soil deep enough for the plow. The work of smooth finishing and assembling of timbers is undertaken at the site of the house to be erected.

FIG. 183.—Rear corner post of "More-back House" which was erected by Chief Skaul. The carved figures at top and bottom represent the bear, a crest often used by members of the Raven clan. At Kasaan, the raven crest was the vogue with adherents of the Eagle clan who lived for the most part in the western half of the village. The carved figure at the center of the post has to do with the adventures of Raven and the moon. See fig. 182.

Erection of a totem pole or a house was the occasion for much feasting and jollification by the Indians, calling to mind those social gatherings which attended barn raisings in colonial days in the United States. The day of the erection of the house was made known to the natives of neighboring villages sometimes months in advance. All those who were invited to attend the event were expected to cooperate in the work of erecting the frame and posts. Rivalry sometimes developed between competing clans, each of which was assigned a different log plate to place into position.

FIG. 184.—All that remains of "Furthest Forward House" at Kasaan. Unusual form of house post carving.

The posts, never more than 10 feet high, are first raised into position by means of rope guys and props and firmly planted in deep holes. The log plates are next put into position with skids and parbuckles. Before the posts are firmly planted in the ground, the head of each is cut out to the shape of a crescent so as to exactly fit the log plate which is to rest on it. The plate is then rolled to within working distance of the posts and parallel to its final position. The posts are braced on the opposite side, while on the near side skids are rested at an angle to form an incline up which the log plate is rolled. Ropes are rove over the top of the post, under and over the log, then back again over the post. The ropes serve as a parbuckle, take the weight of the log and hold it in position. Forked pike poles are rested against the log with their other ends in the ground to help the parbuckle take the weight as the plate is gradually rolled up, the poles being shifted as it rises. The combined efforts of the workers suffice to get the plate to the top of the incline. By means of pushing with poles and by pulling on the parbuckle the log is finally rolled into the rounded notch at the top of the house post. According to old accounts this work was conducted with much confusion, shouting, and sometimes fighting. The clan which finished first taunted their competitors who were still struggling to get the other log into position. The entire procedure, together with the festivities that followed the completion of the task, was the cause of much bad blood between neighboring villages.

The erection of memorial columns or totem poles was accomplished by means of poles, props, and rope guys in much the same manner that construction gangs erect telephone and telegraph poles. The essential difference is in the use of a forked pike pole by the Indians, the more modern method using steel-capped pike poles and the block and tackle.

Long hewn plates are grooved or beveled to receive the upper and lower ends of the split slabs forming the side and front and rear walls of the house. The top purlines form the supports for the roof which is made up of "shakes" or slabs of wood and bark held down by superimposed cross pieces and by rocks. In some of the newer houses at Kasaan the writer found large copper spikes and some made from wrought iron which held the spliced beams and girders in place. Usually there were only wooden pins and pegs placed at strategic points. The method relied on most for holding the framework together was dovetailing or the placing of interlocking mortises at the places of juncture of all beams and girder plates. Mere weight sufficed to keep the two huge main log plates in position.

The smoke hole is surmounted by a shutter which is closed in the direction of the wind. The shutter has a motion about the axle. When the wind changes and blows down the smoke hole a chain or rope is pulled and the shutter revolves to the other position against the wind. As the house faces the beach, and the wind usually blows up or down channel, the shutter faces one side or other rather than the front or rear of the house.

In the more recent years of their occupancy of Kasaan and other villages which are now abandoned, long voyages were undertaken by the natives to Port Simpson, at the head of Dixon Entrance, in British Columbia, for sawed boards and hardware to make a false front for their houses after the fashion of the houses of white men with a door and windows. This custom is reminiscent of the old pioneer store of early villages in the West, when the false store front with its high squared and impressive top section was almost an institution. It must be conceded that the old style of Indian house with its entrance through the base of the totem pole and its huge open smoke hole at the center of the roof, although minus windows, was architecturally more of a unit than the later Indian houses with their hardware and windows from Port Simpson.

During a recent visit by the writer to the village of New Kasaan, rather deplorable conditions with regard to ventilation and sanitation of native houses were observed. The Indian of today in southeast Alaska lives in a house built of sawed boards throughout. There are windows enough, but they are kept closed against the damp air without. There is invariably a stove in the center of the living room which consumes quantities of oxygen. If one were forced to choose between the evils of the cold unlighted slab side native Indian house and the poorly ventilated, unsanitary house of the Indian of today one would not hesitate in preferring the old purely native type of dwelling which was at once both health producing and artistically beautiful.

ARCHEOLOGICAL INVESTIGATIONS IN THE COLUMBIA RIVER VALLEY

During the spring and early summer months of 1926 a regional archeological survey of the middle and upper Columbia River valley was made by H. W. Krieger, curator of ethnology in the U. S. National Museum, on detail to the Bureau of American Ethnology. The project began with a study of the extensive collections obtained by members of the Columbia River Archeological Society from burials and surface finds at various ancient and historic Indian village sites and cemeteries.

Most noteworthy among the collections studied are those of Mr. H. T. Harding, of Walla Walla, Washington; of Messrs. Earl Simmons, Gibson, and Charles Simpson, of Quincy, Washington; and of Messrs. A. H. East, O. B. Brown, Guy C. Browne, Drs. R. T. Congdon, and T. H. Grosvenor, all of Wenatchee, Washington. Other collections studied are those of Dr. F. C. Evertsbusch and others at Pateros, Washington, the Eells collection at Whitman College, and the extensive material collected by Dr. F. S. Hall and others for the State Museum, at Seattle. Enthusiastic interest in the project was shown by members of the Columbia River Archeological Society and others who have done pioneer work in locating many ancient villages and burial sites and in gathering and classifying many different types

Fig. 185.—View along the Columbia River at Vantage Ferry, Grant County, Washington. Many pictographs and petroglyphs appear on the western escarpment of columnar basalt near the water's edge. The low-lying bench land to the left was the site of a large pit house village.

of archeological material. Information as to location of sites and distribution of type specimens was in every instance cheerfully given.

The next step in the survey was the plotting of an archeological map of the Columbia and tributary valleys showing known village sites and cemeteries. A check was made on data already collected, amplified in several instances by a visit to the reported location of an isolated pit house ruin, village site, or cremation burial.

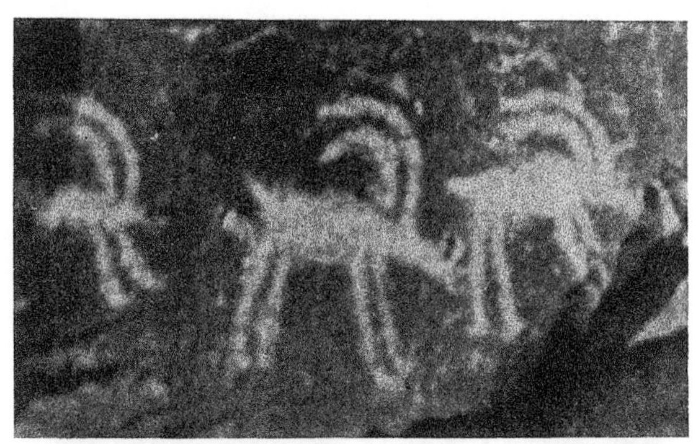

Fig. 186.—Petroglyphs pecked on the columnar basaltic rock escarpment of the Columbia River at Vantage Ferry, Beverly, and Rock Island, Washington.

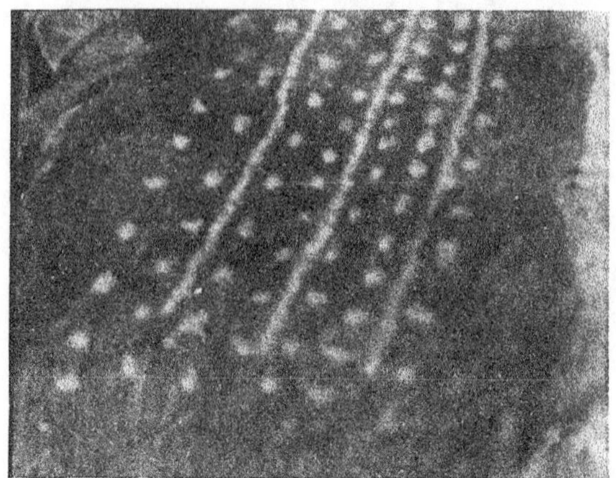

Fig. 187.—Petroglyphs pecked on the columnar basaltic rock escarpment of the Columbia River at Vantage Ferry, Beverly, and Rock Island, Washington.

The necessity for obtaining an archeological map of the valley at this time becomes apparent upon noting that the Indian village site is also the most favored location chosen by the modern orchardist for his planting. The reason for this lies in the need for protection, shelter, and an adequate water supply. A young orchard can best be successfully developed on a narrow level river bench which is high enough to be secure against seasonal flood waters and near enough to the towering escarpment of the river for shelter from the winds which sweep over the plateau above. It was just this type of narrow bench land, situated above danger from floods, and close to the precipitous basaltic or lava capped river escarpment which was selected by the prehistoric occupant of the Columbia valley as a location for his permanent winter home. Here, also, under the well-nigh inaccessible barrier of the cliffs the primitive village group was secure from attack by marauding hostile bands.

FIG. 188.—Gap in the gorge of the Columbia River where the Saddle Mountain range crosses at right angles to the river at the head of Priest Rapids. On the far bank of the river at the left was located the village of Smohalla, a leader in the Ghost Dance cult. Saddle mountains separated in historic times the territory of the Shahaptian Indians from that of the Salish tribes.

As the middle and upper Columbia River valley is semiarid and barren to a degree, an adequate water supply is essential. The bench land selected as a village site must be neither too high nor the banks too steep to preclude easy access to the river. On the sloping beach below the bench were obtained useful varieties of stone pebbles, float bowlders, and drift wood.

The mapping of archeological stations along the middle and upper Columbia and tributary rivers, such as the lower Yakima, Snake,

Walla Walla, Deschutes, Wenatchee, Methow, Okanogan and others began at the Dalles, in the state of Oregon, and continued to the environs of Kettle Falls, near the Canadian border. The falls and gorge of the Columbia River in the vicinity of the Dalles where the Columbia breaks through the Cascades, mark the beginning of the wooded area of the lower river which possessed an equally well marked distinct type of native culture, the Indian tribes there using principally wood in their arts and crafts, and the tribes of the middle and upper river being expert stone cutters and workers in horn and bone. Accompanied by Mr. H. T. Harding, a section of the river was covered as far north as Wenatchee. Mr. Harding's assistance was invaluable as he is intimately acquainted with the archeology of the region due to many years of experience in the field. Traces of Indian occupation are being rapidly obliterated by the plow, which is today the most productive excavator of antiquities. Where land has been brought under the plow, no record of former village sites and cemeteries is available other than that collected by members of the Columbia River Archeological Society.

Of the many sites inspected, excavation was undertaken at eight. The site yielding the largest collection of material such as ceremonial burial offerings and skeletal material was the prehistoric site at Wahluke in Grant County. There was no evidence of the burials there ever having been disturbed. Neither was there any indication of Hudson Bay Company influence in the objects recovered from the graves, such as trade beads of glass or of the shell beads which in historic times were traded to the Indians as a substitute for the *Dentalium indianorum,* or of iron tools and weapons.

The village of Wahluke is located on the east bank of the Columbia in Grant County, at a point where the river, which at this part of its course flows north, strikes a precipitous escarpment formed of yellowish gray volcanic débris and ash, known as White Bluffs. The Columbia here changes its course to a general southeasterly direction and completes the final segment of the course known as the big bend. Wahluke lies well within the territory occupied by the Shahaptian-speaking tribes within historic times, although early accounts and evidence obtained from the nature of the burial offerings indicated that in prehistoric times the entire area on the north and west bank of the Columbia as far south as the Dalles was Salish territory.

White Bluffs, which lies hard against the northern end of Wahluke, is a continuation of a range of hills known locally as Saddle Mountains from the fact that the range extending from west to east lies at right angles to the Columbia where it breaks through the narrow gorge just below the confluence of Crab Creek, about 40 miles up the river from Wahluke. The old Indian term Wahluke is said to mean "where one can see and watch." If this is the correct meaning of the word, it would explain the term Sentinel Mountains, a name still applied to the same range. The flat east bank of the river opposite Wahluke

FIG. 189.—One of the many caves in the lava and vesiculate basalt bluffs of the Columbia River. Such caves were used as temporary habitations by hunting and fishing parties, also by bats. This cave, as may be seen, is half filled with debris composed of charred bat guano, camp refuse, such as fabrics, mats, and cooking stones, together with weapon parts and tools.

was known to the Indians as Yanuke "the place where animals come to the water and drink."

Burials at Wahluke were mostly primary and ceremonial cremation in type. Graves were placed in irregular rows along the river beach up-stream from the village proper. But one site along the river is known where a village had been built on top of a site and cemetery

Fig. 190.—A Wanapum or Columbia River Indian mother and two children in front of their tule mat covered tipi. The dog is of the type formerly kept by these Indians for the use of their shaggy coat of hair in blanket making.

Fig. 191.—Shahaptian Indian sweat house built up of tule reeds with matting of cat-tail rushes and covered over with earth. Still used by the Wanapum or Columbia River Indians. Located at the Lower Falls of the Yakima River, Benton County, Washington.

of an older date. This location, known as Simmon's graveyard is about five miles down-stream from Trinidad, the point where the Great Northern Railway crosses the Columbia. The flood waters of the river had covered the cemetery to a depth of six or more feet. A pit house village is erected over and immediately above it. At many places

FIG. 192.—Looking south from interior of Eagle House, Kasaan, across McKenzie Inlet, Kasaan Bay. Pete Williams, a Tlingit Indian belonging to Eagle clan, standing in foreground. See figs. 177-180.

along the Columbia there are similar strata of sedimentary deposits, each covering the burn and other evidence of the existence of a camp site or burial ground. These strata were formed at intervals of several years, probably generations. Temporary fishing camps which were abandoned at the close of the fishing season were left with camp débris, fragments of shell and other kitchen débris such as charred cooking stones, charcoal, and stone and bone implements. Some of these sites are again exposed in later years when the stream forms a new channel or the flood waters erode the banks. At Pateros, the confluence of the Methow with the Columbia, seven layers of burn with intervening

FIG. 193.—Tipi used by the Wanapum or Columbia River Indians at the Lower Falls of the Yakima River, Benton County, Washington. The habitation is a temporary one and is constructed of tule reeds covered with skins and burlap. The old type of semisubterranean pit house was abandoned before the time of the Lewis and Clark expedition.

sedimentary deposits are exposed on the flanks of a little island formed on the Methow side of the channel.

Cremation graves are usually three or more feet below the surface when undisturbed. A layer of flat stones was invariably placed in oblong or circular form as a protective cover against marauding animals and erosion of the loose sand bench formation by the action of the wind. Loose sand with a slight covering of soil or sometimes dry powdery volcanic débris makes up the formation of the bench above the river bank. This was excavated with crude shovel-shaped paddles of stone, or with the hand aided by the digging stick. The body to be cremated was placed on a piece of matting of Indian hemp, tule

Fig. 194.—A prehistoric Indian grave at Wahluke, Grant County, Washington. The burial is protected with a circular ring of bowlders which have been carried up from the river beach. Among the many ceremonial burial offerings found within this grave were no objects such as glass beads or iron knives which might indicate Hudson Bay Company influence.

reeds, or cat-tail rushes. About the body, which was oriented sometimes with head up-stream, sometimes to the east, or, again apparently haphazard, were arranged the personal belongings of the dead. The pyre was built up of logs of driftwood. Many of the objects placed in the grave are merely charred; others, including the skeletal material, are completely burned or calcined. No indications of burial houses such as were erected by the tribes on the lower river were found at Wahluke or elsewhere on the middle Columbia.

In some instances the skeletons were oriented in such position as to suggest secondary burial, parts of several skeletons in such cases being

FIG. 195.—Stages in the production of a pipe. Beginning at the lower left may be seen a roughly cut and grooved section of steatite. The next step shows the tubular outline formed by crumbling and pecking; the third object is perforate and shows polishing, while the pipe at the right is complete except for the reed mouthpiece which is not shown. The steatite bowl pipe at the top right is modern and is made from a design showing influence of trade pipes. All obtained from graves at various sites along the Columbia River by Mr. A. H. East.

FIG. 196.—Types of stone pipes found in graves at various sites in the Columbia River valley, in the state of Washington. The tubular soapstone pipe at the right indicates contact with Californian tribes, while the three tubular soapstone pipes at the center are the type made by the ancient inhabitants of the Columbia and Fraser river valleys. The bowl pipe with carving of human face at the side is of the type made by the tribes of the northwest Pacific coast. The large catlinite bowl pipe with inlay of lead around the margin may have come from the Sioux or other tribes of the Plains.

jumbled all in a heap. Individual cremation burials were never secondary, the burial being effected with knees flexed and skull facing downward. Incineration was usually so complete as to prevent the securing of any one complete skeleton. Several fragments of charred skeletons including several skulls were recovered from the burn, providing sufficient material to reconstruct later at the Museum. Skulls showed in every case a frontal-occipital deformation. Skulls from graves other than at Wahluke do not always show this deformation,

FIG. 197.—Frame of Salish type of sweat house on west bank of the Columbia River six miles below Pateros.

leading to the assumption that the practice was not general. The physical type is that of the tribes which occupied the upper plateau country within historic times. Skulls found in graves accompanied by ceremonial offerings having a distinctly Hudson Bay Company aspect, such as trade beads of glass, and metal objects of copper, brass, and iron, were in every instance of the same type as those occurring in prehistoric graves.

Many of the objects found in the burn among the charcoal and charred bones were objects of daily use in the life of the Plateau Indian of historic times. Most of the larger pieces as bowls and pestles were intentionally broken at burial. Objects found included basketry, matting, objects of stone, wood, bone, horn, shell, human and dog hair, and hammered nuggets of native copper. No pottery was made by the ancient occupants of the upper Columbia valley, nor were utensils or bowls of wood discovered. This seems strange when it is noted that other objects shaped from driftwood occur in the graves intact; also many ornamental objects formed from jade, dentalium, haliotis, soft slate, and lead inlay, all of which must have been brought from the Pacific coast, British Columbia, or Alaska. It would appear from this that the burials at Wahluke are of a date preceding that of the wood-working tribes of the lower Columbia.

ARCHEOLOGICAL WORK IN LOUISIANA AND MISSISSIPPI

Mr. Henry B. Collins, Jr., assistant curator of ethnology, U. S. National Museum, was engaged from the middle of April to the latter part of June in archeological field-work in Mississippi and Louisiana for the Bureau of American Ethnology. The greater part of this period was given to investigations along the Louisiana Gulf Coast, a region which was practically unknown from an archeological standpoint. This section of Louisiana is but slightly above sea level and consists for the most part of great stretches of marsh, habitable only along the narrow ridges of comparatively high land that border the many lakes and bayous. It is not a region which might be expected to have supported either a large or a very highly developed aboriginal population, and yet unmistakable evidence was found that in pre-Columbian times Indians had lived here in considerable numbers, and that some of them possessed a culture closely allied or identical in general to that found throughout the widespread mound area to the east and north.

The investigations were begun at Pointe a la Hache, on the Mississippi River about 40 miles below New Orleans. About five miles southwest of Pointe a la Hache was found an important group of mounds, nine in number, the largest of which was between 40 and 50 feet high, over a hundred feet in diameter, conical in shape, and with a flat top. The other mounds were lower but were of considerable basal diameter and were all covered with an almost impenetrable growth of palmettos, vines, briars, and other vegetation. These mounds had been built on a ridge of land at the very edge of the marsh and their location was known to only a few fishermen and trappers.

From Pointe a la Hache, Mr. Collins proceeded to Houma, in Terrebonne Parish, and examined a number of mounds and shell heaps.

Here he was fortunate in having the co-operation of Mr. Randolph A. Bazet, who is deeply interested in the local archeology and who was able to supply valuable information on the many earth mounds and shell deposits of Terrebonne Parish. Such remains were found in unexpected numbers along the lakes and bayous, ranging from comparatively small accumulations of shells mixed with charcoal, potsherds, bones and other refuse to huge deposits of the same material, or "islands" as they are locally called, sometimes a hundred yards or more wide, about 10 feet above the marsh level, and extending in some cases for a distance of almost a quarter of a mile. These Terrebonne Parish shell heaps, or kitchen middens, and the others throughout southern Louisiana, are composed almost entirely of the shells of a small brackish water clam, *Rangia cuneata,* which is very common in the bayous and lakes of the Gulf region. They represent merely the accumulated kitchen refuse of the Indians who once lived along these water ways. The clams were eaten, and the shells, along with other trash, were cast aside until in the course of time an extensive heap was formed.

Fig. 198.—Mound on the Fairview Plantation, Berwick, La. About six feet below the surface was a thin stratum of fire-burned earth in which were found seven pits or fireplaces.

After devoting some 10 days to the mounds and shell heaps of Terrebonne Parish, investigation was made of those to the west and north, at Gibson, Lake Palourde, Bayou l'Ours, Berwick, Charenton and Avery Island. Having examined and carried on minor excavations at these localities, Mr. Collins continued westward to Pecan Island in the southern part of Vermillion Parish where he remained for three weeks.

Pecan Island, lying between White Lake and the Gulf, is a long narrow strip of land, averaging less than 200 yards in width, and extending across the marsh for fifteen miles in a general east and west direction. It is the same type of formation as Grand Chenier and other narrow parallel ridges of land found in the marsh region which have resulted from storm wave action throwing up old beach material such as sand and finely crushed shell. Pecan Island today

Fig. 199.—Mound on the property of Mr. Ulysses Veazey, Pecan Island, Vermillion Parish, La.

supports a population of some 400 people, whose only means of communication with the outside world is by the mail boat which makes the 57-mile trip from Abbeville once a week. That it was also inhabited by Indians at an early date is shown by the presence of 21 artificial mounds and considerable quantities of potsherds and other surface refuse on the island. These mounds proved to be of two distinct groups, those on the property of Mr. J. Morgan, four in number, being the larger, with an average basal diameter of over 100 feet and a height ranging from five to twenty-five feet. These four mounds were stratified, with several thick layers of crushed shell and sand (the material of which the island is largely composed) separated by thin strata of soil mixed with charcoal, animal bones and other débris.

FIG. 200.—At its eastern end Pecan Island divides into several long narrow ridges. This view shows the main ridge viewed from Cypress Point, from which it is here separated by a stretch of marsh.

FIG. 201.—Small mound surrounded by marsh at Cypress Point, on Pecan Island, La.

thus revealing successive levels of occupancy and affording a clear picture of the manner in which the mounds were constructed. The mounds of the second type, most of which were on the property of Mr. Ulysses Veazey, were unstratified. Human bones were found in two of the large Morgan mounds and in one mound on the Veazey

FIG. 202.—Detached skulls in Copell burial ground on Pecan Island. The bones when uncovered were soft and usually broken, but all skulls, some fifty in number, were saved and have been repaired.

place. The skulls from the former locality were all of the well-known "flat-head" type, which resulted from the practice of tightly binding the head during infancy. Skulls from the Veazey mound and those from another burial ground on the property of Mr. John Copell, which proved to belong to the same culture, were for the most part undeformed. A similar distinction was observed in the material recovered from these mounds, indicating that the two types represented different cultures. The objects associated with burials in the Veazey mounds and the Copell cemetery, while not especially abundant, were typical of those usually found in mounds in other parts of the country. These included chipped stone and worked bone arrowpoints; the bone end of an atlatl or "spear thrower"; beads and other ornaments of shell; large double-disk or spool shaped ear ornaments of slate covered with native sheet copper; various types of worked stone, shell and bone implements; lumps of galena; hematite; bitumen; and decorated potsherds of a characteristic type. The Morgan mounds, on the other hand, produced no such material, but the invariable flattening of the

Fig. 203.—Large kitchen-midden, or shell heap, on Chenier du Fond, south of Grand Lake, Cameron Parish, Louisiana.

skulls, the different construction of the mounds themselves, and the quite distinctive type of pottery, all point clearly to a different culture.

Pecan Island was the most western point on the Louisiana coast at which artificial mounds were found, but kitchen middens were found to continue westward and there are evidences that these extend well into Texas. After examining the aboriginal remains as far west as Grand Chenier in Cameron Parish, Mr. Collins returned eastward and began excavation of a group of three large mounds on the property of Mr. Adolph Melanson, at Gibson, in northern Terrebonne Parish. These mounds were stratified in much the same manner as the four large mounds on Pecan Island, and although a few burials were found in one of them it was obvious that all three had been erected primarily for habitation purposes.

Considering the evidence thus obtained by reconnoissance and excavation in the Louisiana Gulf Coast region, several points of interest develop. First, there is here found the most southern and western extension thus far recorded of the wide-spread and highly developed mound culture of the Mississippi Valley and Gulf States. The immediate cultural affiliations of this southern Louisiana mound area is seen to be toward the east, forming a direct connection along the Gulf Coast with Florida. The strongest evidence for this deduction is found in the pottery, certain types of which are practically identical from Florida to western Louisiana; these consist of the "checker-board" design, produced by the application of a stamp, and of enlarged rims bearing characteristic incised and punctate decoration. The exact identification of the builders of these Louisiana mounds and shell heaps will, of course, be difficult. The western part of the Louisiana coast, including Pecan Island, was inhabited in historic times by the Attacapa, a cannabalistic tribe of low culture. They could hardly have been the people responsible for either of the two Pecan Island cultures. However, they may have left behind some of the shell heaps here described, although the shell heaps cannot as a class be separated culturally from the earthen mounds. Further to the east were the Chitimacha, much more advanced than the Attacapa, and if the descendants of the prehistoric mound-building tribes of southern Louisiana are to be sought nearby, the Chitimacha might well fit into the rôle. The undoubted cultural connection with the coastal regions of Mississippi, Alabama and Florida, however, suggest the strong possibility that there may have been a direct tribal movement in one direction or the other.

The Louisiana work was completed about the middle of June, when Mr. Collins proceeded to Marion County in southern Mississippi and located seven ancient Indian village sites. From there he went to eastern Mississippi and succeeded in locating the sites of two old Choctaw villages, which were described by French and English explorers as early as 1729, but the present locations of which were unknown. These villages were Chickachae in Clarke County and Okatalaya in Newton County. The potsherds from these village sites were of the same type as those found on other historic Choctaw sites in 1925. This is a ceramic type which is restricted to comparatively recent Choctaw sites and which is quite different from the earlier mound pottery

On July 1 Mr. Collins proceeded to Philadelphia, Miss., to continue the anthropometric studies of the living Choctaw begun in the summer of 1925. The present investigation was made possible by an appropriation from the American Association for the Advancement of Science. Seventy-two adult Choctaws of both sexes were measured, which number, together with those obtained in 1925, affords an adequate idea of the physical type of the modern Choctaw. The number of Choctaw still remaining in Mississippi is around 1,000. Unlike some other Indian tribes they are not mixing to any great extent with whites, nor have they done so for some years past. However, they are far from being full blooded as a group, due to infiltration of white and negro blood at an early date.

FIG. 204.—Group of Choctaw Indians of the Conehatta district, Newton County, Miss. Mr. T. J. Scott of the Choctaw Agency, in back row.

Among those whose interest and co-operation aided the work in Louisiana and Mississippi, particular thanks are due to Senator E. S. Broussard; Father Girault, of Pointe a la Hache, La.; Mr. R. A. Bazet and Dr. Marmande, Houma; Mr. J. A. Pharr and Mr. W. B. Reed, Morgan City; Mr. and Mrs. Sidney Bradford and Miss Sarah McIlhenny, Avery Island; Prof. C. S. Brown, University, Miss.; Mr. T. J. Scott, Philadelphia, Miss.; and Mr. Rufus Terral, Quitman, Miss.

ARCHEOLOGICAL FIELD-WORK IN ARIZONA

During the summer of 1926, from the closing days of May to the end of August, Dr. J. Walter Fewkes, Chief of the Bureau of American Ethnology, excavated and repaired a little known ruin about six and a half miles east of Flagstaff, at the base of Elden Mesa. To this ruin he gave the name Elden Pueblo. He was assisted in this field-work by Mr. J. P. Harrington, ethnologist, and Mr. A. W. Wilding, stenographer, both of whom contributed to the success of the expedition. Valuable assistance was rendered by Mr. J. C. Clarke, Custodian of the Wupatki National Monument, and by Prof. H. S. Colton, of the University of Pennsylvania, both of whom spent much time at the ruin and aided in many ways. These gentlemen called Dr. Fewkes' attention to the clearing in the pine trees that indicated the site of a prehistoric building which had been long suspected by a few local residents. The discovery of Elden Pueblo, or rather the demonstration that the clearing indicated a pueblo site, was an important one, since it opened up an extensive area where many small ruins occur but of which practically nothing is known save obscure sites. The work at Elden Pueblo developed a new type of building characteristic of the region south of the Grand Canyon and west of the Little Colorado, extending west and south to the Lower Gila, the massive ruins of which, known as compounds, attracted attention as far back as the beginning of the seventeenth century.

This pueblo was called by Prof. Colton, Sheep Hill Ruin,[1] but has been given its present name from the mesa which towers over it on the west. The ruin is situated about 200 yards from the National Old Trails Highway, not far from where the road to Tuba City branches from the main thoroughfare. This road is a much used one along which daily, in full view of Elden Pueblo, pass many automobiles with tourists keenly interested in the scenic wonders of the country and in the attractive remains of the former inhabitants. Very few of these tourists passed without stopping to inspect the work the Bureau was doing in bringing to light this remarkable relic of the past.

The mound which covered Elden Pueblo was not very attractive as an archeological site when work began. It was devoid of trees, surrounded by a pine forest, and covered with bushes and stones. Indistinct lines of rocks were visible on the surface of the ground, but even these could not be traced many feet. No standing walls existed above ground, the rooms later excavated having been filled level to the top with fallen stones, earth and sand. The general appearance of the site before excavation is shown in figure 205. Several old residents of Flagstaff claimed they had often herded stock or sheep over

[1] In his manuscript now awaiting publication in the Bureau of American Ethnology. The name Elden Pueblo is there applied to another site.

FIG. 205.—Elden Pueblo before excavation.

FIG. 206.—West wall of Elden Pueblo after excavation. (Photograph by A. W. Wilding.)

the site but never dreamed that under the surface of the earth there existed so many walls and relics of a former pueblo. Taking as guides the few stones in line, the outside walls of the rectangular building (fig. 206) were revealed by a trench dug to the level of the foundation

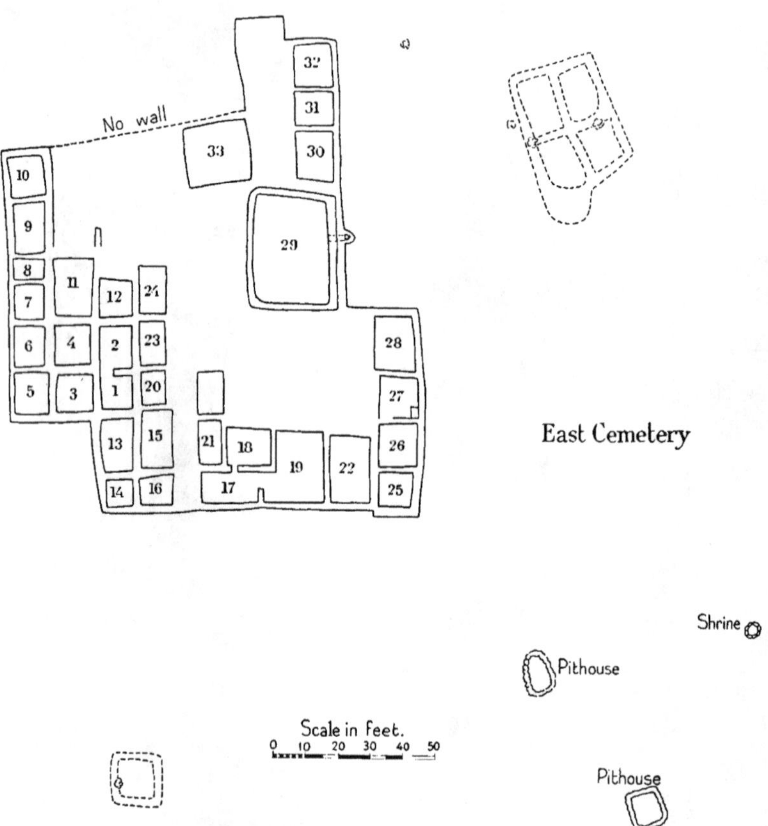

Fig. 207.—Sketch map of Elden Pueblo.

and extending entirely around the ruin. When once this bounding wall had been determined the contents of the rooms, composed of the fallen tops of walls, were dug out and the number of chambers determined. The most laborious part of this work was the removal to a distance from the ruin of the rocks and other material that filled the rooms or had accumulated outside their bounding walls.

FIG. 208.—Elden Pueblo, showing south wall and recess in the southwest corner. (Photograph by J. W. Fewkes.)

FIG. 209.—Elden Pueblo from the northwest; Sheep Hill in the distance. (Photograph by J. W. Simmons.)

Judging from the ground plan shown in the accompanying sketch (fig. 207), Elden Pueblo was a massive walled building composed of large chambers used for domiciles or granaries forming a lower story, upon which a second was built, especially evident on the western end, where, although considerably broken down, fragments of a roof rested on top of the lower rooms, with rudiments of upper walls. As will be seen by an examination of figures 206, 208, and 209, Elden Pueblo was a rectangular structure 145 by 125 feet, oriented approximately east-west. A complete wall was missing on the north side. The standing walls varied in height from about two feet at the lowest to seven feet at the highest point (fig. 210).

Fig. 210.—Recess in the exterior wall on the southwest corner, showing highest point in the wall. (Photograph by J. W. Fewkes.)

This latter occurred at the angular inset of the southwest corner, where there are indications of a shrine near a megalith forming the foundation, under which was a recess containing two or three small undecorated clay vessels and a number of rock concretions. From its position at the southwest angle of the building, facing the sunset point at the winter solstice, this shrine may be supposed to represent the winter solstice house of the sun, or the place where offerings to the sky god were placed, in the Elden Pueblo worship.

The masonry of Elden Pueblo is crude, as would be expected in a ruin of great age. Stone slabs standing on edge and megaliths occur as foundations in several places, and although here and there we find stones laid in courses, this is not a common mural characteristic. Large stones alternating with rubble thrown roughly together and exhibiting no evidence of being shaped by human hands are most abundant. In this respect the walls are quite unlike those of the buildings situated in the Wupatki National Monument, some 25 miles away, or in the middle valley of the Little Colorado.

The secular rooms are square or rectangular in shape, but there is one room, recognized at once as different from the others in size and function, which was not secular but rather communal or ceremonial, later identified as a kiva. This kiva was indicated by a depression in the surface and as is customary was situated half underground. It lies near the northeastern corner of the pueblo, surrounded on three sides by secular rooms and by a wall only on the east. It is large and rectangular with rounded corners, a low seat or banquette extending completely around the inner wall. This banquette has no vestiges of pilasters resting on it for roof supports, the roof evidently having been held up by vertical logs set in the floor. A ventilator shaft opens into the room midway in the east wall at floor level. The vertical portion of this shaft is enclosed by stone masonry bulging from the external wall of the room. The fallen rock and adobe within this ceremonial room rendered it impossible to detect a deflector (windshield) or fire hole in the floor, and the indications are from the amount of charcoal and absence of roof beams that there had been a conflagration in this room before or after the pueblo fell into ruins. A comparison of this kiva with one partially excavated at Wupatki, and with those at Marsh Pass, Arizona, shows a great similarity. This likeness supports an Indian legend that the Marsh Pass people were the same race as the ancient Snake Clan, still represented at Walpi on the East Mesa. Very few objects of Indian manufacture were found in the rooms of Elden Pueblo, and although there were a few bowls and one or two well-preserved human skeletons, these were so few in comparison with mortuary objects from the cemeteries that a consideration of them will be left to a final report on the ruin.

There were many good metates and manos of customary pueblo shape and several stone implements, axes, mortars, spear-points, arrowheads, and the like, a consideration of which would enlarge this preliminary account too much. There should be mentioned, however, a type of paint-grinding stones similar to those from Casa Grande. A spindle whorl may be mentioned because instead of being of stone, thin and perforated like those from the pueblo ruins, it is of clay, thick, almost spherical, like Mexican spindle whorls, identical with several elsewhere figured from Casa Grande. Another form of spindle whorl has a groove around the rim, also suggesting Mexican influence.

Fig. 211.—Incised deer leg bone used, with feathers attached, as a hair ornament. (Length 7½"; width 1".)

Many bone needles were found in the rooms, a few bearing incised decorations. One bone object, figure 211, was a hairpin or ornament worn in the hair. Similar carved bones with attached feathers are still used by the Hopi and Zuñi warriors, as shown in the discussion of the possible use of one of these objects from Youngs Canyon, 12 miles from Elden Pueblo. On the cranium of one of the skeletons, figure 212, found in the Elden Pueblo cemetery there was a similar bone hairpin, still in place.

Two cemeteries were located in the work, one on the eastern, the other on the northern side of the pueblo. Many skeletons and mortuary objects were taken from these burial places. There may have been cemeteries also on the south and west sides, as a burial was found near the west wall a number of years ago when a logging road was constructed through the pine forest. Although a few "test holes" were dug on these sides, no systematic excavations were made there by the Bureau's expedition, which confined itself to the east and north cemeteries of the pueblo. These cemeteries, however, were not completely dug out on account of limitation of funds, but other interments may be brought to light by future archeologists.

There was nothing on the surface of the ground to indicate the position of skeletons in the cemeteries, but the remains of the dead were found by trenching or by probing with iron rods. The number of skeletons located was about 150, many of which were in fairly good condition, but in a much larger number the component bones were either missing or so much broken that they were of very little scientific value. After a skeleton had been located and the mortuary bowls and other objects which accompanied it had been removed with care they were replaced about the remains and a photograph made of the interment *in situ,* showing the relative position of accompanying objects (fig. 213).

There is some variety in the positions in which the dead were placed in their graves. As a rule, however, they lay on their backs, although a few were placed on their sides; but there were no flexed burials. Some of the skeletons were a few inches under the surface, others several feet below it. The latter were accompanied by the same kinds of pottery as that accompanying the shallow burials. As a rule, to which there were exceptions, the skeletons were oriented with the head to the east. Double burials in the same grave, infant interments, and

Fig. 212.—Burial showing position of the effigy of an unknown animal at the pelvis, and carved deer tibia (shown in fig. 211) between vase and the skull. (Photograph by J. P. Harrington.)

a skeleton of a woman with skull fragments of an embryo in place were noted. In one grave was found the skull of a dog, and bones of other mammals and antelope horns occurred in several graves.

One of the most interesting graves was that which had the skeleton encased or covered by a hard crust of adobe (fig. 214), but this seems to have been the only instance of this form of burial. In no instance was the skeleton enclosed in a stone cyst as occurs so often at Wupatki. Individual stones standing on edge were sometimes found, but whether these were placed in that position for a purpose could not be determined. The dead did not appear to have had any wrappings when buried; if so, these wrappings had long since disappeared. They were accompanied by their ornaments, such as bracelets made of shell, necklaces, turquoise ear pendants, finger rings, and other adornments. Shell and stone beads of a necklace of considerable size were sifted out

FIG. 213.—Burial with vertical rock like a head-stone and various mortuary objects, including 6 shell bracelets on left forearm. (Photograph by J. W. Simmons.)

of the earth surrounding the neck of one skeleton. Several paint grinders and one or two mortars and pestles were taken from the graves, but a larger number were found on the floors of the rooms.

There were also clay images of quadrupeds and a small clay effigy of a bird (figs. 217, b, c) with outstretched wings. Miniature vessels, corrugated or painted black on white, often accompanied the dead. A particularly fine small cup of black and white ware with graceful

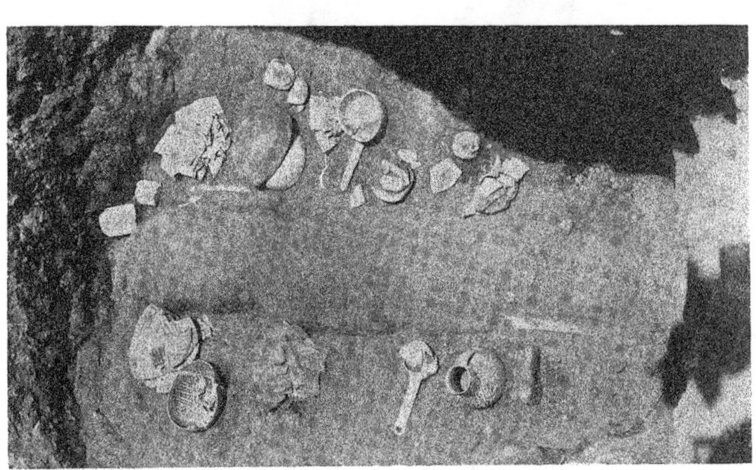

FIG. 214.—Burial with skeleton covered with a hard crust made of adobe, showing disposition of mortuary pottery. (Photograph by J. W. Simmons.)

FIG. 215.—Burial shown in figure 214 with crust of adobe removed. (Photograph by J. W. Simmons.)

neck and handle (fig. 217, a) lay over the breast of a priest. This vessel probably once contained medicine water.

By far the most numerous mortuary objects obtained in the Elden Pueblo graves were pottery vessels or utensils made of burnt clay. Each skeleton was accompanied by several pieces of pottery, differing in form, color, and other features, but all characteristic of the ruin. Among the objects represented were ollas, vases, bowls, jugs, dippers, ladles, effigy jars and other forms. Some of these are painted, others are plain pieces without decorations, but there were no specimens of true glazed ware, leading to the conclusion that, in common with the

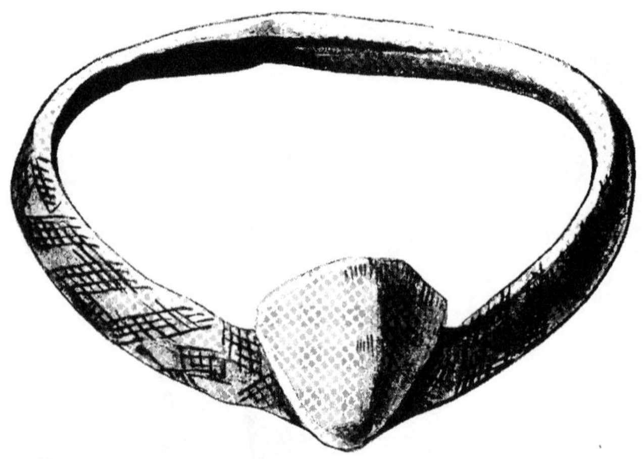

FIG. 216.—Bracelet made of Pacific Coast clam shell (*Pectunculus*), similar to those on the left forearm of fig. 213. (Diam. $3\frac{3}{8}$".)

San Juan region, glazing of ceramics had not yet come into vogue at Elden Pueblo when it was in its prime.

As a rule Elden Pueblo pottery is not of the finest texture—some specimens are very coarse—nor are the decorations carefully made, although variegated. The fact that a relatively large number of pots and bowls were small naturally attracted attention. As several appear too minute for utensils it has been suggested that some of them were specially made as mortuary vessels. None of the vessels was punctured or "killed" before burial, and only one specimen had a "life-line" or break in the surrounding bands.

The corrugated and coiled ware largely represented at Elden Pueblo appears to be contemporaneous with the black and white specimens which are equally abundant. The latter are coarser than

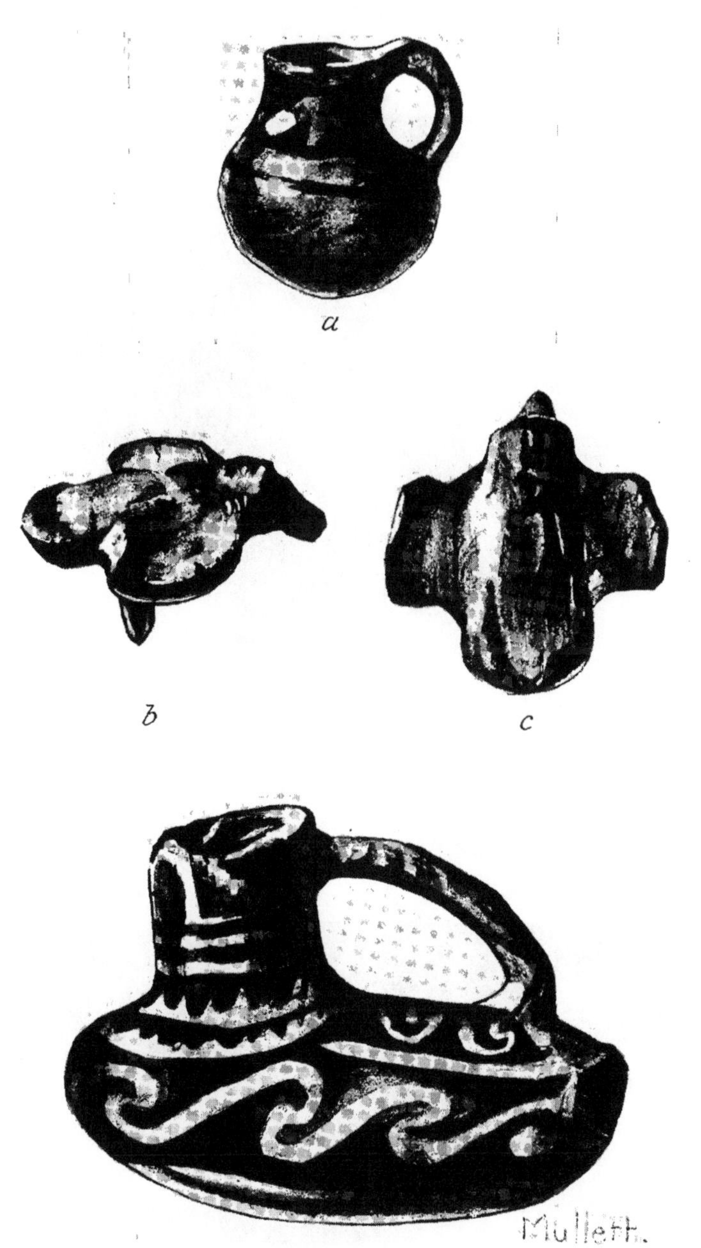

FIG. 217.—Small pottery objects and clay image of a bird. *a*, Medicine cup from breast of a skeleton; *b*, *c*, Lateral and dorsal view of a bird fetish; *d*, Medicine vessel.

a, Largest diam., $1\frac{3}{8}''$; height $1\frac{1}{2}''$. *b*, *c*, Length $1\frac{7}{8}''$; width $1\frac{3}{4}''$. *d*, Length $3\frac{1}{2}''$; height $2\frac{1}{4}''$.

a

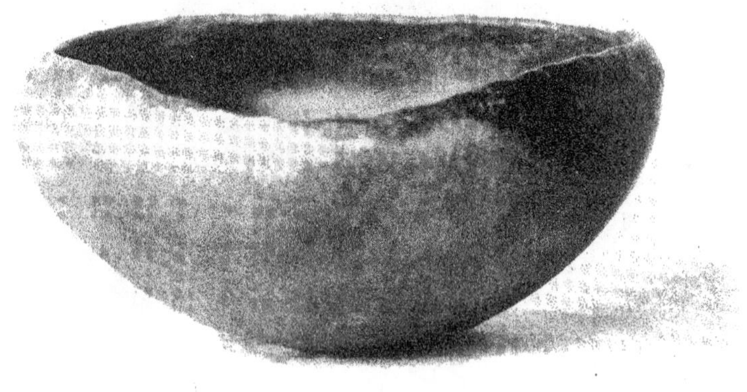

b

FIG. 218.—*a*, Smooth red vase with blackened interior; *b*, red food bowl with lustrous black interior.
a, Diam. 6″; height 5¼″. *b*, Diam. 6½″; height 3″.

those from Mesa Verde and show an imperfect technique, nor are they as boldly drawn and cleverly combined. They show a close resemblance to the black and white ware of the Marsh Pass ruins and those identified as belonging to the San Juan culture. There is no external ornamentation on the black and white food bowls, but jugs and other forms of the same ware show elaborate external decoration.

A prevalent type of pottery obtained in the Elden Pueblo cemetery has a bright red color on the exterior and lustrous black on the interior (fig. 218). Some of these specimens have their outer surface corrugated, but the black interior surface occurs both in many of those which are corrugated and others whose outer surface is smooth. This ware, whether bowls, jars, or vases, is characteristic, and may be known as the Flagstaff ware; although widely distributed all over the Southwest, it is especially abundant in localities where Gila Valley ware prevails. The black color on the inner surface of the jar or bowl was formed in the same way as at the modern Santa Clara pueblo, where it is produced by the action of smoke from a smothered fire. Various substances may be used to create this smoke. The Hopi often use for that purpose a thin piece of corn bread, called paper bread, but other substances are likewise employed. A similar method of using smoke to impart this black color was employed by the ancient Greeks and Egyptians.

The forms of the red pottery here described as Flagstaff type are food bowls, vases, jugs, ladles, ollas, and the like. The several jugs and bowls, some smooth, others with corrugated exterior and lustrous interior surface, would indicate that these two groups were contemporary. Both, however, are supposed to antedate what is called the black on white pottery, which type is decorated and by far the most attractive of all the archaic ware from Elden Pueblo. The designs on this ware are conventionalized figures painted on a white ground.

In order to simplify a study of the variety of pottery found at Elden Pueblo these artifacts are here divided into six groups, classified as follows: 1. Rude and undecorated, with no evidence of coil: 2. Corrugated and coiled; 3. Mat red ware; 4. Smooth red ware; 5. Polychrome; and 6. Black on white, or white on black, decorated with geometrical designs.

1. Rude, undecorated, and uncoiled.—The few objects referred to this type of pottery are very coarse and show no indications that they were made by coiling. They seem to have been fashioned by the hands, working a lump of clay and possibly smoothing it by the aid of a stone or some other implement. Finger prints and other markings occur on the surface of these vessels. They show no attempts at painting a glaze decoration and their shapes recall the very crude products of workmen who were unskilled in the work. Among the varieties of this form may be mentioned small crucible-shaped receptacles, like children's playthings, larger bowls, ollas, cooking pots, vessels with grooved handles like gourd ladles, and others. Images of animals, birds, quadrupeds, and even human effigies were pressed into shape from balls of clay.

FIG. 219.—Red corrugated vessel. Diam. 5½"; height 4½".

2. *Corrugated and coiled ware.*—Elden pottery of this group is very characteristic and is easily distinguished from the corrugated and coiled ware of other pueblos by its red color. We sometimes find examples, mostly bowls, vases and jugs, of gray color with black lustrous interiors, the corrugations being quadrilateral in shape, smooth on the surface, indentations shallow. The quadrangles are arranged in regular rows (fig. 219). Variants of this regular corrugation are many, the differences being mainly in form and relative size but the corrugations are so regular that some form of angular instrument must have been used in making them. There are also specimens of coiled ware in which the corrugations are very small, closely

crowded and apparently made with a stone, a piece of wood, or some pointed implement. The coiling often appears broadly separated by parallel grooves. Vessels made of corrugated ware have not as a rule thin walls like those characteristic of the San Juan, especially Mesa Verde cliff dwellings, and lack appended scrolls.

3. Mat red ware.—There are two types of red ware, both of which may be called abundant. The most common red ware has a bright burnished red color, generally more or less blackened in burning, with a lustrous black color on the interior. It takes the form of globular bowls, flat open food bowls, ladles with short handles, and jugs. The most abundant red ware specimens are globular jars. Some of the red food bowls have the opening pear-shaped, resembling a form of ladle. The lips of red jugs are outcurved and the neck is compressed, generally short. One of the most highly decorated examples of red ware with dull black decoration on the interior (fig. 222, a) has also on the outside white figures representing a circle of human hands or animal paws, which is rare among food bowls from Elden Pueblo. The designs on this bowl recall those on a vessel formerly used in the snake washing of the Hopi snake priests.

4. Smooth red ware.—The most unusual type of pottery and therefore the most instructive is the smooth red ware, commonly without decoration exteriorly but often blackened on the inside. This ware occurs sporadically in the pueblo region but is well known from the ruins in southern Arizona and is most abundant among modern Pima, Kwahadt, and Papago. The smooth red ware of Papago is generally decorated with geometrical patterns in black. Several specimens of both the smooth red ware and the corrugated have their interiors a lustrous black color, but with no designs. The white or gray ware bears black decorations.

We have this same condition of corrugated ware with inner surface black in the cemeteries of Pipe Shrine House on the Mesa Verde, a specimen of which has been figured elsewhere. The smooth red ware which is so abundant at Elden Pueblo is very rare in the Rio Grande and exceptional in the Little Colorado. The decorated red ware of Elden Pueblo is like the ornamented red ware at Homolobi and Chevlon, pueblos near Winslow, and is abundant in the ruins at Tuba City and on the road to Marsh Pass.

5. Polychrome ware.—Polychrome ware is very rare at Elden Pueblo, only a few sherds and but one or two small food bowls having been found in the collection from this ruin. This is astonishing when we bear in mind that collections made at Wupatki and Marsh Pass contain many specimens of this ware. As a general rule, polychrome

a

b

FIG. 220.—Two specimens of black and white ware. *a*, jug; *b*, canteen with glossy black and white decoration.
a, Diam. 5″; height 4½″. *b*, Diam 6″; height 5¼″.

FIG. 221.—Inside decoration of six food bowls of black and white ware showing variety of the geometrical decorations.

a, Diam. 4½″; height 2″. *b*, Diam. 9¼″; height 4¾″. *c*, Diam. 5½″; height 3″. *d*, Diam. 7¾″; height 4¼″. *e*, Diam. 6″; height 3″. *f*, Diam. 5½″ x 5″; height 3″; handle 6½″.

FIG. 222.—Decorations on the interior of food bowls. *a*, interior of dull red ware bowl; *b*, exterior of same bowl; *c-f*, interior decoration of black and white bowls.

a, *b*, Diam. 11½″; height 5″. *c*, Diam. 6¾″; height 4″. *d*, Diam. 7″; height 3¾″. *e*, Diam. 5¼″; height 2½″. *f*, Diam. 6½″; height 4″.

Fig. 223.—Characteristic designs on the interior of black and white food bowls.

a, Diam. 10¼″; height 6″. *b*, Diam. 8½″; height 5″. *c*, Diam. 8¾″; height 3¾″. *d*, Diam. 8½″ x 7⅞″; height 5″. *e*, Diam 8″; height 4½″. *f*, Diam. 3½″; height 2¼″.

ware in the last mentioned localities has the same geometrical symbols as the Elden Pueblo black on white ware.

6. *Black and white ware.*—The common decorated pottery at Elden Pueblo is the well-known black on white. In many instances the

FIG. 224.—Ladle of black and white ware. Bowl, $3\frac{3}{4}'' \times 4\frac{1}{4}''$; handle, $4\frac{1}{4}''$.

figures, which are here always geometrical, are white in color on a black base, but generally the reverse is the case, *i. e.,* the design is in black on a white base. The former may be called "negative." The distribution of the white and black commonly leads to patches or bands of checkerboard areas or mosaic patterns well illustrated in pottery from the middle Little Colorado valley. While the black decoration on white is mainly geometric, some of the designs are very intricate and beautiful. No realistic and few conventionalized figures appear. Decoration appears on the inside of food bowls and ladles (figs. 221 to 224), but are confined to the outside of vases (fig. 225), ollas, jars, and seed bowls. Many of the designs are modifications of the swastika or of the friendship sign. There are no ornamental figures on the outside of bowls and generally no broken lines. The

FIG. 225.—Decorations on two black and white vases.

bowls are hemispherical in shape, generally thick walled with square, round, or outcurved lips. In the last-mentioned, the outcurved inner rim when broad enough is decorated with simple designs differing from those of the body of the bowl.

The globular vessels of black and white ware called seed bowls are represented in the collection by several specimens, all highly decorated with scroll, frets, and other designs. These seed bowls are sometimes designated as globular vessels. They do not always contain seeds, for two specimens of red ware are filled with red and green pigments.

One of the most exceptional forms of Elden Pueblo pottery is an effigy vessel here placed under the black on white ware and shown in figure 226. This specimen was found upon the pelvis of what appeared to be a priest, as shown in figure 212. It is a most remarkable piece of prehistoric pottery, especially when it is borne in mind that it is made of archaic black and white ware and must therefore be very old. Its purpose is unknown, but it may have been a receptacle for medicine liquid employed in ancient rites and ceremonies. It was evidently

Fig. 226.—Vessel of black and white ware in the form of an unknown animal probably used in carrying medicine or sacred meal in ceremonies. Length 8½", height 6¾".

carried by a cord forming a handle attached to a perforated ridge between the stumps of horns or ears and to a similar ridge at the posterior end of the back.

It is not possible to identify what animal this effigy was intended to represent but it was a quadruped with divided hoofs at the extremities of short and stumpy legs, three of which are partially broken. The mouth is wide open and the head bears between the eyes and near the two ears the broken remnants of two horns. The body is almost globular in shape and is covered with designs in which sun emblems predominate. The designs repeat several times a circle with short extensions from the periphery. The circle among the Pueblo people symbolizes several supernatural conceptions in the Hopi art, among which are sun or sky god and earth, but whether the effigy jar represents either of these or some other conception is not evident.

The form of this effigy is not unlike an inferior specimen from the cemetery of Pipe Shrine House, Mesa Verde National Park, Colorado.[1]

The mortuary pottery from Elden Pueblo is allied in several features to that found in the great cliff ruins of the Navajo National Monument and also to that from the Marsh Pass, northern Arizona. It has some points of resemblance to that of an archaic prepueblo of the Mesa Verde culture that antedated the cliff dwellings of the Mesa Verde, as shown by specimens collected by the Bureau from the cemeteries at Far View Tower and at Pipe Shrine House of the Mummy Lake Group. It likewise crops out elsewhere at most unexpected localities in the Southwest, as in ruins in the Walpi Wash and Jedditoh Valley. Dr. Walter Hough has called attention to the similarity of pottery from McDonald Canyon, 22 miles south of Holbrook, to that from Elden Pueblo. It would seem to characterize the oldest culture of the central Little Colorado valley as well as that from the San Juan.

In conclusion, it should be pointed out that although many important contributions have been made in recent times to our knowledge of pueblo pottery and its distribution in prehistoric times, these conclusions are tentative, as there remain many unexplored areas in our Southwest, the pottery of which is unknown. One of these is the region west of the Little Colorado and south of the Grand Canyon, as far west as California and southwest to the Gila Valley. Elden Pueblo lies in the geographical center or heart of this extensive area, of which archeologically and ceramically we knew next to nothing up to the past summer. It is a type ruin strategically placed, adding new facts bearing on several problems of the prehistoric Southwest.

The likeness of Elden Pueblo architecture and ceramics to the oldest ruins in Arizona is very pronounced. It has many points of resemblance to the ruins in the Gila basin, apparently connecting them with the pueblos of the San Juan. The relative age of Gila compounds and San Juan pueblos is a problem we are as yet unable to satisfactorily solve. We now greatly need more information on the region between Elden Pueblo and the mouth of the Gila in order to show intimate connections between compounds and pueblos, but in this re-

[1] Smithsonian Misc. Coll., Vol. 74, No. 5, Fig. 107c.

gion we evidently have a meeting place of the types of prehistoric cultures in the northern and southern parts of Arizona.

RECORDING OF HOPI INDIAN MUSIC

At the request of the Starr Piano Co. of Richmond, Ind., Dr. J. Walter Fewkes superintended the recording of eleven Katcina songs of the Hopi Indians. Permission was received from the Office of Indian Affairs to take four Indians from their reservation to the Grand Canyon, where the recording was done. The singers represented the older generation and remembered Dr. Fewkes from the time of his studies among their tribe 30 years ago. The age of the singers insured the genuineness of the songs, many of the words of which are archaic and cannot be translated.

Dr. Fewkes noted with much interest the improvements that have been made in recording apparatus since his pioneer efforts along this line over 30 years ago. The Indians displayed no hesitancy in singing into the microphone and seemed quite pleased on hearing the result.

The original master records will eventually be deposited with the Smithsonian together with a set of the reproductions.

ARCHEOLOGICAL AND ETHNOLOGICAL RESEARCHES IN CALIFORNIA

The beginning of the year found Mr. J. P. Harrington, ethnologist, engaged in the work of following up what information is still available on the culture and archeology of the Mission Indians of southern California. Work was continued at ruined village sites in the Santa Ines, Ojai, and Simi valleys, and important discoveries were made revealing an earlier and a later coast Indian culture. The archeological sites of the region are being built over at an alarming rate, due to the settling up of the country by Americans, and thus are being lost forever to scientific investigation. The numerous pictographs, legendary stones, and place names were also thoroughly investigated. The rancheria of Misyahu in the Cañada de las Uvas was traced with more than usual success, although it seems that the cemetery has been washed away by the arroyo. The village consisted of 30 or more dome-shaped huts, from 12 to 20 feet in diameter, clustered irregularly on and about a great rocky hill, also of dome shape. Most of the hut circles can still be traced, but little was found under the surface of the floors. The near-by village of Sikutip had an entirely different arrangement, standing on the floor of a meadow beside a little swampy patch of ground that must have supplied the Indians with drinking water. Four large springs with pictographs traced on their rocky walls were located in the vicinity.

Fig. 227.—Housepit at Misyahu.

Fig. 228.—Indian fortification parapet at Santa Maria ranch.

Fig. 229.—Indian sign post.

Fig. 230.—Site of Simomo rancheria.

The country in Indian times was literally populated with "petrified" "first people," who lived at the beginning of the world and were transformed to stone for one reason or another, as special legends tell. There is a rocky pinnacle on a hilltop which used to be a person and evidently still has some life in it, for it is said to change its position at times, being seen by the Indians now erect, now tilted, now reclined. Two more petrified people shot arrows at each other across a canyon, with the result that one of the rocks is badly shattered. Another rock has horns. Even a whole house is petrified; a ghastly magical woodrat described as being some two feet in length is said to live in a rock

FIG. 231.—The old and the new: sheet iron stovepipes emerging from Hopi kivas. (Photograph by J. P. Harrington.)

that looks exactly like a primitive Indian wigwam. Another rock is the home of a magical beaver. Still another rock is a warclub left by the first people. Up another boulder two petrified rattlesnakes are crawling, seen as streaks in the formation of the rock.

The first people also left their barefooted or sandaled tracks. A good photograph was obtained of one of these footprints. It is a perfect human footprint, fourteen inches long. The god who made this print was heading toward the ocean.

The Indians also had the custom of placing a small rock on top of a boulder to mark the trails. One would go along the trail looking for these guides, which are always seen bobbing up ahead in conspicuous places. It is denied that Indians put all of them there; it is said that some of them were there always, even in the time of the first people. They were the Indian sign posts.

At Simomo the cemetery was located. The great shell refuse mound of this rancheria shines white and makes the site conspicuous at a distance. The houses were on this mound and as the mound grew,

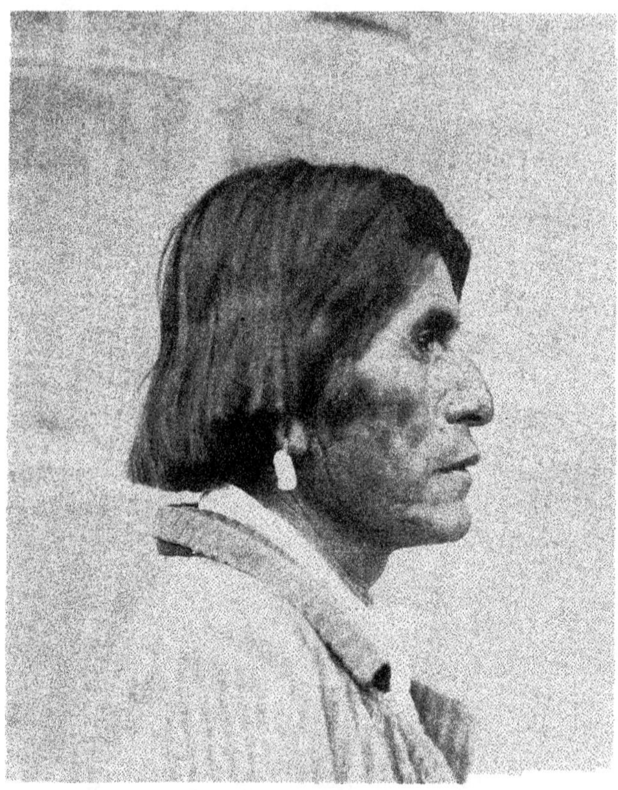

FIG. 232.—Kutqa, Chief of the Walpi Indians, who sang at the Grand Canyon. (Photograph by J. P. Harrington.)

by fresh accumulations of débris, the whole village gradually rose in elevation as the centuries passed. The spring was at the foot of the slope and had a strong flow of excellent water; it is now used as a watering place for cattle. No material of very early date was recovered from the reconnaissance of the Simomo site. At Shisholop site, the shell content of the mound was found to be unusually large and few artifacts were found in it. The shell species contained in these mounds are practically all identified as modern.

At the close of the California work, in May, Mr. Harrington was detailed to visit Walpi, Arizona, for the purpose of bringing a party of Hopi Indians to the Grand Canyon, where records were made of their songs. The Indians whose services were secured were Kutqa, Chief of Walpi Pueblo; Hunawö, head snake chief; Hungi, a leader in the snake ceremonies; and Kakapti, one of the best singers of the tribe. At the Grand Canyon they met their old friend, Dr. Fewkes, and proceeded to give their best renditions of Katcina songs to the strange machine invented by the white man to preserve these priceless songs of the remote Indian past. The Grand Canyon is a sacred place of Hopi mythology, and their visit to it took on almost a religious aspect. Songs were obtained of a dozen different kinds of Katcinas, which may be described as ancestral spirits. On the trip through the Hopi country native place names were gathered from the four aged men.

During the middle of the summer, Mr. Harrington assisted the Chief of the Bureau in the excavation of Elden Pueblo, Arizona. During the excavations, visits were received from a number of Hopi Indians and information of unique character was recorded from some of them. None of the visitors, however, was able to furnish the old Indian name of Elden Pueblo, although they gave without hesitation the name of the near-by Elden Mesa. But they were unanimous in thinking that the Elden inhabitants were ancestors of the Hopi. Practically all the artifacts taken from the ruin could be named in Indian and intelligently discussed by the Hopi.

ETHNOLOGICAL STUDIES AMONG THE IROQUOIS INDIANS

Mr. J. N. B. Hewitt, ethnologist, Bureau of American Ethnology, spent the period from May 25 to June 29, 1926, among the Iroquois Indians living on their reserves in the vicinity of Brantford, Ontario, and at Caughnawaga near Montreal, Quebec, Canada.

While on the Grand River Grant to the Six Nations, near Brantford, Ontario, Mr. Hewitt resumed his intensive researches relating to the content and analytic interpretation of the Onondaga, the Mohawk, and the Cayuga native texts recorded by him in former years, relating to the several institutions of the Federal League of the Five (latterly Six) Iroquois tribes, including the contents and the structure of the noble Chants and Rituals of the impressive Federal Council of Condolence for deceased Chiefs and the Installation of their successors.

With the aid of Chief John Buck, an Onondaga-Tutelo mixed-blood, as an Onondaga informant and interpreter, and Chief (retired)

Alexander G. Smith, a Mohawk speaker and informant, Mr. Hewitt obtained a fine Mohawk version and literal translation of the remarkable Requickening Address of this famous Council.

The psychological insight of the framers of this wonderful ritualistic address is without question unsurpassed in any other composition of its kind in any other literature of the world. Its primary purpose is to thwart the ultimate aim of Death—the ultimate destruction

Fig. 233.—Mr. Joshua Buck, (obiit 1923), Onondaga-Tutelo, Iroquoian stock, ritualist and native physician.

of all living—and to restore the mental equipoise of one who has been stricken with inconsolable grief through the loss of a kinsman or kinswoman, or a beloved ruler, a chief, by the ruthless hand of Death. One so stricken gives vent to extravagant expressions of grief and sorrow, lasting weeks, and months and even years, by self-torture and self-humiliation, by foregoing food and drink to the verge of starvation, by denying himself the ordinary comforts and needs of life, and even by sitting among the ashes of the hearth and casting the ashes and coals over his head and person, thus filling in time the eyes, the ears and the throat with them, in such manner that the environing world of persons and things are no longer objects of sense; and the vitality becomes so reduced in time that the sufferer may well be regarded as moribund. To redeem a grief-tormented sufferer from such a state of collapse is the set task of the Requickening Address.

About the year 1570 of the Christian Era, five linguistically cognate Iroquois tribes, occupying at that time chiefly the central and the eastern portions of the present State of New York, united in establishing the historically well-known League or Confederation of the Iroquois for the avowed purpose of securing Health and Peace, Justice and Righteousness, Order and the Force of Personality (Orenda), as the bases of a beneficent Commonwealth of peoples.

The tribes entering into this organic unity were the Mohawk, the Seneca, the Onondaga, the Oneida, and the Cayuga. At that time these tribes were also united in a more or less close compact with the noted Neutral Nation of the Iroquoian linguistic stock, another federation of tribes, which in the person of the famous Chieftainess, Djigonsa'sen (the Wildcat), took an active part in the conferences and the deliberations resulting in the establishment of the League of the Five Iroquois Tribes and also of the other compact with the Neutral Nation. But for some reason, yet unknown, this alliance was not wrought into a permanent organic institution, and so, in the structure of the Iroquois League as known to us, there appears no concrete indication of this former important alliance; not even nominal recognition of the Neutral Nation exists. There are, however, some essential features of the structure of the Iroquois League which a uniform tradition ascribes to the helpful work of this broadminded stateswoman and Chieftainess of the Neutral Nation.

Political and religious organizations of the Iroquoian peoples function only through the interaction of two important complementary principles which are embodied in definitely organized groups of persons. This is true of the tribal and of the federal organizations. These two principles are the male and the female functions.

The unit of the tribal organization is the clan. The smallest number of clans in any tribe is three. The Mohawk and the Oneida each have this number. And in either tribe the three clans are grouped in two complementary units, the one representing the Male or Father Principle, and the other, the Female or Mother Principle. Each of these units is usually called a Phratry or a Sisterhood of Clans. The nine clans among the Seneca, the Onondaga and the Cayuga are, in like manner, respectively grouped into two complementary units composed of four and five clans each, the one unit embodying the Male, and the other, the Female Principle.

The highest unit of organization of the federal League of the Iroquois was the Sisterhood or Phratry of Tribes. Two Sisterhoods of Tribes composed the League. The five tribes mentioned above composed these Sisterhoods; in the one were three tribes, namely the

Fig. 234.—Chief (*Merit*) John "Smoke" Johnson, Mohawk, Iroquoian stock, annalist and ritualist.

Mohawk, the Seneca and the Onondaga, in the other were two tribes, namely, the Oneida and the Cayuga; the first Phratry embodied the Father or Male Principle, and the second the Mother or Female Principle.

Viewed from the federal standpoint, what affects the integrity or the welfare of a clan of a tribe affects in like manner the entire Sisterhood or Phratry of Tribes to which that tribe belongs. So the death of a Chief of a clan of a tribe makes mourners of the entire

Phratry to which such tribe belongs, and this fact automatically makes the complementary Phratry Redeemers or Restorers. The mourners are described as *Those whose minds are prostrate,* and the Redeemers as *Those whose minds are virgin* or *unaffected,* and so in a position to restore to full life those who mourn.

In all formal public assemblies or councils, the Phratry of Tribes representing the Father or the Male Principle occupies a side of the real, or the imaginary, fire opposite to that occupied by the Phratry of Tribes representing the mother or female principle.

In formal public assemblies the Male or Father side of the tribe or of the League is addressed as a single personality by the pronoun, "thou," and by the terms, "My Father," "My Father's Brothers," "You, Three Brothers." The Female or Mother side of the tribe or of the league is also addressed as a single personality by the pronoun, "thou," and by the terms, "My Offspring" or "My Children," 'My Weanling," and "You, Two (latterly Four) Brothers." These examples apply specifically to the League institutions. It is these figurative appellations which are employed in the Rituals and Chants and Addresses of the Council of Condolence and Installation. To understand these dramatized lyric compositions the terms in the foregoing examples must be kept carefully in mind.

These brief interpretative comments will enable one the better to grasp the significance of the contents of the Requickening Address in question. This formal lyric address is composed of 15 themes or burdens of hurts to life learned from human experience, and the asserted healing of each hurt by the use of the appropriate remedial means by the ritually prescribed agent, the Celebrant.

Each of these Themes or Burdens is set forth through a formula common to all, with one or two exceptions. First, a specific type of hurt arising from grief at the loss by death of kindred is made; then, this type of hurt is directly asserted as affecting the mourner present; and, lastly, the Celebrant executes the symbolic act which at once heals the hurt or removes its cause.

From one to four strings of wampum about four inches long (the proportion of white to purple beads varying with the theme), accompany each of these Themes or Burdens when they are in use, and they are hung upon a horizontal pole immediately in front of the celebrant Orator. As he ends the recital of a Theme he sends by the hand of his assistant the accompanying wampum string or strings to the opposite or mourning side of the Council Fire, where they are hung on another horizontal pole in front of the Speaker for that side.

The Council of Condolence and Installation requires that if the Father Phratry or Sisterhood of Tribes is the mourner then the Mother Phratry must become the unaffected one, and vice versa. It is the duty of the *unaffected* Phratry to act as the Celebrant of the Rites of the Council of Condolence and Installation.

In this remarkable Council five Rituals are employed, and four of these Rituals, when in use, are divided into two portions in such wise that there necessarily results a perplexing interlocking of one Ritual with another. This curious fact, for which the present writer has so far found no satisfactory explanation, has never yet appeared in print, in so far as the writer's knowledge goes.

The Requickening Address opens its first theme with a frank recognition of the Creator and Source of Life and with an expression of solidarity with him.

Specific names are applied to these Themes or Burdens of the Hurts of Life. In their order these Themes are as follows:

I. Tears. II. One's Ears. III. One's Throat. These three constitute the First Section of the Address, and they are used at the "Edge of the Forest," where the Fire of the Welcoming is kindled, and where the mourners first meet their *unaffected* guests, and they intone the Chant of Welcome. But the final Section of twelve Themes of the Requickening Address is not used until the other set Rituals but one have been recited and answered by the mourners, and is virtually the closing Ritual of the Council.

The intent of the Themes of the first Section is to restore to their normal condition the sight, the hearing, and the vocal organs of a grief-stricken mourner such as is mentioned previously in this paper. After this ceremonial cleansing and revivifying, he is prepared to meet the Condolers in the Principal Place of Assembly. Now he can see, he can hear, and he can talk.

The names of the Themes of the final Section are as follows:

IV. "Within the Breast (or Body)." The shock of deep grief has displaced the internal organs and they are awry, inducing much reduced vitality, with impending dissolution.

But the Celebrant gives a draught of the Waters of Pity, pressed from many words of sympathy expressed, to the sufferer, and as these waters reach the parts affected they quickly replace the disturbed organs and normal vitality is restored.

V. "The Trail of Blood from the Death Mat." "Verily, thou dost writhe in the midst of blood."

But the Celebrant does "wipe away the blood-stream from thy mat," using the "soft skin of the spotted fawn," so "that when thou wilt return to thy mat, it will be in the fullness of peace, and it will be spread out in contentment."

VI. "The Thick Darkness of Night" covers one. "Now thou dost not know the Light of Day upon the earth."

But the Celebrant comforts the mourning one, saying, "We cause it to be Daylight again for thee . . . . the daylight will be fine, shining in perfect peace, and thou wilt again look upon the handiwork of the Perfector of our Faculties . . . . spread out richly upon the earth."

VII. "The Sky is Lost" to the mourner. "Now, the Sky is completely lost to thy mind. Thou knowest nothing of what is taking place in the Sky."

But the Celebrant comforts with the words, "We cause the Sky again to be fine for thee; it will be beautiful, and thou wilt think contentedly as thou wilt again look upon the Sky."

VIII. "The Sun is Lost" to the mourner. "Such a person knows nothing about the Sun in its movements, nothing of its drawing nearer and nearer to him."

FIG. 235.—Chief Abram Charles, Cayuga, Iroquoian stock, annalist and ritualist.

But the Celebrant comfortingly says, "We now replace the Sun in the Sky for thee . . . . and when the time for the dawning of a new day comes thou wilt see the Sun perfectly when it will rise . . . . thy eyes will rest upon it as it draws nearer and nearer to thee . . . . When the Sun will place itself in mid-sky, then around thy body rays of sunlight will abundantly surround thee."

IX. "At the Grave—the Heap of Upturned Clay." The shock of grief "thrusts a person aside to the place where arises the mound of earth which covers the one in whom one's mind confided and received support. . . . . Unhappily one thinks, for one's mind lies there beside the grave of uptorn earth. There it is shaken and rolled about on the ground. . . . . It knows nothing else."

But, the Celebrant in comforting terms says: "We now level the uptorn earth over the place where rests the one thou didst trust for words of wisdom. . . . . A fine slab of wood do We, the Three Brothers, place over these; and gathered moss and grasses plucked up; for verily, there are two things which are done by the Day and also by the Night; the one is that, should it so be, that should the Day put forth fierce rays of sunlight, they shall not, therefore, pierce through to him; the other is, that should it so be, that driving rains fall heavily upon them by night, these too shall not go through to where he lies—though nothing save the bones be visible there. So then there will his bones lie peacefully. . . . . So that even for one little day thou shouldst think your thoughts in contentment."

X. "Twenty (Strings of Wampum) are the Penalty for it" (Homicide).

But the Celebrant comforts with these words among others: "Do thou know it. . . . . My Offspring (*i. e.*, the Mother Side) . . . . that now, we, Three Brothers take that up now, and that, let them say it, 'Now, we wrap up thy bones, one and all (as a protection), fixing the penalty of twenty (strings of wampum) on them (for any hurt done them).'"

XI. "The Fire of the Home, Around Which the People are wont to go to and fro."

"Our grandsires, now dead, whom our minds trusted implicitly, decreed, because they failed to perceive the lineaments of its Face, the Face indeed, of that Being that abuses us ceaselessly by day and by night, of that Being of Darkness, crouching hard by the barklodges, goes about with uplifted bludgeon—with its couched weapon at the very top of our heads—eagerly muttering its fell purpose, saying, 'I, I it is, who will destroy all things,' they decreed, I say, that they would name it the Great Destroyer, the Being Without a Face, the Being Malefic in Itself—Death. So, putting forth its sinister power in thy booth of bark, it struck down one therein on whom thou didst depend confidently for words of wisdom and for kindly service, and there is therein now a vacant mat. By this blow It scattered widely the fire-brands of thy fire, and in mocking derision the Great Destroyer has stamped out thy fire."

But the Celebrant orator utters these words of cheer and comfort: "So now, do thou know it, my Offspring, that we, the Three Brothers, having perfected our preparations, say, 'Now, we gather together again the scattered fire-brands of thy home fire, and so indeed, we rekindle thy fire for thee; and the smoke thereof shall rise again; that smoke shall be fine, and it shall even pierce the sky (smoke=the activities of life). . . . . Now, indeed, we raise thee again to full stature.'"

XII. "Woman."

"Now, another thing. It is that wherein the Perfector of our Faculties who dwelleth in the sky, established it, in that He desired that He should have assistants above and even down to the earth; that some shall devote their care to the matters which pertain to the earth 'I have ordained,' He says, 'one and all.' It is that, in fact, that He, therefore, caused the person of our Mother—the Woman—to be of noble worth. He designed that She shall be entrusted with the duties pertaining to the birth and the nurture of mankind, and that She shall circle around the fire in preparing,—that she shall care for,—that by which life is sustained.

"And that, too, is a calamity, that, it may be, the Great Destroyer will make a swift stroke there in the ranks of our mothers, felling one there. The evil is that a long line of (unborn) persons will be blotted out, in the many-fold lines of grandchildren who would have come from her loins. . . . . Now, furthermore, the minds of all those who still remain have fallen and are prostrate."

The Celebrant orator says: "So now, furthermore, the Three Brothers, having perfected their preparations, let them now say, "now, then, we raise up your minds again and cheer and comfort them. This, indeed, shall come to pass; and you shall now again devote yourselves to your several cares and duties."

FIG. 236.—Onondaga woman and infant on cradle-board. Syracuse, N. Y.

XIII. "Hoyā'ne'r: the Federal Chief."

"They whom our minds highly respected made a decree. They strictly forbade what they denominated 'tossing it over the shoulder.'"

The Celebrant continues: "Now, Sayā'ne'r (thou, Federal Chief), we, Three Brothers, having perfected their preparations say to thee: 'Do thou, listening, hear full well what is said to thee by thy Niece (our Mothers) and by thy Nephew (the Men) of thy Clan. The reason that this must be is that to them also has been given mind—the ability to judge right things.'" (Here follow the details of several admonitions to the Federal Chief for derelictions in official duties by the Mothers, and then by the Mothers and the Men of the Clan). The Celebrant continues: "That too is another grave matter, shouldst thou toss over they shoulder the admonition of justice and right conduct by which all the people live. Should this come to pass, then know that the time is near when the feet of thy people shall hang over the abyss of the sundered earth (of impending ruin). Should this misfortune take place there is no One under heaven who is able to draw them out of it save One, the Creator of our bodies, who has the power to do this and to aid all in distress."

XIV. "The Loss of Reason—Suicide."

"Our grandsires made another decree forbidding another thing, and that is, that the mind should not be permitted to lose its reason by nursing the memory of deep sorrow. . . . .

"That verily is a grave matter when the mind loses its reason because a grievous loss has befallen a person; for, it is well-known that there grow things on the earth, over which the other things of the earth have no power, but which have the power to end the days of a human being, should there be any weakening of the mind."

The Celebrant admonishes: "Should this come to pass, it will be a source of death to the people. So, for this reason, we, Three Brothers, forbid this thing to you, so that now you can again think in contentment."

XV. "The Ever-burning Signal Torch" And "The Short Purple Wampum String of Notification."

"When our grandsires, now long dead, conjoined their affairs they made a decree, saying: 'Here we place two rods horizontally side by side, and held fast by them fix the Ever-burning Torch.'

"And here, over the small horizontal pole of partition where it pierces the bark-wall we suspend a Pouch of the Skin of Dji'nho^{n}'do'hyĕ^{n}''ă', of indifferent fur, in which we put the Short Purple Wampum String of Notification.

"And we, so many as our Council Fires number, have an equal right to these two objects. And they shall be of essential use wherever a grevious thing (=death) has taken place, or where instant peril, menacing death to one and all, is seen creeping like a serpent close at hand.

"It matters not on which side of our Council Fire (the Father Side or the Mother Side) the evil will be, these two things shall be vitally important.

"In either case, the one whose mind is *unaffected* (ago'nigo^{n}'ga''te') shall grasp the Ever-Burning torch from its holders and also unhang the Pouch of the Skin of Dji'nho^{n}'do'hyĕ^{n}''ă' containing the Short Purple Wampum String of Notification, and he shall at once start going through the Lodge of the League so that the message he bears shall quickly be carried through the entire Lodge and all the Council Fires be made acquainted with it. . . . . His going shall be done in such manner that there shall be no traces—no 'forms'—of lying down along the path. . . . .

"So, now, verily, with respect to thee, thy Ever-burning Torch is removed from its place and thy Pouch with the Short Purple Wampum String of Notification is unhung for thee, all this because of the grievous calamity which has befallen thee."

The Celebrant orator comforts with the following:

"Now, we, Three Brothers, again suspend the Pouch of the Skin of Dji'nho^{n}'do'hyĕ^{n}''ă', of indifferent fur, over the Small Horizontal Pole where it pierces the Bark-wall of the Lodge, containing the Short Purple Wampum String of Notification which we have replaced and we also set back the Ever-

Burning Torch between the Two Horizontal Poles—these things to which we, so many as our Council Fires number, have an equal right.

"For, it may be that thou thyself will see close at hand the nameless Being of our destruction then verily thou shalt go at once and taking the Ever-burning Torch and the Pouch with the Short Purple Wampum String of Notification, and thou must go quickly along through the Lodge of the League so that in the shortest time possible all shall be notified.

"So, that verily will make it possible that thy two nephew-niece groups of kindred, and thy grandchildren as well, may live and think in contentment. So that, therefore, for one little day, you my weanlings severally may live pleasantly in the days that are coming.

"So, perhaps thus, let the Three Brothers do, so-called ever since they perfected their affairs. So, now then, do thou my weanling know that these are the sum of our words. Now then we kneel before you (thee) reverently. And now we will know presumably that the full number of our words have been realized in deed. We have now completely set your affairs in order again.

"Thus, then, did they whom we greatly revered do when they united their affairs; they made a decree, saying, 'It matters not on which side of the Council Fire which is between us the need be, it shall be possible that they shall again set his face fronting the people, that they shall again raise him up (requicken him), that they shall again name his name, and that then also he shall again stand upright before the people. In this we are following the ceremonial path. So, now furthermore, let them say, 'Do thou now point out to us the one who shall be again a colaborer with us.'

"Now, my weanling, do thou know that we, Three Brothers, have completed the Ceremony.

"And, now then, that which (short purple wampum string) notified us is on its way back to thee.

"Know thou, then, that now soon thy Father's kinsmen will arise to leave for home, and that there, then, at the edge of the forest will they lay their backs."

FIELD STUDIES OF INDIAN MUSIC

In July, 1926, Miss Frances Densmore, collaborator of the Bureau of American Ethnology, went to Neah Bay, Washington, to continue her study of Makah music. Neah Bay is situated on the Strait of Juan de Fuca, near the end of Cape Flattery. Encircled by the Olympic Mountains, it can be reached only by water, and during the winter months it is practically shut off from communication with the outside world. In this isolation the older Indians have kept their former beliefs and traditions with remarkable clearness, though they are anxious to be "civilized" and adopt the best of the white man's ways. Many years ago the Spaniards visited Neah Bay and the location of their fort is near the village. Traces of Spanish ancestry are seen in some of the Indians and occasional songs bear a resemblance to those recorded on the Mexican border. The intermarriage of the

Makah has been chiefly with the Clayoquot on the west coast of Vancouver Island and with the Quileute who are their nearest neighbors on the Pacific coast. Songs of these tribes were recorded as well as those of the Makah, the entire number of songs obtained being 146. The classes of songs comprise those of war, treating the sick, dances, legends, and songs for children. Eight Clayoquot songs for subduing the waves were recorded, as well as songs for fair weather and a plentiful supply of herring. Two aged people, one a Clayoquot and the other a Makah, said they had never heard of anger on the part of any " spirit," nor of any attempt to " propitiate the spirits." The words of the songs addressed to the breakers were said to " make them ashamed." These songs were sung during a storm, one containing the words:

Your teeth that are trying to get these people are long and homely.

Another song contained the words, " Please be still, you have treated us so badly." Another, sung in rough weather, had the words, " What beautiful weather this is! It is as calm as when the dogfish are moving in." The informant said, " Even when we were in the highest breakers, if we sang these songs it seemed as though very soon the water was smoother."

Although the caste system was rigidly maintained among the Makah, it is interesting to note that songs learned from slaves were recorded and that a slave was allowed to sing a song of his own country at the Klokali. A song received from a shell was recorded, as well as songs concerning the crab, chipmunk, shark, and whale.

The Klokali received special attention. This was an important gathering in the old time and terminated with dances imitating the actions of animals; it was understood, however, that the dancers represented the human beings who were the *ancestors* of the various animals and did not represent the animals themselves. Songs with representations of the wolf, deer, blackfish, and the wild white geese were recorded. One of the most important dances was that representing the elk, a dance which had no songs, the pounding on sticks and drum being the only accompaniment.

A remarkable opportunity for seeing native pageantry and hearing Indian songs occurred on Makah Day and at the rehearsals for that event. The dancing on this annual occasion is intended chiefly to transmit to the younger generation a knowledge of the old beliefs which are dramatized and presented by trained performers. Each dance has its costume and other paraphernalia. Spears decorated with hemlock were carried in the war dance, elk antlers by the elk dancers, and robes of

FIG. 238.—Charles Swan in dance costume. (Photograph by Miss Densmore.)

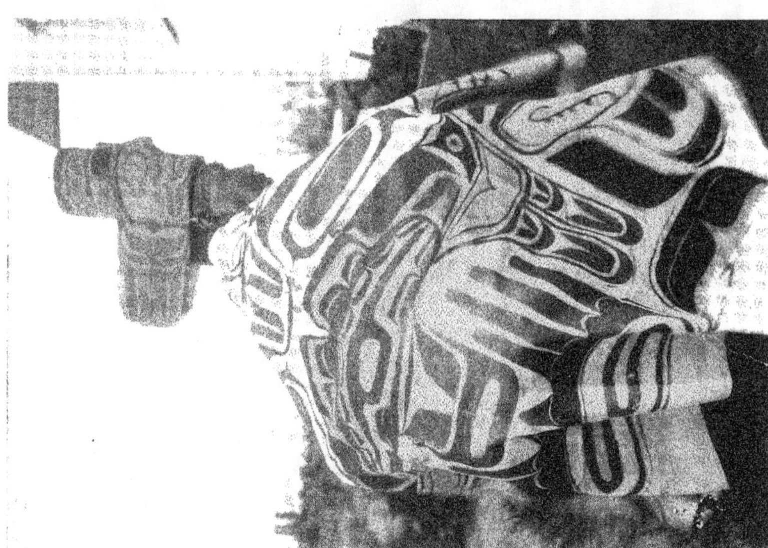

FIG. 237.—James Hunter in dance costume. (Photograph by Miss Densmore.)

bear skin were worn in the "brown bear dance." The thunderbird dancers were followed by women who represented the lightning, known as the "belt of the thunderbird." The whale dancer carried green boughs on his back representing the dead bodies which, according to an old legend, were carried by a man preparing himself to be a whaler. Head-dresses of wood, cedar bark, and feathers were worn,

Fig. 239.—Mrs. Parker. (Photograph by Miss Densmore.)

some being decorated with shells. The legends connected with the dances, as well as their songs, were obtained. James Hunter (fig. 237) in a dance position wears a head-dress representing a duck, and Charles Swan (fig. 238) wears a similar ornament representing a wolf. Many Quileute and Indians from Vancouver Island attended this gathering, and, after the dramatic dances were concluded, each group sang its own songs, making it possible to compare the songs and manner of rendition from these widely separated localities.

The photographs at Neah Bay consisted of 12 portraits, numerous landscape and dance pictures, and a series of 20 posed pictures showing the successive postures in various dances. The 27 specimens obtained included a very old rattle used by a successful whaler, a war knife of whale bone, and a war club of the same material with feathers on the handle " to break the force of the blow " to the arm

FIG. 240.—Albert Irving. (Photograph by Miss Densmore.)

that wielded it. Other specimens were a braided rope of whale sinew and a portion of a whale vertebra used for smoothing the bottom of war and whaling canoes. Old halibut hooks of elk bone were obtained, also an old adze with blade of elk bone, this being the type of implement used in felling trees before the Makah had metal axes. The collection included a small reed instrument commonly called a whistle which, when blown, was concealed in the mouth of a thunderbird dancer, and another wind instrument the sound of which was said to "come through the seams" where the halves of the wooden tube were joined together. An interesting specimen is a "baby-carrier" woven of cedar bark and used only during the first four days of a child's life. In this carrier is placed a doll with pads of cedar bark around its face as formerly used to flatten the forehead and shape the cheeks and chin. A very old wooden cradle used for infants above the age of four days was also obtained.

FIG. 241.—Mrs. Sophie Wilson. (Photograph by Miss Densmore.)

The collecting of plants comprised 58 pressed specimens with descriptions of their economic uses. A majority of these were medicinal, including the entire set of remedies inherited by Mrs. Parker (fig. 239). Her children do not regard the old ways and this knowledge of plants would have perished with her.

The study of general customs included a detailed account of the making of the native canoe (dug-out), the preparation of native dyes, and a description of the whale hunt and its weapons, valuable information on the latter subject being given by Albert Irving (fig. 240).

Numerous legends concerning the mythical hero called Hirokwati were also recorded, with their songs.

On August 24, Miss Densmore arrived at Chilliwack, British Columbia, to record songs among the Indians gathered at the hop-fields where they were employed as pickers. These Indians were more than 1,000 in number. During the next two weeks she obtained more than 120 songs, 28 of which were recorded by old medicine men who use them at the present time in treating the sick. The 20 Indians who recorded songs were from 16 localities distributed over a territory extending about 400 miles north and south and about 150

FIG. 242.—Indians playing Slahal game at Chilliwack, B. C. (Photograph by Miss Densmore.)

miles east and west. Some live in mountain regions whence they go to the railroad on pack horses, while others were from remote settlements on islands or along the Pacific Coast. Among these localities were Cooper Island, Church House, Powell River, Metlakatla, Port Simpson, the Nass, Skeena, Thompson and Fraser rivers; and the west coast of Vancouver Island. Mrs. Sophie Wilson, a singer from Church House on Butte Inlet, is shown in figure 241. Among the interesting songs is one said to have been sung by a seal, another song is about a mountain goat, and another is the song of a doctor who talked with a whale and received from it the power to heal the sick.

An interesting event of each Sunday at the hop-pickers' camp was the playing of the Slahal game, its songs accompanied by pounding on drums and planks. The playing of this game around a huge bonfire at night was a spectacular sight. In figure 242 the leader of one side has the "bones" in his hand and is preparing to give them to a member of his side for hiding. The songs occur while the location of the marked bone is being guessed by the opponents.

The material gathered at Chilliwack affords a remarkable means for comparison among the tribes represented, as well as with songs of tribes previously studied.

Among the important results of this year's work is the obtaining of data on the manner of composing songs by Indians, as distinct from the "receiving of songs in dreams." It appears that the composition of songs was assisted by walking or by the motion of a swing. Two persons sometimes collaborated in the composition of songs. Such persons were interviewed and their songs were recorded.

ARCHEOLOGICAL WORK IN LOUISIANA

Mr. Gerard Fowke, special archeologist, conducted important fieldwork on Indian mounds near Marksville, Louisiana.

It is a matter of actual knowledge that the Natchez Indians built many large mounds along the bluffs bordering the Mississippi on the east, and that this practice continued, though perhaps in a diminishing degree, until the period of French occupation of the territory. But it is not of record that this tribe, or a colony from it, moved permanently to the west of the river until within historic times.

Also, it is now an established fact that the small mounds so numerous over much of Texas, Louisiana and Arkansas, extending in great numbers down the Red River to Alexandria and, sparsely, even beyond that point, are the work of the Caddoans. The latter tribes, as far as we know, did not erect the quadrilateral or flat-topped mounds such as are prevalent to the eastward. This leaves unexplained as yet the comparatively few such structures found along the Red River, always near the stream, reaching up the valley nearly or quite to Texarkana. These may be due to Natchez, or others, who once lived here for a time but left no further traces.

Between the known territory of the Natchez and that of the Caddoans, that is, between the Mississippi and the vicinity of Alexandria, La., is a strip of country which, so far as its ancient remains suggest, did not belong to either of these people, and yet there is some resemblance to both. Whether these works indicate mingling of the two, or an overlapping of boundary lines at different periods, or whether there may have been another people in between them who borrowed somewhat from the customs of both, is not determined.

The low lands subject to overflow from the two rivers are, of course, extremely fertile, but as a rule the soil on the uplands is not productive, is so flat as to be swampy much of the time where not artificially drained, and apparently is not of a nature to invite a primitive people whose sustenance must depend in large measure on agriculture. That there were, nevertheless, settlements of Indians here and there is shown by the tumuli, sometimes more than twenty feet deep; but these are few and far between, and point only to small settlements with much open territory between them. It was a matter

FIG. 243.—Marksville. Trench partly excavated, Mound 4.

of some surprise, then, to find, a mile east of Marksville, a group of earthworks of such extent and character that they would be noticeable even in a region where similar structures are abundant, in Ohio or Georgia, for example. They extend for more than a mile along a bayou known as "Old River," which opens at either end into Red River, and, as its name indicates, is recognized by the present population as having once been the channel followed by that stream. There is little doubt that it flowed here at the time these earthworks were constructed, although at present it is several miles away. This fact, however, has no bearing on the age of the remains; such changes are frequent and extensive.

Most of the works are on the upland, close to the river bluff. Beginning at the south there is first a square, flat-topped mound eighteen feet high, covering a fourth of an acre; the sides are still so steep as to be difficult of ascent. It stands close to a tributary bayou beyond which, toward the south, no mound or other structure exists within three or four miles. Next in order is a nearly circular inclosure measuring somewhat more than three hundred feet across. This has a flat side bordering a moat lying along the outer margin of an embankment which is in the form of a rude semi-circle with each

Fig. 244.—Marksville. Mound 8, showing structure.

end at the top of the bluff; in fact, it seems that part of it has fallen with the caving of the bank. The length of this embankment is thirty-three hundred feet. Inside of it are two flat-topped mounds, each with its highest point thirteen feet above the surrounding surface. One of them covers about three acres, the other being somewhat smaller. It would seem that they were once quadrangular, but their outlines are so altered by cultivation and erosion that this is not certain. There were also within the enclosure a low mound eighty or ninety feet across, one somewhat smaller, and a conical mound twenty feet high.

Beyond this is another enclosure five hundred feet long, forming a fourth of a circle, one end on the river bluff, the other end at the bank of a ravine. Both ends are shortened by the banks caving in. There is a wide, deep moat at the outer side. The space enclosed is about four acres. Within the wall is a flat-topped mound six feet high and one hundred feet across.

Between these two embankments are lodge sites; low, circular embankments with a shallow exterior ditch from which the earth was taken to build them. These do not seem to be a part of the main works, but to pertain to a different period or a different people.

Fig. 245.—Mound inside of enclosure. Marksville.

Over a space of nearly a fourth of a mile to the east of the last mentioned enclosure no structures were erected; beyond this area, close to the margin of the bluff, are six mounds of the so-called "conical" form. Finally, in the bottom, on land subject to overflow, are four large quadrangular, flat-topped mounds from three to thirteen feet high, and three small, low, ordinary mounds, all enclosing a rectangular space of two acres. The three small mounds are on the edge of a slough.

It was a natural supposition that such conditions would be favorable for research, that works of such magnitude would repay investigation.

Six of the mounds were carefully excavated to an extent that disclosed all they had to reveal. Two, the smallest of those opened,

Fig. 246.—Pipe and vase found in Mound 4.

Fig. 247.—Left to right: Vase from Mound 4, vase from Mound 8, and plummet from Mound 10.

were simply piles of earth, containing nothing whatever to signify the reason for their existence. Had it not been for some charcoal, worked flint, and broken pottery, they might have been deemed natural formations. Two others were merely foundations for houses, but each showed plainly that at least two, perhaps three, houses had been erected on the same site, the later ones on new foundations above the old.

The last two mounds, one fifteen feet, the other twenty feet high, were tumuli. In the first a number of bodies had been placed either on the natural surface or in shallow graves, and the mound erected over them, apparently as a continuous operation. There was no evidence of a burial in the body of the structure.

The largest mound was quite different in its arrangement. There were no skeletons at the bottom, the first step having been to build up a platform of earth to an elevation of eighteen or twenty inches. On this a number of bodies had been placed, how many could not be even guessed, as all the bones were completely decayed and had no more consistency than wet chalk or ashes. They entirely covered a space several feet across. The mound had then been carried up over them to a height of about fifteen feet. Scattered through the earth for this entire distance were traces of burials, made as the work progressed, so the construction must have been intermittent or desultory, although there was no stratification or other marks to demonstrate this. Finally, the tumulus in this stage was finished.

Later, though it was impossible to estimate how much time may have elapsed, two large graves were dug in the top of the mound, several bodies placed in each, the graves filled, and the mound then added to until it was five feet higher, with a proportional increase in diameter. But the addition was largely on one side. The point beneath the apex as it was first built, was several feet away from the corresponding point in the mound as completed.

No implements or manufactured objects of any character of stone, shell, or bone were found in any of the mounds. There was some pottery—not much—and that crude, and except for two pots, neither of which is of two ounces capacity, all of it was broken. With a few skeletons were fragments of pots which had been entire when deposited, but had been broken from pressure. Some of them had decorations of incised or impressed lines and figures. A study of these, and comparison with other pottery in the southern territory, may furnish a clue as to the tribal affiliations of these Mound Builders, but unless so, there seems little hope of discovering their identity.

SMITHSONIAN MISCELLANEOUS COLLECTIONS
VOLUME 78, NUMBER 8

THE FLORA OF BARRO COLORADO ISLAND, PANAMA

BY
PAUL C. STANDLEY

(PUBLICATION 2914)

CITY OF WASHINGTON
PUBLISHED BY THE SMITHSONIAN INSTITUTION
MAY 20, 1927

The Lord Baltimore Press
BALTIMORE, MD., U. S. A.

THE FLORA OF BARRO COLORADO ISLAND, PANAMA

By PAUL C. STANDLEY

The logical position of the Republic of Panama as a center for educational work has been recognized throughout the American countries ever since work was begun upon the Panama Canal. It is therefore particularly appropriate that there should be established here in the Canal Zone a laboratory for tropical research in the biological sciences. On April 17, 1923, Barro Colorado Island, in Gatún Lake, was set aside by the Governor of the Canal Zone as a permanent reservation, to preserve in a primitive state the animal and plant life of the region. This result was accomplished largely through the personal interest and effort of Dr. Thomas Barbour and Mr. James Zetek, the latter now resident custodian of the island.

Through the persevering effort of these two persons, also, there has been constructed upon the island a commodious and substantial laboratory with ample living quarters, in which one may enjoy every bodily comfort while carrying on investigations of the highly diversified fauna and flora. Although secluded from the distractions of such towns as Colón and Panama, one is within easy reach of their conveniences. From the windows of the laboratory, situated at the top of a high, steep slope, one may see all day long an ever-changing procession of the world's ships, passing almost before the door.

The laboratory is operated by the Institute for Tropical Research, under the direction of the National Research Council, and a cordial welcome is extended to scientists who wish to make serious use of its facilities. The expenses of administration are borne in part by subscriptions of scientific and educational institutions, and in part by private individuals.

Barro Colorado, the largest island of Gatún Lake, covers approximately six square miles, being about three miles in greatest length and width. It is of artificial origin, and before the water was turned into the lake formed merely a part of the hills along the Chagres River. Near the laboratory site ran one of the cuts of the old French canal, and close at hand was the town of Bohio, now submerged.

The island consists of a mass of hills, steep in places, broken by ravines through which run a few small clear streams. Since the low land along the Chagres was submerged when the lake was filled, there is little swampy land on the island, although about the upper end there is a small amount of aquatic and semi-aquatic vegetation. The shore line, nearly 25 miles long, is very irregular, with innumerable indentations, in some of which are still standing the gray trunks of trees killed when the lower slopes of the hills were inundated. The highest part of the island is 537 feet above sea level, and 450 feet above the main level of Gatún Lake.

Along its north side the island is separated from the mainland by a narrow channel, formed by a diversion canal of the old French days. Toward the south lies the main expanse of Gatún Lake, traversed by the Canal, and on the distant shore is Frijoles, a station of the railroad which crosses the Isthmus from Colón to Panama.

Most of Barro Colorado Island is covered with dense forest and jungle. In a few places there are patches of comparatively level, deforested land, the sites of recently cultivated clearings now abandoned and overgrown with coarse weeds and second-growth shrubs. Within a few years these fields will be invaded by trees.

It is probably true that little of the island is covered with really virgin forest, but the woods have been so long undisturbed that one will hardly recognize the fact. The large number of palms and tree ferns indicates that some of the slopes and ravines have never been wholly cleared, these being plants which disappear when the forest is opened and probably never reestablish themselves. In a region such as the Canal Zone, for over 400 years under European influence and during all this time an important trade route, it is difficult to prove that a given tract of land has not been cleared or put under cultivation at some time during these centuries, of whose detailed history we know so little.

At any rate, the present plant covering of Barro Colorado has every aspect of the typical virgin forest occupying the humid lowlands of Central America, and is so rank and dense that in order to penetrate it a way must be cut with a machete. Many of the trees tower to a vast height, and have massive trunks swathed in a mantle of epiphytic vegetation that is still to be studied. Ropelike vines or lianas dangle from the crowns of the tallest trees, whose branches are loaded with aroids, bromeliads, orchids, and other epiphytes.

Palms are unusually abundant, and many of the 22 genera known from the Canal Zone exist here. Ferns, particularly handsome tree ferns of the genus *Hemitelia,* are plentiful, although in Central America most species of ferns must be sought at much higher elevations. Species of *Piper* are numerous, also Araceae, Rubiaceae, and Bignoniaceae, and most of the important groups of lowland Central American plants are represented. Thus far the cryptogamic plants have been little studied, but there must be a wealth of fungi. The lichens, hepatics, and mosses of the tropics are not highly diversified at so low an altitude.

The present list of the plants known from Barro Colorado Island is based chiefly upon personal collections and notes. I visited the island first on January 17, 1924, and collected that day about 300 numbers of plants. Collecting was then difficult, because there was only a single, inadequate trail; but now trails have been opened upon every hand, and may be extended easily, so there is little limit to one's range of activity.

During November, 1925, I spent a week upon the island as the guest of the laboratory. About 500 specimens of plants were taken, chiefly of the rarer and more interesting species, and notes were made of all the common plants observed. Trips were made each day in some new direction, hence it is believed that the list here offered is fairly representative of the flora. No one familiar with tropical conditions would venture to say that it is nearly complete, for by the very nature of its vegetation, such an area, with its many local or infrequent species, it is almost impossible to exhaust. Probably the next botanist who visits the island will be puzzled by the omission from this list of some plant which to him appears one of the common species.

November did not seem to be an especially favorable period for collecting, and few plants were found in flower. Probably the beginning of the rainy season, in spring, would be the best time for botanizing, although even then one must have good luck to find in proper condition some of the trees and shrubs that flower for only a brief season. The trees are difficult to study, since usually one must guess at their identity from their lofty branches as viewed from the ground, or sort the bits of leaves and flowers strewn upon the soil. There must be several species of trees on Barro Colorado that are not enumerated here, and more than a few shrubs and herbs.

No botanist can fail to be interested by the tropical vegetation so luxuriantly displayed here, and it is to be hoped that many botanical workers will take advantage of the opportunity offered for studying a characteristic area of tropical vegetation, at slight expense. This is an excellent place for making one's first acquaintance with tropical American plants, for no local flora of tropical America is better known, and its variety is equal to that of most localities of similar altitude.

For a study of the ecology of a typical area of lowland tropical vegetation, Barro Colorado offers exceptional advantages, and the morphology of certain groups of plants could be investigated profitably. Few indeed are the Central American localities in which it is possible to find comfortable lodging with the jungle but a few steps from the door. A large number of zoologists have visited Barro Colorado Island, some of them remaining several weeks or months to carry on their studies, and the list of published papers based upon work performed here forms an extensive bibliography.

The botanist also will be interested in the wealth of animal life that may be observed. Freedom from molestation has made the mammals and birds tame, and it is possible to see many kinds that elsewhere are timid and seldom visible. Flocks of chattering parrots and parrakeets fly all day long over the trees, and literally hundreds of other birds may be seen about the forest. Peccaries may be encountered along any trail, and sitting in the evening on the steps of the laboratory, one may watch the monkeys going to their sleeping places. Deer are found in the forest, and jaguars have been seen from the laboratory. In the mud the tracks of tapirs, the largest Central American mammal, are found now and then, and one is likely to meet upon the trail other smaller but interesting animals. Insects are not more plentiful than elsewhere, and I do not remember to have been troubled by anything more disagreeable than ants, the worst pests of tropical forests. Snakes exist here, some of them venomous, but they need occasion only a fair amount of caution. I happened to see none upon the island.

This list is little more than an enumeration of the names of the species of plants now known to occur on Barro Colorado Island. I hope that at some time it may be practicable to prepare a descriptive flora of the island, but it is better to leave such a work until the list is more nearly complete. In the near future there will be published, as volume 27 of the Contributions from the National Herbarium, an account of the plants of the Canal Zone, with keys for their determination, and it is felt that to publish here keys to the species would be an unnecessary repetition.

Besides my own collections, I have had access to a few others made on the island: those of Dr. William R. Maxon, who collected here June 6, 1923; of Prof. F. L. Stevens, of the University of Illinois, who visited the island in September, 1924; and of Prof. C. W. Dodge, of Harvard University, who was engaged in study of the fungi during the summer of 1925, and has furnished a list of the flowering plants he collected at that time. Among the other botanists who have visited the island are Dr. A. S. Hitchcock and Mr. O. F. Cook, of the U. S. Department of Agriculture, and Prof. G. R. Bisby, of Manitoba Agricultural College.

In addition to the species listed there are still on hand some sterile specimens which it has been impossible to identify. Most of these represent additions to the flora of Panama and probably to that of North America. Some of them doubtless will prove of great interest, but their recognition must await the collection of more complete material or a fortunate association with named specimens from other regions.

The Spanish vernacular names given for the species here listed are those used in Panama, and many of them were verified upon the island. Well established English names have been cited when available.

In the present paper there are listed for Barro Colorado Island 611 species of plants. Of these at least 38 species are introduced.

FUNGI

The list of fungi is based partly upon specimens collected by myself and identified by Dr. J. R. Weir of the U. S. Department of Agriculture. There are included also numerous records supplied by Prof. F. L. Stevens and Prof. G. R. Bisby.

Arcyria cinerea Pers.

Auricularia mesenterica Bull. This, like most of the fleshy and woody fungi growing upon logs and tree trunks, is called in Panama as well as elsewhere in Central America "orejas" or "orejitas."

Bagnisiopsis peribebuyensis (Speg.) Theiss. & Syd. On *Miconia argentea.*

Camillea cyclops Mont.

Camillea Sagraeana (Mont.) B. & C.

Cookeina sulcipes (Berk.) Kuntze.

Cookeina tricholoma (Mont.) Kuntze.

Fomes Auberianus Mont.

Fomes ferreus Berk.

Fomes marmoratus Berk.

Ganoderma sp.

Geaster sp.

Gloeoporus conchoides Mont.

Hexagonia tenuis (Hook.) Fr.

Hexagonia variegata Berk.

Hirneola delicata (Fr.) Bres.

Hirneola polytricha Mont.

Hymenochaete damaecornis Link & Lev.

Irenina Shropshiriana Stevens, sp. nov. On *Miconia argentea.*

Laschia auriscalpium Mont.

Laschia pezizoidea Berk.

Lentinus strigellus Berk.

Lentinus velutinus Fr.

Meliola Heliconiae Stevens, sp. nov. On *Heliconia* sp.

Meliola Musae (Kunze) Mont. On *Heliconia* sp.

Meliola palmicola Winter.

Meliola Panici Earle. On *Olyra latifolia.*

Meliola peruviana irregularis Stevens, var. nov. On Bignoniaceae indet.

Meliola Pilocarpi Stevens. On *Zanthoxylum* (?)

Polyporus brachypus Lev.

Polyporus gracilis Kl.

Polyporus infernalis Berk.

Polyporus licnoides Mont.

Polyporus lignosus Kl.

Polyporus subelegans Murr.

Polyporus virgatus B. & C.

Polystictus arenicolor Berk.

Polystictus crocatus Fr.

Polystictus occidentalis (Kl.) Fr.

Polystictus sanguineus (L.) Fr.

Polystictus Steinheilianus Berk. & Lev. "Really a thin form of *Trametes rigida* Berk. & Mont."

Polystictus versatilis Berk.

Polystictus versicolor (Dicks.) Fr.

Poria vincta (Berk.) Cke.

Schizophyllum commune (L.) Fr.

Stereum flabellatum Pat.

Stereum glabrescens Berk.?

Stereum papyrinum Mont.

Thelephora pusiola Pat.?

Trametes caperatus Berk.

Trametes cubensis Mont.

Trametes hydnoides (Sw.) Fr.

Trametes rigida Berk. & Mont.

Xylaria axifera Mont.

Xylaria cubensis Mont.

The records of the following rusts have been supplied by Prof. H. S. Jackson, of Purdue University. The specimens were collected by Prof. F. L. Stevens.

Puccinia Emiliae P. Henn. On *Neurolaena lobata* (L.) R. Br.

Uredo Dioscoreae P. Henn. On *Dioscorea urophylla* Hemsl.

LICHENS

The following species has been determined by Mr. G. K. Merrill. The number of lichens occurring on Barro Colorado is not large, but there are other species besides the one listed.

Leptogium azureum (Swartz) Mont.

MOSSES

The following mosses have been determined by Mr. Edwin B. Bartram:

Bryum coronatum Schwaegr.

Crossomitrium Wallisi C. M.

Lepodipilum polytrichioides (Hedw.) Brid.

Neckeropsis disticha (Hedw.) Fleisch.

Octoblepharum albidum (L.) Hedw.

Pilotrichum ramosissimum Mitt.

Taxithelium planum (Brid.) Mitt.

Thuidium schistocalyx (C. M.) Mitt.

SCHIZAEACEAE. Curlygrass Family[1]

Lygodium polymorphum (Cav.) H. B. K. A slender vine, very hairy, in cut-over places.

Lygodium radiatum Prantl.

CYATHEACEAE. Tree Fern Family

Hemitelia petiolata Hook. Frequent; a very handsome plant, the only tree fern known to occur on the island.

---

[1]An annotated list of the ferns and fern allies of Barro Colorado has been published recently by the writer in the American Fern Journal 16: 112-120; 17: 1-8. 1927. The identifications are by Dr. William R. Maxon.

MARATTIACEAE. Marattia Family

Danaea nodosa (L.) J. E. Sm. Frequent in the forest.

POLYPODIACEAE. Polypody Family

Acrostichum sp. A species of this genus grows in shallow water about the edge of the lake, but specimens have not been collected. It is either *A. aureum* L. or *A. daneaefolium* Langsd. & Fisch., both of which are common in the region.

Adiantum lucidum Swartz. Common in the forest.

Adiantum philippense L. Infrequent.

Adiantum sp. (*Standley* 31330). An unidentified and perhaps undescribed species.

Ananthacorus angustifolius (Swartz) Underw. & Maxon. An epiphytic plant.

Anetium citrifolium (L.) Splitg. Epiphytic.

Asplenium serratum L. Epiphytic. The American birds-nest fern.

Cyclopeltis semicordata (Swartz) J. Sm. Abundant.

Dictyoxiphium panamense Hook.

Diplazium delitescens Maxon. Abundant.

Diplazium grandifolium Swartz.

Dryopteris dentata (Forsk.) C. Chr.

Dryopteris Poiteana (Bory) Urban. Frequent in the forest.

Elaphoglossum Herminieri (Bory & Fée) Moore. Epiphytic.

Eschatogramme furcata (L.) Trev. Epiphytic.

Leptochilus cladorrhizans (Spreng.) Maxon. Common.

Nephrolepis pendula (Raddi) J. Sm. Epiphytic.

Pityrogramma calomelaena (L.) Link. In open places.

Polybotrya caudata Kunze. A creeping and climbing epiphyte.

Polybotrya osmundacea Humb. & Bonpl. A large climbing epiphyte.

Polypodium ciliatum Willd. Epiphytic.

Polypodium crassifolium L. A coarse epiphyte.

Polypodium occultum Christ. Epiphytic.

Polypodium pectinatum L., form. An epiphyte.

Polypodium percussum Cav. Epiphytic.

Saccoloma elegans Kaulf. A common handsome terrestrial plant.

Stenochlaena vestita (Fourn.) Underw. A large creeping epiphyte.

Tectaria euryloba (Christ) Maxon.

Tectaria martinicensis (Spreng.) Copel. Common in the forest.

Vittaria lineata (L.) J. E. Sm. A common epiphyte, with grasslike leaves.

HYMENOPHYLLACEAE. Filmy-fern Family

Trichomanes Godmani Hook. Epiphytic, like the other local species of the genus.

Trichomanes Krausii Hook. & Grev.

Trichomanes sphenoides Kunze.

SALVINIACEAE. Salvinia Family

Salvinia auriculata Aubl. Floating in quiet water.

LYCOPODIACEAE. Clubmoss Family

Lycopodium cernuum L. Reported by Prof. C. W. Dodge.

SELAGINELLACEAE. Selaginella Family

Selaginella conduplicata Spreng. Common in the forest.

Selaginella Fendleri Baker.

Selaginella haematodes (Kunze) Spring. Common; easily recognized by its dark red stems.

Selaginella Schrammii Hieron.

Selaginella sylvatica Baker.

TYPHACEAE. Cattail Family

Typha angustifolia L. CATTAIL. In shallow water at the edge of the lake.

POACEAE. Grass Family

The identifications in this family have been made by Dr. A. S. Hitchcock and Mrs. Agnes Chase.

Andropogon condensatus H. B. K. In clearing; scarce.

Arthrostylidium racemiflorum Steud. A common slender bamboo.

Axonopus compressus (Swartz) Beauv. CARPET GRASS. Common.

Cenchrus viridis Spreng. SANDBUR. In open places.

Chloris radiata (L.) Swartz. In clearings; rare.

Chusquea simpliciflora Munro. A slender bamboo, common in the forest.

Cynodon dactylon (L.) Pers. BERMUDA GRASS. In open places; introduced.

Digitaria sanguinalis (L.) Scop. CRABGRASS.

Eleusine indica (L.) Gaertn. In open places.

Gynerium sagittatum (Aubl.) Beauv. CANE. A tall coarse grass, in wet places.

Hymenachne amplexicaulis (Rudge) Nees. In shallow water.

Ichnanthus nemorosus Doell. Common.

Ichnanthus pallens (Swartz) Munro. Common.

Ischaemum rugosum Salisb. In clearings.

Lasiacis sorghoidea (Desv.) Hitchc. & Chase. A common coarse vine.

Olyra latifolia L. Common in forest.

Oplismenus Burmanni (Retz.) Beauv. Very common.

Oplismenus hirtellus (L.) Beauv. Common.

Orthoclada laxa (Rich.) Beauv. In forest.

Oryza sativa L. RICE. ARROZ. Upland rice has been planted on the island.

Panicum pilosum Swartz. In clearing.

Panicum trichoides Swartz. Common.

Paspalum conjugatum Berg. Common.

Paspalum paniculatum L. In clearing.

Pharus glaber H. B. K. Frequent in forest.

Pharus latifolius L. Frequent.

Polytrias amaurea (Büse) Kuntze. Well established in the lawn at the laboratory.

Saccharum officinarum L. SUGAR CANE. CAÑA. Planted at the laboratory, and about the old clearings.

Setaria geniculata (Lam.) Beauv. Common in open places.

Setaria vulpiseta (Lam.) Roem. & Schult. In a clearing; rare.

Streptochaeta Sodiroana Hack. In the forest.

Streptogyne crinita Beauv. In the forest; occasional.

Zea mays L. MAIZE. MAÍZ. Planted at the laboratory.

CYPERACEAE. Sedge Family

Cyperus caracasanus Kunth. JUNCO. In open places.

Cyperus ferax Rich. JUNCO. Occasional in clearings.

Cyperus giganteus Vahl. A giant plant in water at the edge of the lake; in habit resembling the African papyrus.

Dichromena radicans Schlecht. & Cham. CLAVO. In open places.

Fimbristylis diphylla (Retz.) Vahl. In clearings.

Fuirena umbellata Rottb. In shallow water at the edge of the lake.

Kyllinga pumila Michx. In open places.

Mariscus jamaicensis (Crantz) Britton. SAWGRASS. Common in shallow water at the edge of the lake.

Rynchospora cephalotes (L.) Vahl. PAJA MACHO DE MONTE ("tapir grass"). In open places.

Scleria bracteata Cav. CORTADERA, CUCHILLITO. The Spanish name alludes to the fact that the sharp edges of the leaves cut the skin like a knife.

Scleria melaleuca Schlecht. & Cham.

PHOENICACEAE. Palm Family

Other palms than those listed probably occur here.

Acanthorrhiza Warscewiczii Wendl. NOLÍ, PALMA DE ESCOBA. Scarce. The only fan palm of the region. The leaves are used for brooms and for thatching.

Asterogyne sp. (*Geonoma cuneata* Wendl.?) RABO AHORCADO. A nearly stemless, small plant, the mostly simple leaves deeply lobed at the apex; flowers in simple spikes.

Astrocaryum polystachyum Wendl. A tall plant with spiny trunk.

Bactris sp. (Subgenus *Trichobactris*.) A slender, very spiny palm, in forest; common.

Calyptrogyne sp. A small plant, stemless or with a short trunk; leaves with numerous narrow segments; flowers in simple spikes.

Chamaedorea Wendlandiana (Oerst.) Hemsl. CAÑA VERDE, BOLÁ. A slender graceful palm with smooth green stems.

Cocos nucifera L. COCONUT. COCO. A few trees about the sites of former houses; introduced.

Geonoma sp. Probably two species grow here. Slender plants with pinnate leaves, unarmed stems, and branched inflorescences.

Iriartea exorrhiza Mart. STILT PALM. JIRA. A tall palm with slender smooth green trunk, the trunk supported by stout prop roots, which are covered with very short spines.

Pyrenoglyphis major (Jacq.) Karst. LATA, PALMA BRAVA. A very spiny plant, similar to *Bactris*, but with much larger fruits.

Synechanthus Warscewiczianus Wendl. PALMILLA, BOLÁ. A slender palm, similar in appearance to *Chamaedorea*.

CYCLANTHACEAE. Cyclanthus Family

Carludovica palmata Ruiz & Pav. PANAMA HAT PALM. PORTORRICO, JIPIJAPA, RAMPIRA, IRACA. A stemless plant with numerous long-stalked leaves, the blades cleft so as to resemble a Maltese cross. It is from the young leaves of this plant that the famous "Panama" hats are made, in Ecuador.

Cyclanthus bipartitus Poit. PORTORRICO. A stemless plant, the leaves cleft into two broad divisions. Easily recognized by the fruit, which resembles a large screw.

ARACEAE. Arum Family

Plants of this family are particularly abundant on Barro Colorado. The epiphytic species constitute a large part of the vegetation seen upon tree trunks.

Anepsias Moritzianus Schott.

Anthurium aemulum Schott. A large epiphytic vine with parted leaves.

Anthurium Friedrichsthalii Schott. A small acaulescent epiphyte with linear leaves.

Anthurium Holtonianum Schott. A very showy species, a large vine with huge leaves, digitately parted into several broad segments.

Anthurium maximum (Desf.) Engler. An acaulescent epiphyte, with large broad simple leaves.

Anthurium Schlechtendalii Kunth. An acaulescent epiphyte.

Anthurium scolopendrinum (Ham.) Kunth. Acaulescent, with narrow entire leaves.

Anthurium triangulum Engler. Leaves sagittate.

Dieffenbachia Oerstedii Schott. OTÓ DE LAGARTO. Called "dumb-cane" by the West Indians. A coarse terrestrial herb with erect stems and broad leaves. The crushed plant has a skunklike odor. The juice is very irritant in contact with the skin, and care must be exercised in handling the plant.

Monstera dilacerata Koch. A large and handsome epiphytic vine with deeply pinnatifid, broad leaves.

Monstera pertusa (L.) de Vriese. A coarse vine, recognized at once by the broad leaves perforated with numerous large holes.

Philodendron coerulescens Engler. Epiphytic vine with ovate entire leaves.

Philodendron grandipes Krause. An acaulescent terrestrial plant with rounded-cordate leaves; very common.

Philodendron Karstenianum Schott. An epiphyte with oblong leaves.

Philodendron radiatum Schott. AZOTA CABEZA, CHALDÉ. A large handsome vine, the leaves deeply pinnatifid into narrow segments; very common.

Philodendron rigidifolium Krause. CINCHADORA. Epiphyte with broad ovate leaves.

Philodendron tripartitum (Jacq.) Schott. A common vine, recognized readily by the leaves, which are parted into 3 oblong entire segments.

Philodendron Wendlandii Schott. Epiphytic vine with oblong leaves, cordate at base.

Pistia stratiotes L. WATER-LETTUCE. Floating in quiet water. Very unlike the other members of the family, the plant consisting of a rosette of spongy, broadly wedge-shaped, pale green leaves.

Spathiphyllum Patini (Hogg) N. E. Brown. Acaulescent terrestrial plant.

Stenospermation sessile Engler. Large epiphytic vine with lance-oblong leaves.

Xanthosoma helleborifolium (Jacq.) Schott. PAPAYUELO. Terrestrial plant with a single leaf, this parted into 5 to 13 lobed segments; petiole handsomely blotched with brown.

Xanthosoma violaceum Schott. OTÓ. Called "badú" and "coco" by the West Indians. Planted at the laboratory; cultivated commonly in the lowlands of tropical America for its tuberous roots, which are cooked and eaten much like potatoes. The plant resembles the caladium or elephant-ear cultivated for ornament.

LEMNACEAE. Duckweed Family

Lemna cyclostasa (Ell.) Chev. DUCKWEED. Mr. Zetek reports that he has seen a plant of this family in quiet water about the island. The species listed is the only member of the family known at present from the Canal Zone, but it is possible that others occur here.

BROMELIACEAE. Pineapple Family

Ananas magdalenae (André) Standl. PITA, PIÑUELA. Called "pingwing" by the West Indians. Common in forests. Similar in habit to the pineapple, the red flowers forming a large hard globose head. The long, very spiny leaves furnish one of the best fibers known, the "pita floja." The plants often form dense thickets which are almost impenetrable.

Ananas sativus Schult. PINEAPPLE. PIÑA. Planted at the laboratory.

Billbergia pallidiflora Liebm. An epiphyte with pendent flower spikes, the few long leaves spiny-margined and handsomely blotched with silver.

Catopsis tenella Mez. A small epiphyte with dioecious flowers and broad, thin, bright green leaves.

Guzmania minor Mez. An epiphyte with broad, bright green, thin leaves, the inflorescence short and dense, with showy, red or purple bracts.

Tillandsia bulbosa Hook. An epiphyte with a hard, dark, bulblike base.

Tillandsia digitata Mez. An epiphyte with a cluster of many gray leaves.

COMMELINACEAE. Dayflower Family

Campelia zanonia (L.) H. B. K. An erect herb about a meter high, with conspicuous, dark blue, juicy fruit.

Commelina elegans H. B. K. DAYFLOWER. CODILLO. A fleshy procumbent herb with bright blue flowers, resembling the Wandering Jew of gardens.

Dichorisandra hexandra (Aubl.) Standl. An erect branched herb, about a meter high, with small blue flowers.

Tradescantia geniculata Jacq. An inconspicuous, procumbent, very hairy herb with small white flowers.

PONTEDERIACEAE. Pickerelweed Family

Piaropus azurea (Swartz) Raf. WATER-HYACINTH. I have no record of having seen this plant on Barro Colorado, but it certainly must occur somewhere about the shores, since it is frequent in Gatún Lake. If left to itself it would overgrow the lake, but efforts have been made to exterminate the plant, hence it is not abundant anywhere.

LILIACEAE. Lily Family

Taetsia fruticosa (L.) Merrill. Planted at the laboratory. One of the so-called Dracaenas; much planted for ornament in Panama. A tall plant with green or more commonly red or purple leaves.

SMILACACEAE. Sarsaparilla Family

Smilax mollis Willd. A common small vine with pubescent foliage.

Smilax panamensis Morong. GREENBRIER. ZARZA. A common large vine with very prickly stems and glabrous foliage.

HAEMODORACEAE. Bloodwort Family

Xiphidium caeruleum Aubl. PALMITA. Common in the forest. An herb, marked by its fleshy, vertically 2-ranked leaves, suggesting those of an iris; flowers small and whitish, the fruit a small red berry.

AMARYLLIDACEAE. Amaryllis Family

Hymenocallis americana (L.) Salisb. SPIDERLILY. Called "euchar lily" by the West Indians. I found it in the forest on one of the hills of the island, at the site of a former dwelling. It is normally a seashore plant, but is often grown for ornament because of its handsome white flowers.

DIOSCOREACEAE. Yam Family

Dioscorea alata L. YAM. ÑAME. The common yam, planted at the laboratory.

Dioscorea urophylla Uline. BEJUCO DE SAINA. A native species, growing in the forest.

IRIDACEAE. Iris Family

Marica gracilis Herb. An inconspicuous herb with narrow leaves, occasional in the wet forest.

MUSACEAE. Banana Family

Heliconia acuminata Rich. A small herbaceous plant with small leaves; inflorescence erect, with deep red bracts. The Heliconias are known in Panama as "platanillo," or sometimes as "lengua de vaca." They are conspicuous plants in the forests and in swamps. The bracts hold water in which mosquitoes sometimes breed.

Heliconia latispatha Benth. PLATANILLO, GUACAMAYA. Similar to the last species, but much larger; inflorescence erect, the bracts red, tinged with yellow or orange.

Heliconia Mariae Hook. BEEFSTEAK HELICONIA. PLATANILLO. Called by the West Indians "wild plantain" or "wild banana." The largest and most showy species of the region, often forming dense thickets, the plants several meters high, with leaves as large as those of the banana. Inflorescence very large, thick, and heavy, pendent, with broad, closely crowded, red bracts.

Heliconia pendula Wawra. A medium-sized plant with tomentose, pendent, dark red inflorescence.

Musa paradisiaca L. PLANTAIN. PLÁTANO. Planted at the laboratory and elsewhere.

Musa sapientum L. BANANA. Planted at the laboratory and about the old clearings.

ZINGIBERACEAE. Ginger Family

Costus sanguineus Donn. Smith. The species of *Costus* are common in the forests. They are tall plants with simple leafy stems, the stems formed by the tightly rolled leaf petioles. In this species the flower spikes are fusiform, with closely appressed, unappendaged, red bracts.

Costus spicatus (Jacq.) Swartz. Spikes cylindric or subglobose, the bracts not appendaged, in age loose and spreading.

Costus villosissimus Jacq. CAÑAGRIA, CAÑA DE MICO. Plant very villous; bracts with leafy, green or red appendages.

Dimerocostus uniflorus (Poepp.) Schum. A tall plant, usually 3 to 4 meters high, resembling the *Costus* species; usually growing in water. Flowers white, 7 to 8 cm. long, opening one at a time on each plant.

Renealmia occidentalis (Swartz) Sweet. Stems leafy, in clumps, 1 to 2.5 meters high; inflorescences short, arising from the ground at the base of the plant; berries red or dark blue, with orange pulp.

Renealmia strobilifera Poepp. & Endl. Stems leafy, 1.5 to 3 meters high; inflorescence conelike, bright orange.

MARANTACEAE. Arrowroot Family

Calathea insignis Peters. The Calatheas, common in wet forest and swampy places, are coarse herbs with broad leaves like those of cannas, the flowers in dense spikes. In this species the spikes are strongly compressed, the bracts thin and parchment-like.

Calathea lutea (Aubl.) Meyer. HOJA BLANCA. Leaves whitish beneath; bracts distichous but not strongly compressed, thick and leathery.

Calathea macrosepala Schum. BIJAO. Spikes small and headlike, very dense, not compressed.

Ischnosiphon leucophaeus (Poepp. & Endl.) Koern. Leaves white beneath; Flowers in very slender, terete spikes.

Myrosma panamesis Standl. A stemless plant with broad leaves about a foot long, the flowers in simple spikes.

Pleiostachya pruinosa (Regel) Schum. Easily recognized by the broad leaves, which are dark red or purple beneath. Common in forest.

BURMANNIACEAE. Burmannia Family

Ophiomeris panamensis Standl. Known only from Barro Colorado, where it was collected by Prof. C. W. Dodge. A small delicate whitish saprophyte, the slender stem bearing a single lopsided flower, three of whose lobes end in long filiform appendages.

ORCHIDACEAE. Orchid Family

The identifications have been made chiefly by Mr. Oakes Ames.

Aspasia principissa Reichenb. f. Epiphytic.

Bulbophyllum pachyrrachis (A. Rich.) Griseb. An epiphytic orchid with very small flowers in pendent spikes which have a thick fleshy rachis.

Catasetum viridiflavum Hook. A showy epiphytic species, the green and yellow flowers resembling those of the northern lady's-slippers.

Epidendrum anceps Jacq. Epiphytic.

Epidendrum difforme Jacq. Epiphytic.

Epidendrum Rousseauae Schlechter. Epiphytic.

Epidendrum stenopetalum Hook. An epiphyte.

Maxillaria Macleei Batem. Epiphytic.

Oncidium ampliatum Lindl. BUTTERFLY ORCHID. A handsome plant with large, yellow and brown flowers which suggest butterflies.

Ornithocephalus bicornis Lindl. Epiphytic; easily recognized by its equitant leaves, suggesting those of iris. Flowers very small, resembling in form a bird's head, hence the generic name.

Peristeria elata Hook. DOVE ORCHID or HOLY GHOST FLOWER. ESPÍRITU SANTO. A tall terrestrial species, famed for its handsome white flowers, whose central organs suggest by their form a dove with outspread wings.

Pleurothallis Brighamii Wats. Epiphytic.

Pleurothallis marginata Lindl. Both these species are very small plants with inconspicuous flowers.

Sobralia panamensis Schlechter. A terrestrial plant with tall leafy stems and handsome large purple flowers, which last only part of a single day, closing about noon.

Vanilla pompona Schiede. VANILLA. VAINILLA. A large vine, common nearly everywhere in this part of Panama.

PIPERACEAE. Pepper Family

Peperomia caudulilimba, longependula C. DC. All the species of *Peperomia* occurring on the island are small succulent epiphytic herbs.

Peperomia conjungens Trel. Type from Barro Colorado.

Peperomia gatunensis C. DC.

Peperomia rotundifolia (L.) H. B. K. POLEO. Leaves rounded, very thick and lens-like.

Piper acutissimum Trel. CORDONCILLO. All the species of *Piper* growing here are terrestrial shrubs. They are abundant in wet forest, and often grow in open places. The names given to the species are "cordoncillo," "gusanillo," and "hinojo." The West Indians use the name "cowfoot."

Piper auritum H. B. K. SANTA MARÍA DE ANÍS. A large coarse suffrutescent plant, easily recognized by its very broad, deeply cordate leaves, and by the characteristic odor of the crushed leaves, suggestive of sarsaparilla.

Piper cordulatum C. DC.

Piper culebranum C. DC.

Piper imperiale (Miquel) C. DC. A plant with very large leaves, the petioles with numerous fleshy wartlike protuberances.

Piper laxispicum Trel. Type from Barro Colorado.

Piper paulownifolium C. DC.

Piper pseudo-cativalense Trel.

Piper pseudo-garagaranum Trel. Type from Barro Colorado.

Piper pseudo-variabile Trel.

Piper pubistipulum estylosum Trel. Type from Barro Colorado.

Piper san-joseanum C. DC. HINOJO.

Piper smilacifolium C. DC.

Piper subnudispicum Trel.

Piper viridicaule Trel. Type from Barro Colorado.

Pothomorphe peltata (L.) Miq. SANTA MARÍA. A suffrutescent plant with rounded-cordate leaves, the spikes in umbels.

ULMACEAE. Elm Family

Celtis iguanaea (Jacq.) Sarg. Shrub or small tree, the branches usually pendent or clambering, armed with recurved spines.

Trema micrantha (L.) Blume. Small tree with narrow gray leaves and very small, red fruits.

MORACEAE. Mulberry Family

Artocarpus communis Forst. BREADFRUIT. ARBOL DE PAN, FRUTA DE PAN. Planted at the laboratory.

Castilla panamensis Cook. RUBBER TREE. CAUCHO, HULE, ULE. A common forest tree, the only species of the immediate region.

Cecropia sp. GUARUMO. Three species of *Cecropia* are known from the Canal Zone, and all may occur on Barro Colorado. No specimens suitable for identification have been collected on the island. The species are small trees with prop-roots, and very large, deeply palmate-lobed leaves which are white-tomentose beneath. The hollow branches are inhabited by ants.

Coussapoa panamensis Pittier. A tree, usually epiphytic, at least at first, with large ovate leaves white-tomentose beneath.

Ficus costaricensis (Liebm.) Miquel.? Sterile specimens only, and the determination therefore somewhat doubtful. In Panama the wild figs are usually called "matapalo," "higo," or "higuero." They are large trees, often strangling or epiphytic, and frequently with large buttresses.

Ficus crassiuscula Warb.

Ficus glabrata H. B. K. HIGUERÓN. A common tree, with very large fruits.

Ficus Hemsleyana Standl.

Ficus Tonduzii Standl. Common; leaves very broad, with few coarse nerves.

Helicostylis latifolia Pittier. BERBÁ, CHOYBÁ, QUERENDO. Large tree with oblong to obovate, entire leaves.

Inophloeum armatum (Miquel) Pittier. NAMAGUA, MARAGUA, COGUÁ. Large tree with narrow rough leaves. From the bark of this tree the Panama Indians formerly made a coarse cloth which they used for hammocks, blankets, women's clothes, and sails for boats. The cloth is still made in some parts of the country.

Olmedia aspera Ruiz & Pav. Shrub or small tree with oblong long-cuspidate rough leaves. Common.

Sorocea affinis Hemsl. Shrub or small tree, with small red fruits in racemes.

Trophis racemosa (L.) Urban. Tree of medium or large size.

URTICACEAE. Nettle Family

Boehmeria cylindrica (L.) Swartz. An herb in water about the edge of the lake.

Myriocarpa yzabalensis (Donn. Smith) Killip. Large shrub, the minute whitish flowers in numerous pendent, very slender spikes sometimes 60 cm. long.

Urera baccifera (L.) Gaud. ORTIGA. Shrub or small tree, armed with spine-like hairs that sting the flesh painfully.

Urera elata (Swartz) Griseb. A tree 6 to 9 meters high, in this region known only from Barro Colorado.

PROTEACEAE. Protea Family

Roupala darienensis Pittier. Small tree with a skunklike odor; leaves partly pinnate and partly simple.

OLACACEAE. Olax Family

Heisteria costaricensis Donn. Smith. The species of *Heisteria* are shrubs with alternate entire leaves, and are easily recognized by the saucer-shaped calyx which persists with the fruit and is colored bright red.

Heisteria macrophylla Oerst. AJICILLO.

ARISTOLOCHIACEAE. Birthwort Family

Aristolochia sylvicola Standl. Small slender woody vine.

POLYGONACEAE. Buckwheat Family

Coccoloba acuminata H. B. K. Shrub.

Coccoloba leptostachya Benth. Small tree.

Coccoloba nematostachya (Griseb.) Lindau. HUESO. Small tree.

Triplaris americana L. GUAYABO HORMIGUERO, PALO SANTO. Large tree with dense racemes of purple-red flowers. The flowers appear about the first of February and are very showy, lasting for several weeks. The hollow branches are infested with savage ants, usually a species of *Pseudomyrma*.

AMARANTHACEAE. Amaranth Family

Alternanthera ficoidea (L.) R. Br. A small weedy herb.

Alternanthera sessilis (L.) R. Br.

Celosia argentea L. Rare; a few plants found, probably escaped from cultivation. The cristate form of this species, *C. cristata* L., is the cultivated cockscomb ("abanico").

Cyathula prostrata (L.) Blume. CADILLO. Small herb, introduced from the Old World.

Iresine celosia L. A common herbaceous weed.

NYCTAGINACEAE. Four-o'clock Family

Neea Pittieri Standl. Shrub or small tree.

Pisonia aculeata L. Large shrub or small tree, with long, often clambering branches, armed with hooked spines; fruit small, club-shaped, covered on the angles with small sticky glands.

PHYTOLACCACEAE. Pokeberry Family

Petiveria alliacea L. ANAMÚ. Herbaceous or suffrutescent, the crushed leaves with the odor of garlic; flowers appressed to the rachis of the spike; fruit bearing 4 small hooked bristles.

PORTULACACEAE. Purslane Family

Portulaca oleracea L. PURSLANE. VERDOLAGA. A rare weed.

NYMPHAEACEAE. Waterlily Family

Castalia ampla Salisb. WATERLILY. Called "duckweed" by the West Indians. In quiet water. A plant with handsome white flowers.

MENISPERMACEAE. Moonseed Family

Cissampelos pareira L. A slender vine with rounded hairy leaves, common almost throughout Central America.

Cissampelos tropaeolifolia DC.

Hyperbaena panamensis Standl. Woody vine with ovate to oblong, 3-nerved leaves.

Sciadotenia sp. A woody vine, perhaps of this genus, grows on the island, but only sterile specimens have been collected, hence its identification is uncertain. The broad leaves are closely white-tomentose beneath.

ANNONACEAE. Custard-apple Family

Annona acuminata Safford. CAMARÓN. Shrub, or small tree, the leaves glabrous or nearly so, narrow; fruit small, tuberculate, opening at maturity.

Annona Hayesii Safford. Shrub or small tree; fruit smooth, subglobose, about 5 cm. long.

Annona Spraguei Safford. CHIRIMOYA, NEGRITO. Tree; leaves densely pubescent beneath; fruit small, covered with clawlike tubercles.

Desmopsis panamensis (Robinson) Safford. Shrub or small tree; fruit a cluster of stalked pubescent berries.

Guatteria amplifolia Triana & Planch. Shrub or small tree with large oblong leaves; fruit a cluster of small oval berries.

Xylopia macrantha Triana & Planch. COROBÁ, RAYADO. Small tree.

MYRISTICACEAE. Nutmeg Family

Virola panamensis (Hemsl.) Warb. BOGAMANI, MALAGUETA DE MONTAÑA. Large tree with entire oblong leaves, stellate-tomentose beneath. Common.

MONIMIACEAE. Monimia Family

Siparuna pauciflora (Beurl.) A. DC. Large shrub, strong-scented, with broad pubescent leaves.

LAURACEAE. Laurel Family

Ocotea cernua (Nees) Mez. SIGUA. A frequent tree.

Persea americana Mill. AVOCADO, ALLIGATOR PEAR. AGUACATE. Planted at the laboratory.

CAPPARIDACEAE. Caper Family

Capparis baducca L. Shrub.

ROSACEAE. Rose Family

Rosa sp. One of the common roses, planted at the laboratory.

AMYGDALACEAE. Almond Family

Licania hypoleuca Benth. Tree; leaves small, entire, white-tomentose beneath.

CONNARACEAE. Connarus Family

Cnestidium rufescens Planch. Large woody vine with pinnate leaves; leaflets densely pubescent beneath.

Connarus panamensis Griseb. Woody vine; leaflets 3, glabrous or nearly so.

Rourea glabra H. B. K. Large woody vine; leaflets glabrate.

MIMOSACEAE. Mimosa Family

Acacia Hayesii Benth.? UÑA DE GATO.

Acacia melanoceras Beurl., one of the ant-inhabited bullhorn acacias, may occur here, but the writer has not seen it on the island.

Entada scandens (L.) Benth. JAVILLA. Large woody vine with enormous pods several inches broad.

Inga edulis Mart. GUAVO. Like the other species, a good-sized tree.

Inga Goldmanii Pittier. GUAVO DE MONO

Inga marginata Willd.

Inga panamensis Seem. GUAVO.

Mimosa pudica L. SENSITIVE-PLANT. DORMIDERA, CIÉRRATE, CIERRA TUS PUERTAS. Called by the West Indians "shameweed" and "shame-face." Small herb with round heads of pink flowers.

CAESALPINIACEAE. Senna Family

Bauhinia excisa (Griseb.) Hemsl. BEJUCO DE MONO. Large woody vine with bilobate leaves. The stems are compressed and ribbon-like, and perforated with large holes.

Bauhinia sp. Only sterile material collected. Leaflets 2, very silky beneath, acute.

Cassia bacillaris L. Shrub with showy yellow flowers.

Peltogyne purpurea Pittier. NAZARENO, MORADO. A large tree, reported to exist here.

Prioria copaifera Griseb. CATIVO, AMANSA MUJER. A very common, large tree; leaves with 4 leaflets. The short broad flat fruits are much sought by peccaries.

Tounatea simplex (Swartz) Taub. Shrub or small tree.

FABACEAE. Bean Family

Aeschynomene americana L. PEGA-PEGA. Herb with buff flowers.

Aeschynomene sensitiva Swartz.

Andira inermis H. B. K. CABBAGE-BARK. COCÚ. Large tree; leaflets 7 to 13, opposite, oblong, glabrous; flowers purple, in panicles. The wood is of good quality and is much used locally.

Cajanus bicolor DC. PIGEON-PEA. GUANDÚ, FRIJOL DE PALO. Shrub; much cultivated in this region for its edible seeds, and also naturalized.

Clitoria arborescens Ait. An erect or scandent shrub; one of the most beautiful plants of Central America, bearing clusters of shell-pink flowers about 7 cm. long.

Coumarouna panamensis Pittier. ALMENDRO. Common. A large tree; leaves pinnate, the leaflets 5 to 8 pairs, large, oblong, the costa close to the margin; flowers pink, in panicles. The fresh fruit is filled with an oily fragrant liquid that crystallizes when dry.

Dioclea reflexa Hook.? Large woody vine.

Erythrina panamensis Standl. Shrub or small tree with narrow, bright red flowers and red seeds.

Machaerium marginatum Standl.

Machaerium microphyllum (Meyer) Standl. Spiny woody vine with purple flowers.

Machaerium purpurascens Pittier.

Machaerium Seemanni Benth.

Meibomia adscendens (Swartz) Kuntze. A frequent weed.

Meibomia axillaris (Swartz) Kuntze. The pods are sometimes called "guavitas."

Meibomia cana (Gmel.) Blake. PEGA-PEGA, PEGADERA. Known among the Jamaicans as "strong-back," and used by them in domestic medicine.

Meibomia purpurea (Mill.) Vail.

Meibomia scorpiurus (Swartz) Kuntze.

Mucuna urens (L.) DC. CHOCHO. Large vine; pods covered with stiff bristles that penetrate the skin easily.

Phaseolus peduncularis H. B. K. Small herbaceous vine.

Phaseolus vulgaris L. BEAN. FRIJOL. Planted at the laboratory.

Platymiscium polystachyum Benth. QUIRA. Large tree with racemes of small yellow flowers. The wood is of good quality, being known in commerce as Panama redwood.

Platypodium Maxonianum Pittier. CARCUERA. Large tree; fruit 1-seeded, winged, samara-like.

Pterocarpus officinalis Jacq. Large tree with small thin winged fruits. The sap turns red upon exposure to the air.

Rhynchosia pyramidalis (Lam.) Urban. A herbaceous vine with red and black seeds.

ERYTHROXYLACEAE. Coca Family

Erythroxylon amplum Benth. Shrub with entire leaves.

Erythroxylon panamense Turcz.

RUTACEAE. Rue Family

Citrus aurantifolia (Christm.) Swingle. LIME. LIMÓN. Naturalized in the forest.

Citrus sinensis (L.) Osbeck. ORANGE. NARANJO. Planted at the laboratory.

Zanthoxylum panamense P. Wilson. ARCABÚ, ACABÚ, ALCABÚ. Large tree; trunk covered with large pyramidal prickles.

SIMAROUBACEAE. Simaruba Family

Quassia amara L. QUASSIA. GUAVITO AMARGO, PUESILDE, CRUCETA. Shrub or small tree with pinnate leaves and showy red flowers. The leaves and bark are as bitter as quinine.

BURSERACEAE. Torchwood Family

Protium asperum Standl. CARAÑO. A large tree. From wounds in the trunk there are distilled large quantities of a fragrant resin or balsam, which collects upon the ground. Leaflets very rough.

Protium sessiliflorum (Rose) Standl. ANIME. Large tree; common; leaflets smooth.

Tetragastris panamensis (Engler) Kuntze? Large tree; common.

MALPIGHIACEAE. Malpighia Family

Hiraea faginifolia (DC.) Juss. Woody vine, the leaves densely silky beneath.

Stigmaphyllon Humboldtianum Juss. Woody vine with yellow flowers. The broad leaves bear numerous stalked glands along the margins.

TRIGONIACEAE. Trigonia Family

Trigonia floribunda Oerst. Woody vine with entire leaves, densely white-tomentose beneath; flowers small and white.

POLYGALACEAE. Polygala Family

Securidaca diversifolia (L.) Blake. Large woody vine with small entire leaves; flowers pink, showy.

EUPHORBIACEAE. Spurge Family

Acalypha diversifolia Jacq. A common shrub.

Acalypha macrostachya Jacq. Shrub.

Acalypha villosa Jacq. Common shrub.

Alchornea costaricensis Pax & Hoffm. Small tree with ovate crenate leaves; staminate flowers in long slender drooping spikes.

Codiaeum variegatum (L.) Blume. A shrub with colored leaves; one of the tropical "crotons," planted at the laboratory.

Croton Billbergianus Muell. Arg. Large shrub or small tree growing in the wet forest.

Dalechampia panamensis Pax & Hoffm. Vine with 3-parted leaves; inflorescense subtended by 2 green bracts; calyx furnished with stiff hairs which penetrate the skin easily.

Euphorbia hirta L. HIERBA DE POLLO. Called "milkweed" by the West Indians. A small annual herb.

Euphorbia hypericifolia L. HIERBA DE POLLO. A small glabrous annual.

Hura crepitans L. SANDBOX. JAVILLO. A giant forest tree, the trunk covered with small sharp spines. The milky sap causes blisters upon the skin.

Hieronyma alchorneoides Allem. PANTANO. Large tree with broad entire leaves bearing minute stellate scales.

Mabea occidentalis Benth. Shrub or small tree with oblong leaves; flowers in raceme-like terminal panicles.

Manihot esculenta Crantz. CASSAVA. YUCA. Much cultivated in Panama for its edible roots. Planted at the laboratory.

Phyllanthus conami Swartz. Shrub or small tree with small distichous ovate leaves.

Phyllanthus niruri L. Called by the West Indians "seed on the leaf." A small annual herb.

Phyllanthus nobilis (L. f.) Muell. Arg. Shrub or small tree with oblong-elliptic leaves.

ANACARDIACEAE. Cashew Family

Anacardium excelsum (Bert. & Balb.) Skeels. ESPAVÉ. A common large tree with entire leaves. The bark is used in some parts of Panama as a fish poison.

Astronium graveolens Jacq. ZORRO. A common tree with pinnate leaves having serrate or entire leaflets.

Mangifera indica L. MANGO. Naturalized and planted.

Spondias mombin L. HOGPLUM. JOBO. Tree with pinnate leaves and a juicy yellow edible fruit.

HIPPOCRATEACEAE. Hippocratea Family

Hippocratea volubilis L. Large woody vine, on the highest trees. The capsule is large, vertically compressed and nearly flat, and deeply 3-lobed.

Salacia praecelsa (Miers) Griseb. GARROTILLO. Large woody vine with globose fruit.

SAPINDACEAE. Soapberry Family

Allophylus psilospermus Radlk. Shrub or small tree with 3-foliolate leaves and winged fruit.

Cupania cinerea Poepp. GORGOJO, GORGOJERO. Shrub or small tree with pinnate leaves, whitish beneath.

Cupania fulvida Triana & Planch. CANDELILLO, GORGOJO, GORGOJERO. Shrub or small tree, often simple, densely brown-hirsute. The leaves are pinnate, but on young plants they are simple.

Cupania latifolia Kunth. Leaflets glabrous, rounded or retuse at apex.

Cupania Seemanni Triana & Planch. Leaflets glabrous, acuminate.

Paullinia alata Don. All the species of *Paullinia* are woody vines. They are used in tropical America as fish poisons.

Paullinia bracteosa Radlk.

Paullinia glomerulosa Radlk.

Paullinia turbacensis H. B. K.

Serjania trachygona Radlk. Woody vine.

Talisia nervosa Radlk. Small tree with very large, pinnate leaves.

RHAMNACEAE. Buckthorn Family

Gouania lupuloides (L.) Urban. Woody vine.

Gouania polygama (Jacq.) Urban. JABONCILLO. Called "chewstick" in the West Indies. The stems when chewed produce lather.

VITACEAE. Grape Family

Cissus salutaris H. B. K. Woody vine with 3-foliolate leaves and small red flowers.

Cissus sicyoides L. Vine with simple leaves. The inflorescences of this species are frequently distorted by a smut, *Mycosyrinx Cissi.*

Vitis tiliaefolia Humb. & Bonpl. GRAPE. UVA, BEJUCO DE AGUA. The fruit is small and very sour.

TILIACEAE. Basswood Family

Apeiba aspera Aubl. Tree with entire leaves. Fruit resembling a sea-urchin, and covered with stiff spines.

Apeiba tibourbou Aubl. PEINE DE MICO, CORTEZO. Leaves finely dentate.

Belotia panamensis Pittier. Tree with very showy flowers, the sepals pink, the petals violet; fruit compressed, obcordate, 2-celled.

Heliocarpus popayanensis H. B. K. MAJAGÜILLO. Tree, the small flowers panicled; fruits very small, compressed, the margin bearing a row of stiff radiating hairs.

Luehea Seemannii Triana & Planch. GUÁCIMO. A common, very large forest tree; leaves tomentose beneath; fruit small, woody, obtusely 5-angled.

Triumfetta lappula L. CADILLO, CEPA DE CABALLO. Shrub bearing small globose spiny burs.

MALVACEAE. Mallow Family

Hibiscus rosa-sinensis L. CHINESE HIBISCUS. PAPO, TAPO. Planted at the laboratory.

Pavonia dasypetala Turcz. Shrub with showy pink flowers 4 to 6 cm. long; leaves broad and velvety.

Pavonia rosea Schlecht. Herbaceous or suffrutescent, with small pink flowers; fruit armed with barbed spines.

Sida rhombifolia L. ESCOBILLA. One of the most common weedy plants of tropical America.

BOMBACACEAE. Cotton-tree Family

Bombacopsis Fendleri (Seem.) Pittier. CEDRO ESPINOSO. Large tree with spiny trunk, flowering in winter when leafless.

Bombacopsis sessilis (Benth.) Pittier. CEIBO. Trunk unarmed.

Cavanillesia platanifolia H. B. K. CUIPO, BONGO, QUIPO. Large tree with smooth swollen trunk; leaves deciduous, 5 or 7-lobed; flowers small, with red petals. The trees are conspicuous when in flower, in late March and early April. The wood is very soft and light.

Ochroma limonensis Rowlee. BALSA. Large or medium-sized tree, the cordate leaves 3-angled or shallowly 3-lobed, pale beneath; flowers large and whitish. The balsa trees have one of the lightest woods known.

STERCULIACEAE. Cacao Family

Buettneria aculeata Jacq. ESPINO HUECO, ZARZA, RABO DE IGUANA. Prickly shrub, often scandent; young leaves often blotched with silver.

Sterculia apetala (Jacq.) Karst. PANAMA. Large tree with 3 or 5-lobed leaves, stellate-tomentose beneath; flowers without petals, the large calyx 5-lobed, reddish; fruit of 5 carpels, the large brown seeds resembling chestnuts. It is from the Indian name of this tree that the Republic of Panama derives its name.

Theobroma cacao L. CACAO. Planted and also naturalized in the forest.

Theobroma purpureum Pittier. CACAO CIMARRÓN, CHOCOLATILLO. Shrub or small tree; leaves digitately compound, with 5 large leaflets; fruit small, covered with stiff hairs which penetrate the skin readily.

DILLENIACEAE. Dillenia Family

Davilla rugosa Poir. Woody vine with rough, obovate, nearly entire leaves and yellow flowers.

Dillenia indica L. Planted at the laboratory. A handsome tree with large toothed obovate leaves, very large white flowers, and a huge globular green fruit.

Doliocarpus major Gmel. Woody vine with glabrous but punctate leaves.

OCHNACEAE. Ochna Family

Ouratea Wrightii (Van Tiegh.) Riley. Shrub with narrow lustrous leaves; flowers yellow, in terminal panicles; fruits several, black, borne on a red disk.

HYPERICACEAE. St. Johnswort Family

Vismia ferruginea H. B. K. SANGRE DE PERRO. Shrub with ovate entire leaves, brownish beneath. The sap turns red upon exposure to the air.

CLUSIACEAE. Clusia Family

Calophyllum longifolium Willd. MARÍA. Large tree with very handsome, narrow, oblong leaves, 30 cm. long or larger; sap yellowish.

Clusia rosea L. COPEY. Tree; leaves thick, nearly as broad as long; flowers pink, waxy; fruit a leathery fleshy capsule; sap milky, sticky.

Rheedia madruno (H. B. K.) Planch. & Triana. CERILLO, TOMÉ, MACHARI. Tree with oblong to elliptic, acuminate leaves.

Symphonia globulifera L. f. CERILLO. Tree with small oblong-lanceolate leaves.

Tovomitopsis nicaraguensis (Oerst.) Triana & Planch. Shrub or small tree; flowers small, whitish.

VIOLACEAE. Violet Family

Hybanthus anomalus (H. B. K.) Standl. Shrub with alternate leaves.

Rinorea squamata Blake. MOLENILLO. Shrub with opposite leaves.

Rinorea sylvatica (Seem.) Kuntze.

FLACOURTIACEAE. Flacourtia Family

Casearia arguta H. B. K. RASPA-LENGUA. Shrub.

Casearia guianensis (Aubl.) Urban. PALO DE LA CRUZ.

Casearia nitida (L.) Jacq. RASPA-LENGUA.

Casearia sylvestris Swartz. Shrub with entire leaves.

Hasseltia floribunda H. B. K. RASPA-LENGUA. Small tree with oblong to elliptic, coarsely serrate, glabrate leaves, and small white flowers.

Oncoba laurina (Presl) Warb. GUAVO CIMARRÓN, CARBONERO. Small tree with spiny globose fruit.

TURNERACEAE. Turnera Family

Turnera panamensis Urban. Shrub with lance-oblong leaves and yellow flowers.

PASSIFLORACEAE. Passionflower Family

Passiflora auriculata H. B. K. Leaves ovate-lanceolate, 3-lobed or subentire.

Passiflora vitifolia H. B. K. GUATE-GUATE. A very showy species, a woody vine, with large, deep red flowers.

CARICACEAE. Papaya Family

Carica papaya L. PAPAYA. Planted at the laboratory; also wild or naturalized.

BEGONIACEAE. Begonia Family

Begonia filipes Benth. A small and inconspicuous plant.

CACTACEAE. Cactus Family

Epiphyllum phyllanthus (L.) Haw. An epiphytic spineless plant with large white flowers.

LYTHRACEAE. Loosestrife Family

Adenaria floribunda H. B. K. FRUTA DE PAVO. Shrub with entire, opposite, nearly sessile leaves.

LECYTHIDACEAE. Brazilnut Family

Grias Fendleri Seem. Tree with large sessile leaves, entire or nearly so.

Gustavia superba Berg. MEMBRILLO. Medium-sized tree with few branches; leaves 30 to 100 cm. long, serrate; flowers about 10 cm. broad, white; fruit edible. Common.

RHIZOPHORACEAE. Mangrove Family

Cassipourea elliptica Poir. HUESITO, LIMONCILLO. Shrub or small tree with glabrous entire opposite leaves.

COMBRETACEAE. Combretum Family

Terminalia Hayesii Pittier. AMARILLO REAL. A common large tree; leaves obovate, entire; flowers minute, green, in long spikes.

MYRTACEAE. Myrtle Family

Calycolpus Warscewiczianus Berg. GUAYABILLO. Slender shrub with pink or whitish flowers.

Eugenia uniflora L. SURINAM CHERRY. A South American shrub, planted at the laboratory.

Psidium guajava L. GUAVA. GUAYABA. Frequent in open places.

MELASTOMACEAE. Meadowbeauty Family

Clidemia petiolata (Rich.) DC. Shrub.

Conostegia bracteata Triana. Shrub.

Conostegia speciosa Naud. DOS CARAS, RASPA-LENGUA, FRUTA DE PAVA. Shrub.

Heterotrichum octonum (Bonpl.) DC. Shrub with 7 or 9-nerved, broadly ovate leaves.

Miconia argentea (Swartz) DC. DOS CARAS, CANILLO, PAPELILLO. Common shrub or small tree, with large broad leaves very white beneath.

Miconia Beurlingii Triana.

Miconia lacera (Humb. & Bonpl.) Naud. Common shrub.

Miconia nervosa (Smith) Triana.

Miconia impetiolaris (Swartz) Don. DOS CARAS, OREJA DE MULA. Leaves large, brownish beneath, sessile.

Miconia lonchophylla Naud.

Mouriria parvifolia Benth. ARRACHECHE. Shrub, glabrous throughout, with sessile entire ovate leaves and small axillary flowers.

Ossaea diversifolia (Naud.) Cogn. FRUTA DE PAVA. Shrub with pink or reddish flowers and small, black or purple fruit.

Ossaea micrantha (Swartz) Macfad.

Tibouchina longifolia (Vahl) Benth. Herb with small white flowers.

ONAGRACEAE. Evening-primrose Family

Jussiaea suffruticosa L. A common herb with yellow flowers, growing in wet places.

ARALIACEAE. Ginseng Family

Dendropanax arboreum (L.) Decaisne & Planch. VAQUERO. Small tree with entire or 3-lobed leaves.

Didymopanax Morototoni (Aubl.) Decaisne & Planch. MANGABÉ, GARGORÁN, PAVA. Large tree; leaves digitately compound, with 7 to 10 long-stalked entire acuminate leaflets, pale-tomentose beneath. Common.

Nothopanax Guilfoylei (Cogn. & Marché) Merrill. Planted at the laboratory. Shrub with pinnate white-margined leaves.

MYRSINACEAE. Myrsine Family

Ardisia compressa H. B. K. Shrub with white or pinkish flowers and black juicy fruit.

Ardisia myriodonta Standl. Described from Barro Colorado. A small shrub.

Stylogyne laevis (Oerst.) Mez. Glabrous shrub with thick entire leaves; flowers white or pinkish, the branches of the panicle bright red.

Stylogyne ramiflora (Oerst.) Mez.

SAPOTACEAE. Sapodilla Family

Chrysophyllum cainito L. STAR-APPLE. CAIMITO. Large tree; leaves covered beneath with silky golden-brown hairs; fruit edible. Common forest tree.

LOGANIACEAE. Logania Family

Spigelia Humboldtiana Cham. & Schlecht. A small herb.

Strychnos darienensis Seem. Woody vine.

Strychnos panamensis Seem. CANJURA, FRUTA DE MURCIÉLAGO. Woody vine with nearly glabrous, entire leaves; fruit large, globose, with hard shell.

Strychnos toxifera Benth. A very hairy vine. This species furnishes curare poison, used by the Indians of Panama and elsewhere for poisoning their arrows.

GENTIANACEAE. Gentian Family

Leiphaimos simplex (Griseb.) Standl. A small saprophyte, without any green coloration, the slender stem bearing a single pale blue flower; common in dark wet forest.

APOCYNACEAE. Dogbane Family

Odontadenia speciosa Benth. NEGRILLO. Woody vine with large yellow flowers.

Prestonia obovata Standl. Woody vine with glabrous obovate leaves.

Tabernaemontana grandiflora Jacq. HUEVO DE GATO, LECHUGA, VENENILLO. Glabrous shrub with yellow flowers about 5 cm. long.

Thevetia nitida (H. B. K.) A. DC. COJÓN DE GATO, LAVAPERRO, HUEVO DE TIGRE. Shrub or small tree with thick obovate leaves, yellow flowers, and bright red fruit.

ASCLEPIADACEAE. Milkweed Family

Asclepias curassavica L. NIÑO MUERTO, PASORÍN. Herb with red and orange flowers. The only species of *Asclepias* found in the region.

Vincetoxicum pinguifolium Standl. A herbaceous vine, known only from Barro Colorado.

CONVOLVULACEAE. Morning-glory Family

Maripa panamensis Hemsl. Large glabrous woody vine with oblong to ovate leaves.

Rivea campanulata (L.) House. BATATILLA. Large vine with broadly cordate leaves; flowers pink, 7 to 8 cm. long.

The genus *Ipomoea* must occur on the island, but I have no record of it.

BORAGINACEAE. Borage Family

Cordia alliodora (Ruiz & Pav.) Cham. LAUREL. Tree with stellate-pubescent leaves and small but showy, white flowers. The nodes are often swollen, and inhabited by ants.

Tournefortia obscura A. DC. Small woody vine.

VERBENACEAE. Verbena Family

Petrea volubilis Jacq. VILDA, FLOR DE MAYO, FLOR DE LA CRUZ. Large woody vine with very showy racemes of purple-blue flowers.

MENTHACEAE. Mint Family

Coleus Blumei Benth. COLEUS. POMPOLLUDA, CHONTADURA. Called by the West Indians "Jacob's-coat." Planted at the laboratory.

Hyptis capitata Jacq. SUSPIRO DE MONTE. A weedy herb.

Salvia occidentalis Swartz. A small weedy herb with minute blue flowers

SOLANACEAE. Potato Family

Capsicum annuum L. PEPPER. CHILE, AJÍ. Planted at the laboratory.

Capsicum macrophyllum (H. B. K.) Standl. PINTAMORA DE MONTE. Large coarse herb with bright red, cherry-like fruit.

Cestrum panamense Standl. Small tree with pale green, tubular flowers.

Lycianthes Maxonii Standl. Nearly glabrous, erect shrub with small violet flowers. The typical form of the species is known only from Barro Colorado, but a variety occurs in the forests beyond Panama City.

Solanum allophyllum (Miers) Standl. HIERBA DE GALLINAZO, HIERBA GALLOTA. Herb with entire or lobed leaves.

Solanum bicolor Willd. Unarmed shrub with long-stalked cymes of white flowers.

Solanum diversifolium Schlecht. FRIEGA-PLATO, HUEVO DE GATO. Called by the Barbadians "susumba." Erect prickly shrub.

Solanum parcebarbatum Bitter. Nearly glabrous, unarmed shrub. For this I have been given in Panama the name "sauco," but that name belongs properly to the genus *Sambucus*.

Solanum scabrum Vahl. FRIEGA-PLATO, ARAÑA-GATO. Large, very prickly, woody vine.

Solanum sp. Only imperfect material is available; probably an undescribed species.

SCROPHULARIACEAE. Figwort Family

Scoparia dulcis L. ESCOBILLA AMARGA. Called "sweet broom" by the West Indians. Herbaceous or suffrutescent weed, with very small, white flowers.

Stemodia parviflora Ait. Small herb with blue flowers.

Torenia crustacea (L.) Cham. & Schlecht. Small weedy herb with blue-purple flowers.

BIGNONIACEAE. Bignonia Family

Adenocalymna flos-ardeae Pittier. Woody vine, the leaflets with large yellow glands on the lower surface.

Amphilophium paniculatum (L.) H. B. K. Leaflets covered with minute scales; flowers pink and white.

Arrabidaea pachycalyx Sprague. Leaflets white-tomentose beneath; flowers purple, small but in large panicles and very showy.

Cydista aequinoctialis (L.) Miers. Reported by Dodge. Nearly glabrous vine with pale purple flowers 5 to 8 cm. long.

Jacaranda copaia Don. PALO DE BUBA. Tree with twice-pinnate leaves and large, bluish, very showy flowers.

Macfadyena uncinata (Meyer) DC. Easily recognized by the sharp hooks terminating the tendrils; flowers pale yellow.

Paragonia pyramidata (Rich.) Bur. Woody vine; leaflets minutely lepidote beneath; flowers pink, 6 to 7.5 cm. long.

Petastoma patelliferum (Schlecht.) Miers. Reported by Dodge. Glabrate vine with purple flowers about 4 cm. long.

Phryganocydia corymbosa (Vent.) Bur. Nearly glabrous vine with handsome, bright pink flowers 6 to 9 cm. long.

Tabebuia guayacan (Seem.) Hemsl. GUAYACÁN. Tree with digitately compound leaves and large yellow flowers.

Tabebuia pentaphylla (L.) Hemsl. ROBLE, ROBLE DE SABANA. Tree with digitately compound, minutely lepidote leaves; flowers varying from pale to deep pink. When in full flower, this is one of the most beautiful of Central American trees.

GESNERIACEAE. Gesneria Family

Achimenes panamensis (Seem.) Hemsl. Small herb with white flowers.

Columnea purpurata Hanst. Rare. Coarse suffrutescent plant with large, very hairy leaves, and bright red, axillary inflorescence.

Drymonia spectabilis (H. B. K.) Mart. Epiphytic shrub; corolla dull dark red.

Kohleria tubiflora (Cav.) Hanst. Herb with scarlet flowers.

Tussacia Friedrichsthaliana Hanst. Small herb with large orange flowers.

PINGUICULACEAE. Butterwort Family

Utricularia mixta Barnh. Small floating aquatic plant, in quiet water; flowers yellow.

ACANTHACEAE. Acanthus Family

Aphelandra Sinclairiana Nees. Showy shrub with bright red flowers in dense bracted spikes, the bracts orange-red.

Aphelandra tetragona (Vahl) Nees. Shrub with red flowers, the bracts small and green.

Blechum pyramidatum (Lam.) Urban. Common herbaceous weed with small purple flowers.

Blechum panamense Lindau. Herb with purple flowers.

Chaetochlamys panamensis Lindau. Erect herb with showy purple flowers.

Mendoncia retusa Turrill. Vine; flowers white, with purple veins; fruit a black plumlike drupe.

RUBIACEAE. Madder Family

Alibertia edulis (L. Rich.) A. Rich. LAGARTILLO, TROMPITO. Called "wild guava" by the West Indians. Shrub with sessile glabrous lance-oblong leaves; flowers small, clustered, sessile; fruit globose, 2.5 cm. in diameter. Young seedling plants, which are very common on the island, have the leaves handsomely striped or mottled with purple and pink.

Bertiera guianensis Aubl. Shrub with small, bright blue fruit.

Borreria laevis (Lam.) Griseb. Small weedy herb.

Borreria latifolia (Aubl.) Schum. Reported by Dodge. An herb with small white flowers.

Borreria ocymoides (Burm.) DC. Small annual herb.

Cephaelis ipecacuanha (Brot.) Rich. IPECAC. RAICILLA. A small glabrate plant, about 30 cm. high; leaves oblong; flowers small, white, in a single terminal head. Ipecac is obtained from the thickened roots.

Cephaelis tomentosa (Aubl.) Vahl. Shrub with very hairy leaves; flowers in a dense head, subtended by showy, bright red bracts.

Faramea occidentalis (L.) Rich. HUESITO. Shrub with white flowers. In its general appearance and in its fruit this plant sugests coffee, to which it is related.

Geophila herbacea (L.) Schum. Creeping herb with heart-shaped leaves and small white flowers; fruit juicy, red or purple-black.

Guettarda foliacea Standl. Shrub with globose red fruit.

Hamelia nodosa Mart. & Gal. Shrub with orange-red flowers.

Hemidiodia ocimifolia (Willd.) Schum. A weedy herb.

Isertia Haenkeana DC. CANELITO. Showy shrub with large leaves and dense panicles of bright yellow and red, tubular flowers.

Ixora coccinea L. BUQUET DE NOVIA. Shrub with red flowers, planted at the laboratory.

Oldenlandia corymbosa L. Small herb with linear leaves and white or pinkish flowers.

Palicourea guianensis Aubl. Shrub; flowers yellow, in a terminal thyrse, its branches red or orange.

Pentagonia macrophylla Benth. HOJA DE MURCIÉLAGO. Shrub with very large, obovate leaves.

Pentagonia pubescens Standl. Common shrub.

Posoqueria latifolia (Rudge) Roem. & Schult. BOCA VIEJA, BORAJÓ, FRUTA DE MONO. Large shrub or small tree with broad thick leaves; flowers tubular, very slender, 12 to 16 cm. long.

Psychotria brachiata Swartz. The species of this genus are common shrubs of the forest.

Psychotria calophylla Standl. Reported by Dodge.

Psychotria chagrensis Standl.

Psychotria cuspidata Bredem.

Psychotria emetica L. f. RAICILLA MACHO, RAICILLA. Small shrub, with axillary white flowers and blue fruit. The roots yield a kind of ipecac.

Psychotria granadensis Benth.

Psychotria grandis Swartz.

Psychotria horizontalis Swartz.

Psychotria involucrata Swartz.

Psychotria limonensis Krause.

Psychotria marginata Swartz.

Psychotria micrantha H. B. K.

Psychotria patens Swartz. GARRICILLO.

Psychotria Pittieri Standl. Fruit blue.

Psychotria racemosa (Aubl.) Willd. Fruit 5-celled; it is 2-celled in the other species.

Randia armata (Swartz) DC. ROSETILLO. Spiny shrub with large white flowers.

Rudgea fimbriata (Benth.) Standl. Shrub with subsessile leaves.

Uncaria tomentosa (Willd.) DC. Woody vine armed with hooked spines; flowers yellowish, fragrant, in dense globose heads.

CUCURBITACEAE. Gourd Family

Anguria Warscewiczii Hook. f. Glabrous vine with 3-foliolate leaves; flowers small, with bright red petals.

Cayaponia Poeppigii Cogn. Fruit small, globose, 6-seeded.

Cucurbita pepo L. SQUASH. CALABAZO, SAPUYO. Planted at the laboratory.

Gurania Seemanniana Cogn.? BEJUCO PICADOR. Large herbaceous vine with red inflorescence.

Gurania suberosa Standl. Type from Barro Colorado. Large woody vine, climbing on high trees; stems covered with corky bark; flowers small, red, borne on the naked stems near the ground. The leaves have not been collected.

Luffa cylindrica (L.) Roem. SPONGEGOURD. CALABAZO. Vine with large yellow flowers. The interior of the fruit resembles a sponge, and may be used in the same manner.

Melothria guadalupensis (Spreng.) Cogn. SANDILLITA. Slender vine with small yellow flowers. The fruit resembles a small watermelon, and has the odor of cucumber.

Posadaea sphaerocarpa Cogn. BRUJITO. Herbaceous vine with a globose gourdlike fruit.

Sicydium tamnifolium (H. B. K.) Cogn. Leaves nearly entire, softly pubescent; flowers minute, in large panicles; fruits small, black.

ASTERACEAE. Aster Family

Baccharis trinervis (Lam.) Pers. Shrub with dirty-white flowers.

Bidens pilosa L. CADILLO, SIRVULACA. Called "Spanish needles" by the West Indians.

Chaptalia nutans (L.) Polak. Leaves white-tomentose beneath, in a basal rosette; rays short, white to red-purple.

Eclipta alba (L.) Hassk. Small weedy herb.

Emilia sonchifolia (L.) DC. Small weedy herb with pale purple or pink, discoid heads.

Erechtites hieracifolia (L.) Raf. TABAQUILLO. Coarse herb with greenish discoid heads.

Erigeron bonariensis L. TABAQUILLO. Weedy herb with linear leaves.

Erigeron spathulatus Vahl. Weedy herb with spatulate or obovate leaves.

Eupatorium macrophyllum L. Coarse herb with greenish white heads.

Eupatorium microstemon Cass. Small annual with purple heads.

Eupatorium odoratum L. PALECA, HIERBA DE CHIVA. Called "Christmas-bush" by the West Indians. Large herb or shrub with lavender flowers.

Eupatorium Sinclairii Benth. Weedy herb with purplish flowers.

Mikania leiostachya Benth. Herbaceous vine; heads in spikes.

Mikania micrantha H. B. K. Herbaceous vine; heads small, fragrant, whitish, in cymes.

Neurolaena lobata (L.) R. Br. CONTRAGAVILANA. Coarse herb with yellow heads.

Pluchea purpurascens (Swartz) DC. Viscid herb with purple heads; growing in shallow water at edge of lake.

Porophyllum ruderale (Jacq.) Cass. Glabrous annual with discoid heads of bronze flowers; rare here.

Pseudelephantopus spicatus (Juss.) Rohr. ESCOBILLA BLANCA, CHICORIA. Weedy herb with pale purple heads.

Rolandra fruticosa (L.) Kuntze. Coarse herb with 1-flowered white heads; leaves white-tomentose beneath.

Tridax procumbens L. Weedy procumbent annual herb with pale yellow heads.

Vernonia canescens H. B. K. HIERBA DE SAN JOSÉ. Heads pink or white.

Vernonia cinerea (L.) Less. Small weedy herb with purple heads; naturalized from the Old World.

Vernonia patens H. B. K. LENGUA DE VACA, LENGUA DE BUEY. Shrub with white heads.

Wulffia baccata (L. f.) Kuntze. Arching shrub with rough leaves; heads 2 cm. broad, with small yellow rays.